SpringerWienNewYork

T0181148

Tansel Özyer
Keivan Kianmehr
Mehmet Tan

Editors

# Recent Trends in Information Reuse and Integration

SpringerWienNewYork

*Editors*

Tansel Özyer
Department of Computer
Engineering
Tobb University
Söğütözü Caddesi No. 43, Söğütözü
Ankara, Turkey
ozyer@etu.edu.tr

Mehmet Tan
Department of Computer Engineering
Tobb University
Söğütözü Caddesi No. 43, Söğütözü
Ankara, Turkey
mtan@etu.edu.tr

Keivan Kianmehr
Department of Electrical and Computer
Engineering
University of Western Ontario
Building 373
London, ON, N6A 5B9, Canada
kkianmeh@uwo.ca

© 2012 Springer-Verlag/Wien
Softcover reprint of the hardcover 1st edition 2012

SpringerWienNewYork is part of Springer Science + Business Media
springer.at

Typesetting: SPi Publisher Services

Printed on acid-free paper and chlorine-free bleached paper
SPIN: 80032388

With 134 (partly coloured) Figures

ISBN 978-3-7091-1689-0    ISBN 978-3-7091-0738-6 (eBook)
DOI 10.1007/978-3-7091-0738-6
SpringerWienNewYork

# Foreword

We are delighted to see this edited book as the result of our intensive work over the past year. We succeeded in attracting high quality submissions of which we could only include 19 papers in this edited book. The present text aims at helping the reader whether researcher or practitioner to grasp the basic concept of reusability which is very essential in this rapidly growing information era. The authors emphasize the need for reusability and how it could be adapted in the right way. Actually reusability leads to satisfy a multiobjective optimization process, that is, to minimize the time, cost and effort spent to develop new products, technologies, information repositories, etc. Instead of always starting from scratch and reinventing the wheel in a process that consumes and wastes time, effort, and resources, practitioners and developers should always look into the possibility of reusing some of the existing entities to produce new ones. In other words, reuse and integration are essential concepts that must be enforced to avoid duplicating the effort. This problem is investigated from different perspectives. In organizations, high volumes of data from different sources form a big threat for filtering out the information for effective decision making. To address all these vital and serious concerns, this book covers the most recent advances in information reuse and integration. It contains high quality research papers written by experts in the field. Some of them are extended versions of the best papers which were presented at IEEE International Conference on Information Reuse and Integration, which was held in Las Vegas in August 2010.

Chapter 1 by Udai Shanker, B. Vidya Reddi, and Anupam Shukla studies a real time commit protocol to improve the performance based on two approaches. They are: 1. Some of the locked data items which are unused after completion of processing of transaction can be unlocked immediately after end of processing phase to reduce data contention. 2. A lending transaction can lend its dirty data to more than one cohorts by creating only a single shadow in case of write–read conflicts to reduce the data inaccessibility. Its performance has been compared with existing protocols.

Chapter 2 by Ladjel Bellatreche estimates the complexity of the problem of selecting dimension table(s) to partition a fact table. It also proposes strategies to perform their selection that take into account the main characteristics of queries such as access frequencies, size of tables, size of intermediate results of joins, etc. Experimental studies using a mathematical cost model and the obtained results have been executed on Oracle11G DBMS for validation, and a tool has been implemented to assist data warehouse administrators in their horizontal partitioning selection tasks.

Chapter 3 by Shabnam Pourdehi, Dena Karimipour, Navid Noroozi, and Faridoon Shabani addresses an adaptive fuzzy controller for a large class of nonlinear systems in the presence of uncertainties, input nonlinearities, and unknown time-delay. Based on the combination of the sliding mode control (SMC) with fuzzy adaptive control, it presents a design algorithm to synthesize a robust fuzzy sliding mode controller. Later, an adaptive fuzzy observer-based SMC scheme is proposed for stabilization. The unknown nonlinear functions have been approximated with fuzzy logic systems in both proposed control schemes and its asymptotical stability according to the corresponding closed-loop system is shown with Lyapunov–Krasovskii approach. The synchronization of two nonidentical time-delayed chaotic systems is investigated as an application of control schemes with simulated examples to show the effectiveness of the proposed techniques.

Chapter 4 by Ladjel Bellatreche, Guy Pierra, and Eric Sardet presents a technique for integrating automatically ontology-based data sources in a materialized architecture and a framework dealing with the asynchronous versioning problem. Existing solutions proposed in traditional databases is adapted to manage instance and schema changes. The problem of managing ontology changes is overcome by introducing a distinction between ontology evolution and ontology revolution. Finally, it proposes floating version model for ontology evolution. It fully automates the whole steps of building an ontology-based integration systems (OBIS). It is validated by a prototype using ECCO environment and the EXPRESS language.

Chapter 5 by Iyad Suleiman, Shang Gao, Maha Arslan, Tamer Salman, Faruk Polat, Reda Alhajj, and Mick Ridley suggests a a computerized assessment tool that can learn the user's skills and adjust the assessment tests for assessment of school readiness. The user plays various sessions from various games, while the Genetic Algorithm (GA) selects the upcoming session or group of sessions to be chosen for the user according to his/her skills and status. It describes the modified GA and the learning procedure that is integrated with a penalizing system into the GA and a fitness heuristic for best choice selection. Two methods for learning are proposed, a memory system and a no memory system. Furthermore, it includes several methods for the improvement of the speed of learning. In addition, learning mechanisms that are based on the social network paradigm to address further usage of assessment automation is used.

Chapter 6 by Grégory Claude, Gaël Durand, Marc Boyer, and Florence Sèdes defends the idea that a defect carries information by the simple fact that it exists, the characteristics of the detected incident (the problem) and of the applied protocol to resolve it (the solution). The various actors who took part in its description and its resolution collect this information. This knowledge is essential to achieve for assistance in corrective actions for future defects and prevention of their emergence. Taking the advantage of this knowledge by working out a model of defect makes it possible to define a set of grouping criteria of defects that were solved in the past. These groups are the cornerstone of the corrective and preventive processes for new defects.

Chapter 7 by Abdullah M. Elsheikh, Tamer N. Jarada, Taher Naser, Kelvin Chung, Armen Shimoon, Faruk Polat, Panagiotis Karampelas, Jon Rokne, Mick Ridley, and Reda Alhajj presents a comprehensive approach for the mapping between the object database language (ODL) and XML. It includes both structure specification and database content. It concentrates on deriving a separate set of transformation rules for each way mapping. For the first mapping (from ODL to XML).

Chapter 8 by Taghi M. Khoshgoftaar, Kehan Gao, and Jason Van Hulse proposes a novel approach to feature selection for imbalanced data in the context of software quality engineering. This process follows a repetitive process of data sampling followed by feature ranking and finally aggregating the results generated during the repetitive process. It is compared against filter-based feature ranking technique alone on the original data, and data sampling and feature ranking techniques with two different scenarios.

Chapter 9 by Zifang Huang and Mei-Ling Shyu presents a k-nearest neighbors (k-NN)-based least squares support vector machine (LS-SVM) model with multi-value integration to tackle the long-term time series prediction problem. A new distance function, which incorporates the Euclidean distance and the dissimilarity of the trend of a time series, is deployed.

Chapter 10 by Carlos M. Cornejo, Iván Ruiz-Rube, and Juan Manuel Dodero discusses their approach to provide a complete set of services and applications to integrate diverse web-based contents of the cultural domain. It is purposed to extend over the web the knowledge base of cultural institutions, build user communities around it, and enable its exploitation in several environments.

Chapter 11 by Abdelghani Bakhtouchi, Ladjel Bellatreche, Chedlia Chakroun, and Yamine Aït-Ameur proposes an ontology-based integration system with media-tor architecture. It exploits the presence of ontology referenced by selected sources to explicit their semantic. Instead of annotating different candidate keys of each ontology class, its set of functional dependencies are defined on its properties.

Chapter 12 by Mark McKenney gives an efficient algorithm for the map construction algorithm (MCP). The algorithm is implemented and experiments show that it is significantly faster than the naive approach, but is prone to large memory usage when run over large data sets. An external memory version of the algorithm is also presented that is efficient for very large data sets, and requires significantly less memory than the original algorithm.

Chapter 13 by Jairo Pava, Fausto Fleites, Shu-Ching Chen, and Keqi Zhang proposes a system that integrates storm surge projection, meteorological, topograph-ical, and road data to simulate storm surge conditions. The motivation behind the system is to serve local governments seeking to overcome difficulties in persuading residents to adhere to evacuation notices.

Chapter 14 by Richa Tiwari, Chengcui Zhang, Thamar Solorio, and Wei-Bang Chen describes a framework to extract information about co-expression relationships among genes from published literature using a supervised machine learning approach. Later it rank those papers to provide users with a complete specialized information retrieval system with Dynamic Conditional Random Fields (DCRFs) for training the model. They show its superiority against Bayes Net, SVM, and Nave Bayes.

Chapter 15 by Brandeis Marshall identifies similar artists, which serves as a precursor to how music recommendation can handle the more complex issues of multiple genre artists and artist collaborations. It considers the individual most similar artist ranking from three public-use Web APIs (Idiomag, Last.fm, and Echo Nest) as different perspectives of artist similarity. Then it aggregates these three rankings using five rank fusion algorithms.

Chapter 16 by Ken Q. Pu and Russell Cheung presents a query facility called tag grid to support fuzzy search and queries of both tags and data items alike. By combining methods of Online Analytic Processing from multidimensional databases and collaborative filtering from information retrieval, tag grid enables users to search for interesting data items by navigating and discovering interesting tags.

Chapter 17 by Stefan Silcher, Jorge Minguez, and Bernhard Mitschang introduces an SOA-based solution to the integration of all Product Lifecycle Management (PLM) phases. It uses an Enterprise Service Bus (ESB) as service-based integration and communication infrastructure. Three exemplary scenarios are used to illustrate the benefits of using an ESB as compared to alternative PLM infrastructures. Furthermore, it describes a service hierachy that extends PLM functionality with value-added services by mapping business processes to data integration services.

Chapter 18 by Awny Alnusair and Tian Zhao describes an ontology-based approach for identifying and retrieving relevant software components in large reuse libraries. It exploits the use of domain-specific ontologies to enrich a knowledge base initially populated with ontological descriptions of API components.

Chapter 19 by Du Zhang proposes an algorithm for detecting several types of firewall rule inconsistency. It also defines a special type of inconsistency called setuid inconsistency and highlights various other types of inconsistencies in the aforementioned areas. Finally, it concludes that inconsistency is a very important phenomenon, and its utilities can never be underestimated in information security and digital forensics.

Last but not the least, we would like to mention the hard workers behind the scene who have significant unseen contributions to the successful task that produced this valuable source of knowledge. We would like to thank the authors who submitted papers and the reviewers who produced detailed constructive reports which improved the quality of the papers. Various people from Springer deserve large credit for their help and support in all the issues related to publishing this book. In particular, we would like to thank Stephen Soehnlen for his dedications, seriousness, and generous support in terms of time and effort; he answered our emails on time despite his busy schedule, even when he was traveling.

*Tansel Özyer, Keivan Kianmehr, Mehmet Tan*

# Contents

# Contributors

**Yamine Aït-Ameur** LISI/ENSMA – Poitiers University Futuroscope, France, yamine@ensma.fr

**Reda Alhajj** Computer Science Department, University of Calgary, Calgary, AB, Canada

Department of Computer Science, Global University, Beirut, Lebanon

Department of Information Technology, Hellenic American University, Manchester, NH, USA, alhajj@ucalgary.ca

**Awny Alnusair** Department of Science, Math & Informatics, Indiana University Kokomo, 2300 S. Kokomo, IN 46904, Washington, aalnusai@iuk.edu

**Maha Arslan** Haifa University, Haifa, Israel

**Abdelghani Bakhtouchi** National High School for Computer Science (ESI), Algiers, Algeria, a_bakhtouchi@esi.dz

**Ladjel Bellatreche** LISI/ENSMA – Poitiers University, 1, avenue Clément Ader, 86960 Futuroscope, France, bellatreche@ensma.fr

**Marc Boyer** Université de Toulouse, Université Paul Sabatier, Inserm UMR 825 Place du Docteur Baylac, Pavillon Riser 31059 Toulouse cedex 9, France, Marc.Boyer@iut-tlse3.fr

**Chedlia Chakroun** LISI/ENSMA – Poitiers University Futuroscope, France, chakrouc@ensma.fr

**Shu-Ching Chen** Distributed Multimedia Information Systems Laboratory, Florida International University, Miami, FL 33199, USA, chens@cs.fiu.edu

**Wei-Bang Chen** Department of Computer and Information Sciences, The University of Alabama at Birmingham, 115A Campbell Hall, 1300 University Boulevard, Birmingham, Alabama 35294, wbc0522@cis.uab.edu

**Russell Cheung**  University of Ontario Institute of Technology, 2000 Simcoe Street N, Oshawa, ON, Canada, cheugn@mycampus.uoit.ca

**Kelvin Chung**  Computer Science Department, University of Calgary, Calgary, AB, Canada

**Grégory Claude**  Université de Toulouse, Université Paul Sabatier, IRIT UMR 5505 118, route de Narbonne 31062, Toulouse cedex 9, France, Gregory.Claude@irit.fr

**Carlos M. Cornejo**  Computer Languages and Systems Department, University of Cadiz, C\Chile. 1, 11002 Cadiz, Spain, carlos.cornejo@uca.es

**Juan Manuel Dodero**  Computer Languages and Systems Department, University of Cadiz, C\Chile. 1, 11002 Cadiz, Spain, juanma.dodero@uca.es

**Gaël Durand**  Intercim LLC, 1915 Plaza Drive, Eagan, MN 55122, USA

Intercim, 32, rue des Jeûneurs 75002, Paris, France, GDurand@intercim.com

**Abdullah M. Elsheikh**  Computer Science Department, University of Calgary, Calgary, AB, Canada

**Fausto Fleites**  Distributed Multimedia Information Systems Laboratory, Florida International University, Miami, FL 33199, USA, fflei001@cs.fiu.edu

**Kehan Gao**  Eastern Connecticut State University, Willimantic, CT 06226, USA, gaok@easternct.edu

**Shang Gao**  Computer Science Department, University of Calgary, Calgary, Alberta, Canada

**Zifang Huang**  Department of Electrical and Computer Engineering, University of Miami, Coral Gables, FL 33124, USA, z.huang3@umiami.edu

**Tamer N. Jarada**  Computer Science Department, University of Calgary, Calgary, AB, Canada

**Panagiotis Karampelas**  Department of Information Technology, Hellenic American University, Manchester, NH, USA

**Dena Karimipour**  School of Electrical and Computer Engineering, Shiraz University, Shiraz, Iran

**Taghi M. Khoshgoftaar**  Florida Atlantic University, Boca Raton, FL 33431, USA, taghi@cse.fau.edu

**Brandeis Marshall**  Computer and Information Technology, Purdue University, West Lafayette, IN 47907, USA, brandeis@purdue.edu

**Mark McKenney**  Department of Computer Science, Texas State University, 601 University Drive, San Marcos, TX 78666-4684, USA, mckenney@txstate.edu

**Jorge Minguez** Graduate School for advanced Manufacturing Engineering, GSaME, Institute for Parallel and Distributed Systems, IPVS, University of Stuttgart, Universitaetsstrasse 38, 70569 Stuttgart, Germany, jorge.minguez@ipvs. uni-stuttgart.de

**Bernhard Mitschang** Graduate School for advanced Manufacturing Engineering, GSaME, Institute for Parallel and Distributed Systems, IPVS, University of Stuttgart, Universitaetsstrasse 38, 70569 Stuttgart, Germany, bernhard. mitschang@ipvs.uni-stuttgart.de

**Taher Naser** Department of Computing, School of Computing Informatics and Media, University of Bradford, Bradford, UK

**Navid Noroozi** School of Electrical and Computer Engineering, Shiraz University, Shiraz, Iran

**Jairo Pava** Distributed Multimedia Information Systems Laboratory, Florida International University, Miami, FL 33199, USA, jpava001@cs.fiu.edu

**Guy Pierra** LISI/ENSMA – Poitiers University, 1, avenue Clément Ader, 86960 Futuroscope, France, pierra@ensma.fr

**Faruk Polat** Dept of Computer Engineering, Middle East Technical University, Ankara, Turkey

**Shabnam Pourdehi** School of Electrical and Computer Engineering, Shiraz University, Shiraz, Iran, Shabnamp_el83@yahoo.com

**Ken Q Pu** University of Ontario Institute of Technology, 2000 Simcoe Street N, Oshawa, ON, Canada, ken.pu@uoit.ca@mycampus.uoit.ca

**Mick Ridley** Department of Computing, School of Computing Informatics and Media, University of Bradford, Bradford, UK

**Jon Rokne** Computer Science Department, University of Calgary, Calgary, AB, Canada

**Iván Ruiz-Rube** Computer Languages and Systems Department, University of Cadiz, C\Chile. 1, 11002 Cadiz, Spain, ivan.ruiz@uca.es

**Tamer Salman** Dept of Computer Science, Technion, Haifa, Israel

**Eric Sardet** CRITT Informatique CRCFAO, 86960 Futuroscope, France, sardet@ensma.fr

**Florence Sèdes** Université de Toulouse, Université Paul Sabatier, IRIT UMR 5505 118, route de Narbonne 31062, Toulouse cedex 9, France, Florence.Sedes@irit.fr

**Faridoon Shabani** School of Electrical and Computer Engineering, Shiraz University, Shiraz, Iran, Shabani@shirazu.ac.ir

**Udai Shanker** Dept. of CSE, M. M. M. Engg. College, Gorakhpur-273010, India, udaigkp@gmail.com

**Armen Shimoon** Computer Science Department, University of Calgary, Calgary, AB, Canada

**Anupam Shukla** ICT Stream, ABV IIITM Gwalior, India, dranupamshukla@gmail.com

**Mei-Ling Shyu** Department of Electrical and Computer Engineering, University of Miami, Coral Gables, FL 33124, USA, shyu@miami.edu

**Stefan Silcher** Graduate School for advanced Manufacturing Engineering, GSaME, Institute for Parallel and Distributed Systems, IPVS, University of Stuttgart, Universitaetsstrasse 38, 70569 Stuttgart, Germany, stefan.silcher@ipvs.uni-stuttgart.de

**Thamar Solorio** Department of Computer and Information Sciences, The University of Alabama at Birmingham, 115A Campbell Hall, 1300 University Boulevard, Birmingham, Alabama 35294, solorio@cis.uab.edu

**Iyad Suleiman** Department of Computing, University of Bradford, Bradford, UK

**Richa Tiwari** Department of Computer and Information Sciences, The University of Alabama at Birmingham, 115A Campbell Hall, 1300 University Boulevard, Birmingham, Alabama 35294, rtiwari@cis.uab.edu

**Jason Van Hulse** Florida Atlantic University, Boca Raton, FL 33431, USA, jvanhulse@gmail.com

**B. Vidyareddi** ICT Stream, ABV IIITM Gwalior, India, vidyareddi b@gmail.com

**Chengcui Zhang** Department of Computer and Information Sciences, The University of Alabama at Birmingham, 115A Campbell Hall, 1300 University Boulevard, Birmingham, Alabama 35294, zhang@cis.uab.edu

**Du Zhang** Department of Computer Science, California State University, Sacramento, CA 95819-6021, USA, zhangd@ecs.csus.edu

**Keqi Zhang** International Hurricane Research Center, Florida International University, Miami, FL 33199, USA, zhangk@fiu.edu

**Tian Zhao** Department of Computer Science, University of Wisconsin-Milwaukee, 2200 E. Kenwood Blvd, P.O. Box 413, Milwaukee, WI 53211, USA, tzhao@uwm.edu

# Chapter 1
# PERDURABLE: A Real Time Commit Protocol

**Udai Shanker, B. Vidyareddi, and Anupam Shukla**

## 1.1 Introduction

Real Time Systems (RTS) are those systems, for which, correctness depends not only on the logical properties of the produced results but also on the temporal properties of these results [27]. Typically, RTS are associated with critical applications in which human lives or expensive machineries may be at stake. Examples include telecommunication systems, trading systems, online gaming, chemical plant control, multi point fuel injection system (MPFI), video conferencing, missile guidance system, sensor networks etc. Hence, in such systems, an action performed too late (or too early) or a computation which uses temporally invalid data may be useless and sometimes harmful even if such an action or computation is functionally correct. As RTS continue to evolve, their applications become more and more complex, and often require timely access and predictable processing of massive amounts of data [21]. The database systems especially designed for efficient processing of these types of real time data are referred as distributed real time database system (DRTDBS). Here, data must be extremely reliable and available as any unavailability or extra delay could result in heavy loss. Business transactions being used in these applications in the absence of real time could lead to financial devastations and in worst case cause injuries or deaths [20].

The applications listed here using DRTDBS require distributed transaction executed at more than one site. The transaction can be of read, blind write or update type having timing constraints. A commit protocol ensures that either all the effects of the transaction persist or none of them persist despite the failure of

U. Shanker (✉)
Dept. of CSE, M.M.M. Engg. College, Gorakhpur-273010, India
e-mail: udaigkp@gmail.com

B. Vidyareddi · A. Shukla
ICT Stream, ABV IIITM Gwalior, India
e-mail: vidyareddi_b@gmail.com; dranupamshukla@gmail.com

T. Özyer et al. (eds.), *Recent Trends in Information Reuse and Integration*,
DOI 10.1007/978-3-7091-0738-6_1, © Springer-Verlag/Wien 2012

site or communication link and loss of messages [5]. Hence, the commit processing should add as little overhead as possible to transaction processing. Therefore, design of a better commit protocol is very much important for DRTDBS.

## 1.2   Related Work

A distributed system is one where data are located on several computers that are linked together by a heterogeneous network. The advantages of such system are increased availability of resources, increased reliability and increased execution speed in less time. The coordination of activities among these computers is a complex task and deadlines make them more complex [6]. If a transaction runs across more than one site, it may commit at one site and may fail at another site. Hence, uniform commitment is needed to ensure that all participating sites agree on a final outcome. The two phase commit protocol (2PC) referred to as the Presumed Nothing 2PC protocol (PrN) is the most widely used commit protocol for maintaining atomicity in distributed real time systems [25]. It ensures that sufficient information is force-written on the stable storage to reach a consistent global decision about the transaction. In case of faults, however, it is not possible to provide such guarantees and an exception state is allowed which indicates the violation of the deadline. The blocking nature of 2PC can make the executing-committing conflict a serious problem to the system performance. A number of 2PC variants [13] have been proposed and can be classified into following four groups [2].

- Presumed Abort/Presumed Commit Protocols
- One Phase Commit Protocols [12]
- Group Commit Protocols
- Pre Commit/Optimistic Commit Protocols

Presumed commit (PC) and presumed abort (PA) [14] are based on 2PC. PA, as originally designed and implemented in R*, does not support coordinator migration. [26] have proposed a protocol that allows individual site to unilaterally commit. If later on it is found that the decision is not consistent globally then compensation transactions are executed to rectify errors. The problem with this approach is that many actions are irreversible in nature. Gupta et al. proposed optimistic commit protocol and its variant in [8]. Enhancement has been made in PROMPT [9] commit protocol, which allows executing transactions to borrow data in a controlled manner only from the healthy transactions in their commit phase. However, it does not consider the type of dependencies between two transactions. The impact of buffer space and admission control is also not studied. Another problem with this protocol is that, in case of sequential transaction execution model, the borrower is blocked for sending the WORKDONE message and the next cohort can not be activated at other site for its execution. It will be held up till the lender completes. If its sibling is activated at another site anyway, the cohort at this new site will not get the result of previous site because previous cohort has been blocked for sending of WORKDONE message due to being borrower [24].

Lam et al. proposed deadline-driven conflict resolution (DDCR) protocol which integrates concurrency control and transaction commitment protocol for firm real time transactions [11]. DDCR resolves different transaction conflicts by maintaining three copies of each modified data item (before, after and further) according to the dependency relationship between the lock requester and the lock holder. This not only creates additional workload on the systems but also has priority inversion problem. The serializability of the schedule is ensured by checking the before set and the after sets when a transaction wants to enter the decision phase. The protocol aims to reduce the impact of a committing transaction on the executing transaction which depends on it. The conflict resolution in DDCR is divided into two parts (a) resolving conflicts at the conflict time; and (b) reversing the commit dependency when a transaction, which depends on a committing transaction, wants to enter the decision phase and its deadline is approaching.

If data conflict occurs between the executing and committing transactions, system's performance will be affected. Pang Chung-leung and Lam K. Y. proposed an enhancement in DDCR called the DDCR with similarity (DDCR-S) to resolve the executing-committing conflicts in DRTDBS with mixed requirements of criticality and consistency in transactions [4]. In DDCR-S, conflicts involving transactions with looser consistency requirement, the notion of similarity are adopted so that a higher degree of concurrency can be achieved and at the same time the consistency requirements of the transactions can still be met. The simulation results show that the use of DDCR-S can significantly improve the overall system performance as compared with the original DDCR approach.

Based on PROMPT and DDCR protocols, B. Qin and Y. Liu proposed double space commit (2SC) protocol [15]. They analyzed and categorized all kind of dependencies that may occur due to data access conflicts between the transactions into two types commit dependency and abort dependency. The 2SC protocol allows a non-healthy transaction to lend its held data to the transactions in its commit dependency set. When the prepared transaction aborts, only the transactions in its abort dependency set are aborted and the transactions in its commit dependency set execute as normal. These two properties of the 2SC reduce the data inaccessibility and the priority inversion that is inherent in distributed real-time commit processing. 2SC protocol uses blind write model. Extensive simulation experiments have been performed to compare the performance of 2SC with that of other protocols such as PROMPT and DDCR. The simulation results show that 2SC has the best performance. Furthermore, it is easy to incorporate it in any current concurrency control protocol.

Ramamritham et al. [18] have given three common types of constraints for the execution history of concurrent transactions. The paper [16] extends the constraints and gives a fourth type of constraint. Then the weak commit dependency and abort dependency between transactions, because of data access conflicts, are analyzed. Based on the analysis, an optimistic commit protocol Two-Level Commit (2LC) is proposed, which is specially designed for the distributed real time domain. It allows transactions to optimistically access the locked data in a controlled manner, which reduces the data inaccessibility and priority inversion inherent and undesirable in DRTDBS. Furthermore, if the prepared transaction is aborted, the transactions in its weak commit dependency set will execute as normal according to 2LC. Extensive simulation experiments have been performed to compare the performance of 2LC with that of the base protocols PROMPT and DDCR. The simulation results show that 2LC is effective in reducing the number of missed transaction deadlines. Furthermore, it is easy to be incorporated with the existing concurrency control protocols.

Many existing commit protocols try to improve system performance by allowing a committing cohort to lend its data to an executing cohort, thus reducing data inaccessibility [7]. These protocols block the borrower when it tries to send WORKDONE/PREPARED message thus increasing the transactions commit time. The paper [22] first analyzes all kind of dependencies that may arise due to data access conflicts among executing-committing transactions when a committing cohort is allowed to lend its data to an executing cohort. It then proposes a static two-phase locking and high priority based, write-update type, ideal for fast and timeliness commit protocol i.e. SWIFT. In SWIFT, the execution phase of a cohort is divided into two parts, locking phase and processing phase and then, in place of WORKDONE message, WORKSTARTED message is sent just before the start of processing phase of the cohort. Further, the borrower is allowed to send WORKSTARTED message, if it is only commit dependent on other cohorts instead of being blocked as opposed to [9]. This reduces the time needed for commit processing and is free from cascaded aborts. To ensure non-violation of ACID properties, checking of completion of processing and the removal of dependency of cohort are required before sending the YES-VOTE message. Simulation has been done for main memory resident as well as disk resident databases. Simulation results show that SWIFT improves the system performance in comparison to earlier protocol. The performance of SWIFT is also analyzed for partial read-only optimization, which minimizes intersite message traffic, execute-commit conflicts and log writes resulting in a better response time. The impact of permitting the cohorts of the same transaction to communicate with each other [11] has also been analyzed. The SWIFT commit protocol is beneficial only if the database is main memory resident [22].

The concept of Shadow was proposed by Azer Bestavros. In this approach, we make a copy of original transaction known as Shadow of the original transaction. The original transaction continues to run uninterrupted, while the shadow transaction is started on a different processor and allowed to run concurrently. In other words, two versions of the same transaction are allowed to run in parallel, each one being at a different point of its execution. Obviously, only one of these two transactions will be allowed to commit; the other will be aborted. The concept of Shadow was first used for commit process in shadow PROMPT [9] which is a modification of PROMPT commit protocol. Here, a cohort forks off a replica of the transaction, called a shadow, independent of the type of dependency. Thus, in shadow PROMPT, shadow is created in case of Abort as well as Commit dependency. However in case of Commit dependency, there is no need to create a shadow because an abort dependent transaction will complete its execution irrespective of final outcome of lender transaction. Thus, in Shadow PROMPT [9], unnecessary shadow is created. The problem has been solved in DSS-SWIFT commit protocol [23], where a shadow is created whenever an abort dependent borrower cohort uses dirty value of a data item updated by an uncommitted lender cohort. Here, each site maintains abort dependency set (ADSS), which is the set of shadow of those cohorts that are abort dependent on lender. Also, the DSS-SWIFT creates the non beneficial shadows in some cases. The SPEEDITY commit protocol [1] reduces non beneficial shadows that are created in DSS-SWIFT by the introduction of $T_{shadow\ Creatlon\ Time}$ where the abort dependent cohort having deadline beyond a specific value ($T_{shadow\ Creatlon\ Time}$) can only forks off a replica of itself called shadow, whenever it borrows dirty value of a data item.

The SWIFT commit protocol typically uses S2PL mechanism to acquire locks [24]. Due to several conditional statements in the code of transaction [3], some of the locked data items may not be used at all. Hence, here in our work, these unused locked data items are being unlocked immediately after completion of processing phase. Also, a lending transaction can lend its dirty data to more than one borrower cohort in case of write-read conflicts by creating a single shadow only.

The remainder of this chapter is organized as follows. Section 1.3 introduces the distributed real time database system model. Section 1.4 presents PERDURABLE and its pseudo code. Section 1.5 discusses the simulation results and Sect. 1.6, finally concludes the chapter.

## 1.3  DRTDBS Model

In distributed database system model, the global database is partitioned into a collection of local databases stored at different sites. A communication network interconnects the sites. There is no global shared memory in the system, and all sites communicate via message exchange over the communication network. We assume that the transactions are firm real time type. Each transaction in this model exists in the form of a coordinator that executes at the originating site of the transaction and a collection of cohorts that execute at various sites, where the required data items reside. If there is any local data in the access list of the transaction, one cohort is executed locally. Before accessing a data item, the cohort needs to obtain lock on data items. Sharing of the data items in conflicting modes creates dependencies among the conflicting local transactions/cohorts, and constraints their commit order. We also assume that:

- The processing of a transaction requires the use of CPU and data items located at local site or remote site.
- Arrival of transactions at a site is independent of the arrivals at other sites and uses Poisson distribution.

- Each cohort makes read and update accesses.
- Each transaction pre-declares its read-set (set of data items that the transaction will only read) and write-set (set of data items that the transaction will write).
- Cohorts are executed in parallel.
- A distributed real time transaction is said to commit, if the coordinator has reached to the commit decision before the expiry of the deadline at its site. This definition applies irrespective of whether cohorts have also received and recorded the commit decision by the deadlines or not.
- Studies have been made for both main memory resident and disk resident database.
- Communication delay considered is either 0 or 100 ms.
- In case of disk resident database, buffer space is sufficiently large to allow the retention of data updates until commit time.
- The updating of data items is made in transaction own memory rather than in place updating.

## 1.4  PERDURABLE Protocol

In conventional databases, the execution time and data to be accessed is not known at the time of transaction submission. However, for most well defined real time transactions, an estimate of execution time and the locks required by the transaction are known prior. Actually, many real-time concurrency control protocols have made these assumptions or even stronger assumptions. For example, priority ceiling protocols [17, 19] explicitly assume the availability of such information. The real-time S2PL (RT-S2PL) protocols do possess desirable features making them suitable for RTDBS, especially for distributed real-time database systems (DRTDBS), in which remote locking is required and distributed deadlock is possible. At the same time, the problem of prior knowledge on the required data objects of a transaction is easy to address in DRTDBS as it is generally agreed that the behavior and the data items to be accessed by real-time transactions, especially hard real-time transactions, are much more well-defined and predictable.

In S2PL, each transaction pre-declares it's read-set and write-set. If a transaction contains conditional statements, a pessimistic approach is taken and data used in all branches are included and locked irrespective of their use. Here, all the locked data items will never be used. Hence, all the locked data items by the transaction, which are still unused after completion of processing of transaction, can be unlocked after end of processing phase to reduce data contention. This is achieved by relaxing the notion of 2PL but still maintaining the database consistency. A major aim in developing a database system is to allow several users to access shared data concurrently. Concurrent access is easier if users are only reading data. Also, a lending transaction can lend its dirty data to more than one borrower cohort in read mode in case of write-read conflicts by creating only a single shadow. In case of commit of lender, the entire borrower cohorts will run independently without any conflict because of being read type. However, in case of abort of lender, the single shadow will be used for the survival of all the borrower cohorts. Based on these concepts, we introduce our new protocol which is enhanced version of base protocol SWIFT.

Sharing of data items in conflicting modes creates dependencies among conflicting transactions and constraints their commit order. We assume that a cohort requests an update lock if it wants to update a data item x. The prepared cohorts called as lenders lend uncommitted data to concurrently executing transactions known as borrower. Here, the borrower is further divided into two categories.

- Commit Dependent Borrower
- Abort Dependent Borrower

Therefore, modified definitions of dependencies used in this chapter are given below.

- *Commit dependency (CD)*. If a transaction $T_2$ updates a data items read by another transaction $T_1$ a commit dependency is created from $T_2$ to $T_1$. Here, $T_2$ is called as commit dependent borrower and is not allowed to commit until $T_1$ commits.
- *Abort dependency (AD)*. If $T_2$ reads/updates an uncommitted data item updated by $T_1$, an abort dependency is created from $T_2$ to $T_1$. Here, $T_2$ is called as abort dependent borrower. $T_2$ aborts, if $T_1$ aborts and $T_2$ is not allowed to commit before $T_1$.

Each transaction/cohort $T_i$ that lends its data while in prepared state to an executing transaction/cohort $T_j$, maintains two sets.

- Commit Dependency Set CDS $(T_i)$: set of commit dependent borrower $T_j$, that are borrowed dirty data from lender $T_i$.
- Abort Dependency Set ADS $(T_i)$: the set of abort dependent borrower $T_j$ that are borrowed dirty data from lender $T_i$.

These dependencies are required to maintain the ACID properties of the transaction. Each lender is associated with a health factor defined as follows.

$$\text{HF (health factor)} = \text{TimeLeft/MinTime}$$

where TimeLeft is the time left until the transaction's deadline, and MinTime is the minimum time required for commit processing. The health factor is computed at the time when the coordinator is ready to send the YES-VOTE messages. MinHF is the threshold that allows the data held by committing transaction to be accessed. The variable MinHF is the key factor to influence the performance of the protocol. In our experiments, we have taken MinHF as 1.2, the value of MinHF used in PROMPT [9]. Here, each transaction also maintains shadow set (DS), which is the set of shadows of those cohorts that are abort dependent on lender.

DS $(T_i)$: Shadow of those cohorts which are in the abort dependency set of $T_i$.

Hence, if $T_2$ is abort dependent on $T_1$ then Shadow of $T_2$, $S_2$ is created in DS ($T_1$). If $T_2$ accesses a data item already locked by $T_1$, the lock manager processes the data item accesses in conflicting mode as follows.

if($T_1$ has not lent the data item)
```
    {
        If (T₂ CD T₁)
            {
            CDS (T₁) = CDS (T₁ {T2};
            T₂ is granted Update lock;
            }
        else if ((T₂ AD T₁ AND (HF(T₁ > MinHF))
            {
            ADS (T₁) = ADS (T₁ {T2};
            T₂ is granted the requested lock;
            Add shadow of T₂ in DS(S);
            }
            else T₂ will be blocked;
    }
else
        {
        if (T₂ is read type)
                {
                if ((T₂ AD T₁ AND (HF(T₁ > MinHF))
                        {
                                ADS (T₁ = ADS (T₁ {T2};
                                T₂ is granted the requested lock;
                        }
                else T₂ will be blocked;
                }
        }
```

## 1.4.1  Type of Dependency in Different Cases of Data Conflicts

When data conflicts occur, there are three possible cases of conflicts.

**Case 1: Read-Write Conflict**

If $T_2$ requests update-lock while $T_1$ is holding a read-lock, a commit dependency is defined from $T_2$ to $T_1$. The transaction id of $T_2$ is added to the CDS ($T_1$). Then, $T_2$ acquires the update-lock.

**Case 2: Write-Write Conflict.**

If both locks are update-locks and HF($T_1 >$ MinHF, an abort dependency is defined from $T_2$ to $T_1$. The transaction id of $T_2$ is added to ADS ($T_1$). $T_2$ acquires the update-lock; otherwise, $T_2$ is blocked. In case of getting update-lock, the shadow of $T_2$ is created and added to DS ($T_1$).

**Case 3: Write-Read Conflict**

When data conflicts occur, there are two possible situations of conflict.

*Situation 1*: $T_1$ *has not lent the data items.*

If $T_2$ requests a read-lock while $T_1$ is holding an update-lock and HF ($T_1 >$ MinHF), an abort dependency is defined from $T_2$ to $T_1$. The transaction id of $T_2$ is added to ADS ($T_1$). $T_2$ acquires the read-lock; otherwise, $T_2$ is blocked. In case of getting read-lock, the shadow of $T_2$ is created and added to DS ($T_1$).

*Situation 2*: $T_1$ *has lent the data items.*

If $T_2$ requests a read-lock while $T_1$ is holding an update-lock and HF ($T_1 >$ MinHF), an abort dependency is defined from $T_2$ to $T_1$. The transaction id of $T_2$ is added to ADS ($T_1$). $T_2$ acquires the read-lock; otherwise, $T_2$ is blocked.

## 1.4.2 Mechanics of Interaction Between Lender and Borrower Cohorts

If $T_2$ accesses a data item already locked by $T_b$ one of the following three scenarios may arise.

**Scenario 1:** $T_1$ receives decision before $T_2$ has completed its local data processing:

If the global decision is to commit, $T_1$ commits. All cohorts in ADS ($T_1$) & CDS ($T_1$) will execute as usual;

Sets of ADS ($T_1$), CDS ($T_1$) & DS ($T_1$) will be deleted;

If the global decision is to abort, $T_1$ aborts, cohorts in the dependency set of $T_1$ will execute as follows:

Shadows of all cohorts abort dependent on $T_1$ in DS(S) will be activated and sends YES-VOTE piggy bagged with the new result to their coordinator, only if they can complete execution; otherwise, discarded; Transactions in CDS ($T_1$) will execute normally; Delete Set ADS ($T_1$), DS ($T_1$) and CDS ($T_1$) CDS ($T_1$);

**Scenario 2:** $T_2$ is about to start processing phase after getting all its locks before $T_1$ receives global decision.

$T_2$ sends WORKSTARTED message to its master and start processing phase. After end of processing phase, all the unused data items are released.

**Scenario 3:** $T_2$ **aborts before,** $T_1$ **receives decision**

In this situation, $T_2$'s updates are undone and $T_2$ is removed from the dependency set of $T_1$.

## 1.4.3 Algorithm

On the basis of above discussions, the complete pseudo code of the protocol is given below.

```
if (T₁ receives global decision before T₂ ends processing)
    {
        One: if (T₁'s global decision is to commit)
```

{
    $T_1$ enters in the decision phase;
    Delete Set ADS $(T_1)$, DS $(T_1)$ and CDS $(T_1)$;
}
else if ($T_1$'s global decision is to abort)
{
$T_1$ aborts;
Transactions in CDS $(T_1)$ will execute as usual.
For all the abort dependent cohorts
{
if (shadow)
{
if (Shadow of cohort can complete its execution)
    Execute Cohorts Shadow in DS $(T_1)$ and send YES-VOTE piggy
    bagged with the new result;
else discard shadow;
}
}
Delete Set ADS $(T_1)$, DS $(T_1)$ and CDS $(T_1)$;
}
}
else
{
    if ($T_2$ aborted by higher transaction before $T_1$ receives decision OR $T_2$
    expires its deadline)
    {
    Undo the computation of $T_2$;
    Abort $T_2$;
    Delete $T_2$ from ADS $(T_1)$, CDS $(T_1)$;
    if (shadow)
    Delete $T_2$ from DS $(T_1)$;
    }
    else if ($T_2$ ends executing phase before $T_1$ receives global decision)
        $T_2$ sends WORKSTARTED message & start processing phase. After
        end of processing phase, all the unused data items are released.
}

## Main Contributions

- A lender already lent data to abort dependent borrower cohort can lend the same data item in read mode to more than one borrower cohort.
- Locked data items, which are not used during processing phase, can be released even before commit phase.

## Correctness

To maintain ACID property of database, cohort sends the YES-VOTE in response to its coordinator's VOTE-REQ message only when its dependencies are removed and it has finished its processing [20, 22]. An atomic commitment protocol ensures that the cohorts in a transaction agree on the outcome, i.e. ABORT or COMMIT. Each participant vote, YES or NO, on whether they can guarantee the local ACID properties of the transaction. All participants have a right to *veto* the transaction, thus causing it to abort. The proposed protocol has these properties:

**AC1** ⟨*uniform agreement*⟩ All cohorts that decide reach the same decision.

**AC2** ⟨*integrity*⟩ A cohort cannot reverse its decision after it has reached one.

**AC3** ⟨*uniform validity*⟩ COMMIT can only be reached if *all* cohort voted YES.

**AC4** ⟨*non-triviality*⟩ If there are no failures and no cohort voted NO, then the decision will be to COMMIT.

**AC5** ⟨*termination*⟩ Every cohort eventually decides.

## 1.5  Performance Measures and Evaluation

Transactions are characterized by the amount of data they access, the nature of the accesses (read or write), the priority assigned to them, their deadline and so on. Since no meaningful comparison of the algorithms from the different camps is possible and even within the single camp, it is often difficult to compare different algorithms; they were chosen to be in accordance with those used in earlier published simulation studies. The default values of different parameters for simulation experiments are same as in [27] and are given in Table 1.1. The transaction arrival process is assumed to be Poisson with parameter inter-arrival time. Also, to evaluate the Cubic effect of transaction length, range was taken from 3–20 being much more variable in nature which is really a difficult task. Transaction have extensive data resource requirements which are either pre-declared before execution or more often acquired at the record level as the transaction progress. In conventional database systems, the major performance measures are mean system throughput and mean system response time. On the contrary, minimizing the number of missed deadlines is the main concern in DRTDBS. Research work on real time locking is mainly based on D2PL. But, there is no strong evidence till now that the performance of real-time D2PL based protocols is better than those based on

**Table 1.1**  Default values for model parameters

| Parameters | Meaning | Default setting |
|---|---|---|
| $N_{site}$ | Number of site | 4 |
| AR | Transaction arrival rate | 4 Transactions/ Second |
| $T_{com}$ | Communication delay | 100 ms (Constant) |
| SF | Slack factor | 1–4 (Uniform distribution) |
| $N_{oper}$ | No. of operations in a transaction | 3–20 (Uniform distribution) |
| PageCPU | CPU page processing time | 5 ms |
| PageDisk | Disk page processing time | 20 ms |
| DBsize | Database size | 200 Data objects/Site |
| $P_{write}$ | Write operation probability | 0.60 |

S2PL. As a result of the better defined nature of real time transactions, it is not unreasonable to assume that the locking information of transactions is known before processing. Hence, the concurrency control scheme used is static two phase locking with higher priority. Miss Percentage (MP) is the primary performance measure used in the experiments and is defined as the percentage of input transactions that the system is unable to complete on or before their deadlines [9].

Since mathematical modeling makes several crude approximations so it reasonable required verifying the performance of transaction behavior with help of discrete event simulation. Since there were no practical benchmark programs available in the market or with research communities to evaluate the performance of protocols and algorithms, an event driven based simulator was written in C language. In our simulation, a small database (200 data items) is used to create high data contention environment. This helps us in understanding the interaction among the different policies [10]. Transactions make requests for data items and concurrency control is implemented at data item level. A small database means that degree of data contention in the system can easily be controlled by the sizes of the transactions. A small database also allows us to study the effect of hot-spots, in which a small part of the database is accessed very frequently by most of the transactions. However, we have not considered the hot spot issue. The write operation probability $P_{write}$ parameter determines the probability that a transaction operation is a write. This is very important as real time database systems have applications in highly critical environment where the performance of system is important especially in highly loaded environment. For each set of experiment, the final results are calculated as an average of 10 independents runs. In each run, 100,000 transactions are initiated.

### 1.5.1   Simulation Results

In this section we present the performance results from our experiments comparing the various commit protocols using the simulation model. The model consists of a non replicated firm deadline database system that is distributed over a set of sites connected by a network. Simulation was done for both the main memory resident and the disk resident databases at communication delay of 0ms & 100ms. Also, the write probability is 0.60% and approximately 20%–25% data items are not being use by the transaction during processing phase. The settings of the workload parameters were chosen with the objective of having significant data contention thus helping to bring out the performance differences between the various commit protocols. Our experiments consider only data contention because resource contention can usually be addressed by purchasing more and/or fast resources while it is not easily possible with data contention.

We compared PERDURABLE with PROMPT, 2SC and SWIFT in this experiment. Figures 1.1–1.5 show the Miss Percent behavior under normal and heavy load conditions with/without communication delay for main memory resident database system and disk resident database system. In these graphs, we first observe that there is a noticeable difference between the performances of the various commit protocols throughout loading range. This is due to increased availability of data items and allowing multiple cohort/transactions simultaneously with single backup shadow. Here, cohorts are able to obtain lock sooner than base protocols. Hence, the cohorts, who are being aborted due to deadline expiry, tend to make further progress. In this way, we can reduce the data contention and increase system throughput by allowing more cohort/transactions to be carried out at a time.

Sharing of data items in conflicting mode creates dependencies among conflicting transactions and constraints their commit order. It is assumed that a cohort requests an update lock if it wants to update a data item $x$. The prepared cohorts, called as lenders, lend uncommitted data to concurrently executing transactions known as borrower. If a cohort fork off a replica of the transaction, it is called as shadow. The original incarnation of the transaction continues its execution, while the shadow is blocked after finishing its execution. If the lender finally commits, the borrower continues its on-going execution and the shadow is discarded; otherwise, borrower is aborted due to abort of lender and shadow is activated. Here, we maintain a single backup shadow for multiple transactions compared to creating a separate shadow for each cohort/transaction in previous real time commit protocols with still limiting the transaction abort chain to one only and reducing the data inaccessibility. Also, the resources in creating these backup shadows for each cohort are saved. The overhead incurred due to shadow creation is also minimizes. This combines the advantages of both pessimistic as well as optimistic approaches. Here, transaction is restarted from the intermediate conflict point (not from beginning) giving it more chance to meet its deadline. Although transaction class pre-analysis is considered too restrictive in many conventional environments, it is perfectly valid in time critical environments in which all application code is pre-analyzed. Hence, by releasing the locked data items which are unused at the end of processing phase, we are making them available all through commit phase which is usually one-third of total processing time. In fact, the behaviour of all the protocol at normal load is almost similar. However, this is basically beneficial in overload situation and also this is particularly true for multiprocessor environment, in which the data not the CPU is the limiting resource. With this approach, the proposed protocol overcomes the demerits of S2PL to some extents by not locking the data items for longer period with deadlock free environment in overload situation.

A prototype system was developed over which we have implemented our protocol and other protocols (PROMPT, 2SC, SWIFT, etc.) to view the performance of these protocols in the real life applications also. For testing the system, we have used total 15 nodes (2.2 Core 2 Due processor, 1 GB RAM, 160 GB HDD, Windows-XP and three servers). Still, the PERDURABLE commit protocol provides a performance that is significantly better than other commit protocols and be implemented easily by integrated with 2PC protocol by allowing transactions to access uncommitted data-item held by prepared transactions in the optimistic belief that this data-item will eventually be committed.

## Main Memory Resident Database

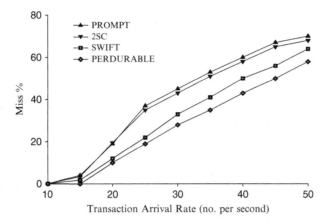

**Fig. 1.1** Miss % with (RC+DC) at communication delay = 0 ms normal and heavy load

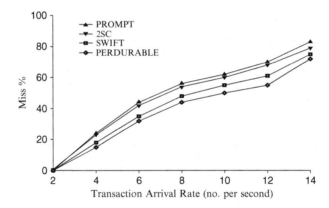

**Fig. 1.2** Miss % with (RC+DC) at communication delay = 100 ms normal and heavy load

**Disk Resident Database**

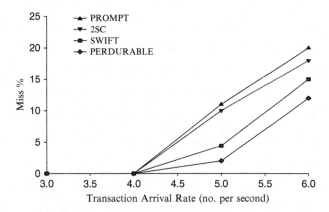

**Fig. 1.3**  Miss % with (RC+DC) at communication delay = 0 ms normal load

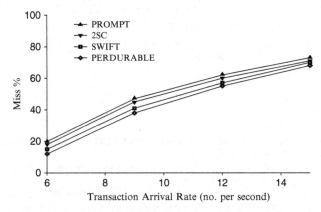

**Fig. 1.4**  Miss % with (RC+DC) at communication delay = 0 ms heavy load

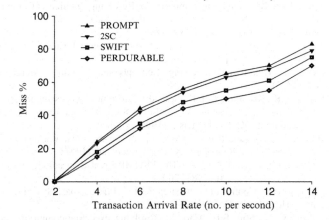

**Fig. 1.5**  Miss % with (RC+DC) at communication delay = 100 ms normal and heavy load

## 1.6    Conclusion

While real time scheduling of transactions are widely studied in a real time databases, the issues of conditional statements and shadowing has not yet been addressed more. Motivated by this observation, we introduced a new commit protocol PERDURABLE, where a lender already lent data to abort dependent borrower cohort can lend the same data item in read mode to more than one borrower cohort and locked data items, which are not used during processing phase, can be released even before commit phase. To the best of our knowledge, such study has not been attempted before. The main conclusion of our study can be summarized as follow.

We compared the proposed protocol with base protocols. Our performance results clearly establish the superiority of PERDURABLE significantly over the standard algorithms for all the workloads and system configurations considered in this study. Here, the new notion has solved the problem of transactions holding locks for a longer time with static locking, because transactions must acquire all their locks before they can begin execution.

As future work, it is desirable to study the performance of the proposed protocol in real environment more extensively.

## References

1. Agrawal, S., Singh, A.N., Anand, A., Shanker, U.: SPEEDITY-A real time commit protocol. Int. J. Comput. Appl. **3**(13) (2010)
2. Attaluri, G.K., Kenneth, S.: The presumed-either two-phase commit protocol. IEEE Trans. Knowl. Data. Eng. **14**(5), 1190–1196 (2002)
3. Buchman, A.P., McCarthy, D.R., Hsu, M., Dayal, U.: Time Critical Database Scheduling: A Framework for Integrating Real Time Scheduling and Concurrency Control. IEEE Fifth International Conference on Data Engineering, pp. 470–480, Los Angeles, California (1989)
4. Chung-Leung, P., Lam, K.Y.: On Using Similarity for Resolving Conflicts at Commit in Mixed Distributed Real-Time Databases. Proceedings of the 5th International Conference on Real-Time Computing Systems and Applications (1998)
5. Gray, J.: Notes on database operating systems: an advanced course. Lecture Notes in Computer Science, Vol. 60, pp. 397–405, Springer, Verlag (1978)
6. Gray, J., Reuter, A.: Transaction Processing: Concepts and Technique. Morgan Kaufman, San Mateo, CA (1993)
7. Gupta, R., Haritsa, J.R., Ramamritham, K., Seshadri, S.: Commit processing in distributed real time database systems. In: Proceedings of Real-time Systems Symposium, IEEE Computer Society Press, Washington DC, San Francisco (1996)
8. Gupta, R., Haritsa, J.R., Ramamritham, K.: More optimistic about real time distributed commit processing. In: Proceedings of Real-Time Systems Symposium (1997)
9. Haritsa, J., Ramamritham, K., Gupta, R.: The PROMPT real time commit protocol. IEEE Trans. Parallel Distr. Syst. **11**(2), 160–181 (2000)
10. Lam, K.-Y.: Concurrency Control in Distributed Real – Time Database Systems, PhD Thesis, City University of Hong Kong, Hong Kong (1994)
11. Lam, K.Y., Pang, C.-L., Son, S.H., Cao, J.: Resolving executing-committing conflicts in distributed real-time database systems. Comput. J. **42**(8), 674–692 (1999)

12. Lee, I., Yeom, H.Y.: A Single Phase Distributed Commit Protocol for Main Memory Database Systems. 16th International Parallel & Distributed Processing Symposium (IPDPS 2002), Ft. Lauderdale, Florida, USA (2002)
13. Misikangas, P.: 2PL and its variants. Seminar on Real-Time Systems, Department of Computer Science, University of Helsinki (1997)
14. Mohan, C., Lindsay, B., Obermarck, R.: Transaction management in the R* distributed database management system. ACM Trans. Database Syst. 11(4), 378–396 (1986)
15. Qin, B., Liu, Y.: High performance distributed real time commit protocol. J. Syst. Software, Elsevier Science Inc, 1–8 (2003)
16. Qin, B., Liu, Y., Yang. J.C.: A commit strategy for distributed real-time transaction. J. Comput. Sci. Tech. 18(5), 626–631 (2003)
17. Rajkumar, R.: Task Synchronization in real time systems. Ph.D. Thesis, Carnegie-Mellon University (1989)
18. Ramamritham, K., Chrysanthis, P.K.: A taxonomy of correctness criteria in database applications. VLDB Journal, 5, 85–97 (1996)
19. Sha, L., Rajkumar, R., Lehoczky, J.P.: Concurrency control for distributed real time data bases. ACM SIGMOD Record 17(1), 82–98 (1988)
20. Shanker, U., Misra, M., Sarje, A.K.: Dependency sensitive distributed commit protocol. In: Proceedings of the 8$^{th}$ International Conference on Information Technology (CIT 05), pp. 41–46, Bhubaneswar, India (2005)
21. Shanker, U., Misra, M., Sarje, A.K.: Some performance issues in distributed real time database systems. In: Proceedings of the VLDB PhD Workshop (2006), the Convention and Exhibition Center (COEX), Seoul, Korea (2006)
22. Shanker, U., Misra, M., Sarje, A.K.: SWIFT-A new real time commit protocol. Int. J. Distrib. Parallel Databases 20(1), 29–56 (2006)
23. Shanker, U., Misra, M., Sarje, A.K., Shisondia, R.: Dependency sensitive shadow SWIFT. In: Proceedings of the 10$^{th}$ International Database Applications and Engineering Symposium (IDEAS 06), pp. 373–276, Delhi, India (2006)
24. Shanker, U., Misra, M., Sarje, A.K.: Distributed real time database systems: background and literature review. Int. J. Distrib. Parallel Databases 23(2), 127–149 (2008)
25. Skeen, D.: Nonblocking commit protocols. In: Proceedings of the 1981 ACM SIGMOD International Conference on Management of Data, pp. 133–142, New York (1981)
26. Soparkar, N., Levy, E.H., Korth, F., Silberschatz, A.: Adaptive Commitment for Real-Time Distributed Transaction. Technical Report TR-92-15, Department of Computer Science, University of Texax, Austin (1992)
27. Ulusoy, O.: Concurrency Control in Real Time Database Systems. PhD Thesis, Department of Computer Science, University of Illinois Urbana-Champaign (1992)

# Chapter 2
# Dimension Table Selection Strategies to Referential Partition a Fact Table of Relational Data Warehouses

**Ladjel Bellatreche**

## 2.1 Introduction

Enterprise wide data warehouses are becoming increasingly adopted as the main source and underlying infrastructure for business intelligence (BI) solutions. Note that a data warehouse can be viewed as an *integration system*, where data sources are duplicated in the same repository. Data warehouses are designed to handle the queries required to discover trends and critical factors are called Online Analytical Processing (OLAP) systems. Examples of an OLAP query are: *Amazon* (www.amazon.com) company analyzes purchases by its customers to come up with an individual screen with products of likely interest to the customer. Analysts at *Wal-Mart* (www.walmart.com) look for items with increasing sales in some city. Star schemes or their variants are usually used to model warehouse applications. They are composed of thousand of dimension tables and multiple fact tables [15, 18]. Figure 2.1 shows an example of star schema of the widely-known data warehouse benchmark *APB-1 release II* [21]. Here, the fact table *Sales* is joint to the following four dimension tables: *Product, Customer, Time, Channel. Star queries* are typically executed against the warehouse. Queries running on such applications contain a large number of costly joins, selections and aggregations. They are called mega queries [24]. To optimize these queries, the use of advanced optimization techniques is necessary.

By analyzing the most important optimization techniques studied in the literature and supported by the most important commercial Database Management Systems (DBMS) like Oracle, SQL Server, DB2, etc. we propose to classify them into two main categories: (1) *optimization techniques selected during the creation of data warehouses* and (2) *optimization techniques selected after the creation of the warehouses*. The decision of choosing techniques in the first category is

L. Bellatreche (✉)
LISI/ENSMA – Poitiers University, 1, avenue Clément Ader, 86960 Futuroscope, France
e-mail: bellatreche@ensma.fr

T. Özyer et al. (eds.), *Recent Trends in Information Reuse and Integration*,
DOI 10.1007/978-3-7091-0738-6_2, © Springer-Verlag/Wien 2012

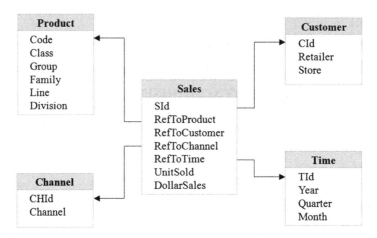

**Fig. 2.1** Logical schema of the data warehouse benchmark *APB-1 release II*

taken before creating the data warehouse. Horizontal partitioning of tables and parallel processing are two examples of the first category. Techniques belonging to the second category are selected during the exploitation of the data warehouse. Examples of these techniques are materialized views and indexes, data compression, etc. Based on this classification, we note that the use of optimization techniques belonging to the first category is more sensitive compared to those belonging to the second one, since the decision of using them is usually taken at the beginning stage of the data warehouse development. To illustrate this sensitivity, imagine the following two scenarios: (1) a database administrator (DBA) decides to partition her/his warehouse using a native data definition language (DDL) proposed by his/her favorite DBMS, later on, DBA realizes that data partitioning is not well adapted for his/her data warehouse. Therefore, it will be costly and time consuming to reconstitute the initial warehouse from the partitions. (2) DBA selects indexes or materialized views (techniques belonging to the second category) and if she/he identifies their insufficiencies to optimize queries, she/he can easily drop or replace them by other optimization techniques. This sensitivity and the carefulness of optimization techniques belonging to the first category motivate us to study one of them, known as referential horizontal partitioning.

Horizontal partitioning is one of important aspect of physical design [23]. It is a divide-and-conquer approach that improves query performance, operational scalability, and the management of ever-increasing amounts of data [28]. It divides tables/views into disjoint sets of rows that are physically stored and accessed separately. Unlike indexes and materialized views, it does not replicate data, thereby reducing storage requirement and minimizing maintenance overhead. Most of today's DBMSs (*Oracle, DB2, SQL Server, PostgreSQL, Sybase* and *MySQL*) offer native Data Definition Language support for creating partitioned tables (*Create Table ... Partition*) and manipulating horizontal partitions (*Merge* and *Split*) [9, 20, 23]. Partitioning is not restricted to data tables. Indexes may also be partitioned. A local index may be constructed so that it reflects the structure of an underlying table [23]. Notice that when horizontal partitioning is applied on materialized views and indexes, it will be considered as a technique of the second category. Its classification depends on the nature of the used access method (table, materialized view or index).

Horizontal partitioning improves performance of queries by the mean of pruning mechanism that reduces the amount of data retrieved from the disk. When a partitioned table is queried, the database optimizer eliminates non relevant partitions by looking the partition information from the table definition, and maps that onto the Where clause predicates of the query. Two main pruning are possible: partition pruning and partition-wise join pruning [9]. The first one prunes out partitions based on the partitioning key predicate. It is used when only one table is partitioned. The second one prunes partitions participating in a join operation. It is applicable when the partitioning key is same in the two tables.

Two main types of horizontal partitioning exist [5]: mono table partitioning and table-dependent partitioning. In the mono table partitioning, a table is partitioned using its own attributes. It is quite similar to the primary horizontal partitioning proposed in traditional databases [22]. Several modes exist to support mono table partitioning: *Range, List, Hash, Round Robin* (supported by Sybase), *Composite* (Range-Range, Range-List, List-List), etc.), Virtual Column partitioning recently proposed by *Oracle11G*, etc. Mono table partitioning may be used to optimize selections, especially when partitioning key matches with their attributes (partition pruning). In the data warehouse context, it is well adapted for dimension tables. In table-dependent partitioning, a table inherits the partitioning characteristics from other table. For instance a fact table may be partition based on the fragmentation schemes of dimension tables. This partitioning is feasible if a parent-child relationship among these tables exists [7, 9]. It is quite similar to the derived horizontal partitioning proposed in 1980s in the context of distributed databases [7]. Two main implementations of this partitioning are possible: native referential partitioning and simulated referential partitioning. The native referential partitioning is recently supported by Oracle11G to equi-partition tables connected by a parent child referential constraint. A native DDL is given to perform this partitioning [9] (*Create Table ... Partition by Reference ...*). It optimizes selections and joins simultaneously. Therefore, it is well adapted for star join queries.

Simulated referential partitioning is a manual implementation of referential partitioning, i.e., it simulates it using the mono table partitioning mode. This simulation may be done as follows: (1) a dimension table is first horizontally partitioned using its primary key, then the fact table is decomposed based on its foreign key referencing that dimension table. This partitioning has been used for designing parallel data warehouses [10], but it is not well adapted for optimizing star join queries [9]. Figure 2.2 classifies the proposed partitioning modes.

By analyzing the most important works done on referential partitioning in the data warehousing environment, we draw the two following points: (a) most of proposed algorithms control the number of generated fragments of the fact tables in order to facilitate the manageability of the data warehouse [2, 16, 25] and

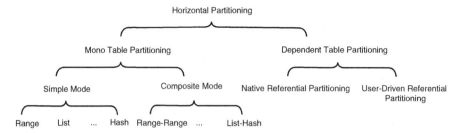

**Fig. 2.2** A classification of the proposed partitioning modes

(b) the fragmentation algorithms start from a bag containing all selection predicates involved by the queries. Therefore, dimension tables referenced by these predicates have the same chance to be used to referential partition a fact table of a warehouse. This situation is not always the best choice to optimize queries. This finding is based on the following observation: To execute a star join query involving several tables in a non partitioned data warehouse, the query optimizer first shall establish an order of joins [26]. This order should reduce the size of intermediate results. To execute the same query on a horizontally partitioned warehouse, the query optimizer will do the same task as in the non partitioned scenario. The join order is somehow partially imposed by the fragmentation process, since fragments of the fact table will be first joined with the fragments of a dimension table used in the referential partitioning process in order to perform partition-wise join pruning. Based on this analysis, we notice that the choice of dimension table(s) used to partition the fact table is a crucial performance issue that should be addressed. Contrary to the existing horizontal partitioning approaches that consider all dimension tables involved by the queries, we propose to identify relevant dimension table(s) and then generate fragmentation schemes using existing algorithms. This partitioning manner will reduce the complexity of the partitioning problem which is known as a NP-complete problem [3], since selection predicates belonging to non desirable dimension table(s) will be discarded. Note that the complexity of horizontal partitioning is proportional to the number of predicates [22].

For data warehouse applications managing a reasonable number of dimension tables, DBA may identify manually dimension tables based on her/his experience, but for BI applications, where hundred of dimension tables are used, it will be hard to do it manually. Therefore, the development of automatic techniques offering DBA the possibility to select dimension tables to partition her/his fact table is necessary. In this paper, we propose a comprehensive procedure for automatically selecting dimension table(s) candidate to referential partition a fact table.

This paper is divided in six sections: Section 2.2 reviews the existing works on horizontal partitioning in traditional databases and data warehouses. Section 2.3 proposes selection strategies of dimension table(s) to support referential horizontal partitioning. Section 2.4 gives the experimental results done on the data set of the APB-1 benchmark using a mathematical cost model and on Oracle11. Section 2.5, a tool supporting our proposal is given. It is connected to *Oracle11G* DBMS. Section 2.6 concludes the paper by summarizing the main results and suggesting future work.

## 2.2 Related Work

Many research works dealing with horizontal partitioning problem were proposed in traditional databases and data warehouses. In the traditional database environment, researchers concentrate their efforts on proposing algorithms for the mono table partitioning. These algorithms are guided by a set of selection predicates of queries involving the table to be partitioned. A selection predicate is defined as follows: *attribute θ value*, where *attribute* belongs to the table to be fragmented, $θ \in \{>, >, =, \leq, \geq\}$ and *value* ∈ domain of attribute. In [3], a classification of these algorithms is proposed: *minterm generation-based approach* [22], *affinity-based approach* [13] and *cost-based approach* [4].

Online analysis applications characterized by their complex OLAP queries motivate the data warehouse community to propose table-dependent horizontal partitioning methodologies and algorithms. [19] proposed a methodology to partition a relational data warehouse in a distributed environment, where the fact table is referentially partitioned based on queries defined on all dimension tables. [17] proposed a fragmentation methodology for a multidimensional warehouse, where the global hypercube is divided into sub cubes. Each one contains a sub set of data. This process is defined by slice and dice operations (similar to the selection and the projection in relational databases). Relevant dimensions to partition the hyper cube are chosen manually. In [3], a methodology and algorithms dealing with partitioning problem in relational warehouses are given. In this work, the authors propose a fragmentation methodology of the fact table based on the fragmentation schemas of dimension tables. Each dimension table involved in any selection predicate is a candidate to referential partition the fact table. Three algorithms selecting fragmentation schemes reducing query processing cost were proposed [3]: *genetic*, *simulated annealing* and *hill climbing*. These algorithms control of generated fragments of a fact table. The main characteristic of these algorithms is that they use a cost model to evaluate the quality of the obtained fragmentation schemes. In [5], a *preliminary work* on selecting dimension table(s) to partition a relational fact table is presented.

In [16], an approach for fragmenting XML data warehouses is proposed. It is characterized by three main steps: (1) extraction of selection predicates from the query workload, (2) predicate clustering and (3) fragment construction with respect to predicate clusters. To form clusters of predicates, the authors proposed to use the k-means technique. As in [3], this approach controls the number of fragments. [23] showed the need of integrating horizontal and vertical partitioning in physical database design, but they did not present algorithm to support referential partitioning of a data warehouse.

To summarize, most important research efforts on horizontal partitioning in the traditional database are focused on the mono table partitioning, whereas in data warehouses, the efforts are concentrated on dependent table partitioning. Proposed algorithms on fragmenting a fact table consider whether all dimension tables of a data warehouse schema or a set of these tables identified by DBA. The choice of these dimension tables to partition a fact table is a crucial issue for optimizing OLAP queries. Unfortunately, the problem of selecting dimension table(s) to partition the warehouse is not well addressed; except the work done by [17] that pointed out the idea of choosing dimension to partition a hyper cube, without giving strategies to perform this selection.

## 2.3 Selection of Dimension Tables

In this section, we propose first a formalization of the problem of selecting dimension tables and then strategies aiding DBA in choosing her/his relevant dimension tables used to referential partition a fact table.

### 2.3.1 Formalization of the Problem

Before proposing this methodology, we formalize the referential horizontal partitioning problem as follows:

Given a data warehouse with a set of $d$ dimension tables $D = \{D_1, D_2, \ldots, D_d\}$ and a fact table $F$, a workload $Q$ of queries $Q = \{Q_1, Q_2, \ldots, Q_m\}$ and a maintenance constraint $W$ fixed by DBA that represents the maximal number of fact fragments that he/she can maintain. The referential horizontal partitioning problem consists in (1) identifying candidate dimension table(s), (2) splitting them using single partitioning mode and (3) using their partitioning schemes to decompose the fact table into $N$ fragments, such that: (a) the cost of evaluating all queries is minimized and (b) $N \leq W$.

Based on this formalization, we note that referential partitioning problem is a combination of two difficult sub problems: *identification of relevant dimension tables* and their *fragmentation schema selection* (which is known as a NP-complete problem [2]). In the next section, we study the complexity of the first sub problem (identification of relevant dimension tables).

### 2.3.2 Complexity of Selecting Dimension Tables

The number of possibilities that DBA shall consider to partition the fact table using referential partitioning is given by the following equation:

$$\binom{d}{1} + \binom{d}{2} + \ldots + \binom{d}{d} = 2^d - 1 \tag{2.1}$$

*Example 1.* Let us suppose a warehouse schema with three dimension tables *Customer*, *Time* and *Product* and one fact table *Sales*. Seven scenarios are possible to referential partition *Sales*: *Sales(Customer), Sales(Time), Sales(Product), Sales (Customer, Time), Sales (Customer, Product), Sales (Product, Time)* and *Sales (Customer, Time, Product)*. For a data warehouse schema with a large number of dimension tables, it will be hard for DBA to choose manually relevant dimension table(s) to partition the fact table and guarantying a reduction of query processing cost. For instance, the star schema of *Sloan Digital Sky Survey database*, which is a real-world astronomical database, running on SQL Server has 18 dimension tables [12]. In this context, DBA needs to evaluate $2^{18} - 1$ (=262 143) possibilities to partition her/his fact tables.

This formalization and real partitioning scenario motivate us to propose strategies aiding DBA to select relevant dimension tables.

### 2.3.3   Dimension Table Selection Strategies to Split the Fact Table

Referential partitioning is proposed essentially to optimize join operations and facilitate the data manageability [9, 23]. Therefore, any selection strategy should hinge on join operation which well present in OLAP queries.

1. *Largest dimension table:* a simplest way to select a dimension table to referential partition the fact table is to choose the largest dimension table. This selection is based on the following observation: joins are typically expensive operations, particularly when the relations involved are substantially larger than main memory [14].
2. *Most frequently used dimension table:* the second naive solution is to choose the most frequently used dimension table(s). The frequency of a dimension table Di is calculated by counting its appearance in the workload.
    These two naive strategies are simple to use. But, they ignore several logical and physical parameters like size of tables, buffer size, size of intermediate results of joins that has a real impact on join cost, etc. Generally, when query involves multiple joins (which is the case of star join queries), the cost of executing the query can vary dramatically depending on the query evaluation plan (QEP) chosen by the query optimizer. The join order and access methods used are important determinants of the lowest cost QEP [27]. The query optimizer estimates the intermediate results and uses this information to choose between different join orders and access methods. Thus, the estimation of join result sizes in a query is an important problem, as the estimates have a significant influence on the QEP chosen. Based on these observations, the choice of dimension table(s) should take into account these the sizes of intermediate results.
3. *Minimum share:* in data warehousing, every join operation involves the fact table. The query graph of such a query contains a root table (representing the fact table)

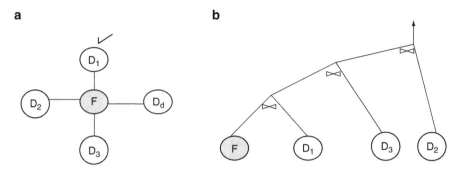

**Fig. 2.3** (a) Star Query Graph, (b) Star query execution using left deep tree

and peripheral tables (representing dimension tables) (Fig. 2.3a). To optimize the execution of a star join query, an optimal sequence in which the tables are joined shall be identified. This is problem is known as *join ordering* [26]. Intuitively, this problem is quite similar to dimension table selection problem. Join ordering is specific to one query (Fig. 2.3b). An important difference between these two problems is that join order concerns only one query, but dimension table selection problem concerns several queries running on data warehouse. As in join ordering problem [26], the best choice of dimension table should reduce the size of intermediate results of the most important join star queries. Based on this analysis, we propose a strategy to referential partition the fact table reducing the size of intermediate results. The size (in terms of number of tuples) of join between the fact table F and a dimension table $D_i$ (which is a part of a star join query), denoted by $\|F \bowtie D_i\|$ is estimated by:

$$\|F \bowtie D_i\| = \|D_i\| \times Share(F.A_i) \tag{2.2}$$

where $\|D_i\|$ and $Share(F.A_i)$ represent respectively, the cardinality of dimension table $D_i$ and the average number of tuples of the fact table $F$ that refer to the same tuple of the dimension table $D_i$ via the attribute $A_i$. To reduce the intermediate result of a star join query, we should choose the dimension table with a *minimum share*.

If we have a star join query involving several dimension tables, a simplest join order is to use the dimension table with a minimum share to ensure a reduction of the intermediate result. Therefore, we propose to choose the dimension table with a minimum share to referential partition the fact table. Several studies were proposed to estimate the share parameter, especially in object oriented databases [8].

### 2.3.4  Referential Partitioning Methodology

Now, we have all ingredients to propose our methodology offering a real utilization of data partitioning in data warehouse application design. It is subdivided into three steps described in Fig. 2.4:

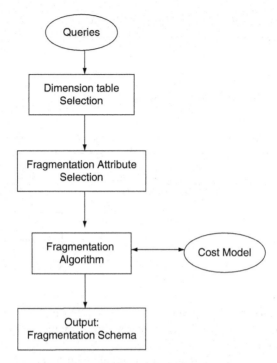

**Fig. 2.4** A methodology to support referential partitioning

1. Selection of dimension table(s) used to decompose the fact table. This selection may be done using one of the previous criteria.
2. Extraction of candidate fragmentation attributes (key partitioning). A fragmentation attribute is an attribute used by a selection predicate defined on a chosen dimension table.
3. Partition each dimension table using single table partitioning type. To generate partitioning schemes of chosen dimension tables, DBA may use his/her favourite algorithm. The complete generation of fragmentation schemes (of fact and dimension tables) is ensured by using a particular coding [2]. Each fragmentation schema is represented using a multidimensional array, where each cell represents a domain partition of a fragmentation schema.

   To illustrate this coding, let us suppose the following scenario: DBA choose dimension table *Customer* and Time to referential partition the fact table Sales. Three attributes: Age, Gender, Season, where Age and Gender belong to *Customer* dimension table, whereas Season belongs to Time. The domain of these attributes are: $Dom(Age) = [0, 120]$, $Dom(Gender) = \{\text{'M'},$ 'F'\}, and $Dom(Season) = \{\text{"Summer"}, \text{"Spring"}, \text{"Autumn"}, \text{"Winter"}\}$. DBA proposes an initial decomposition of domains of these three attributes as follows: $Dom(Age) = d_{11} \cup d_{12} \cup d_{13}$, with $d_{11} = [0, 18]$, $d_{12} = ]18, 60[$, $d_{13} = [60, 120]$. $Dom(Gender) = d_{21} \cup d_{22}$, with $d_{21} = \{\text{'M'}\}$, $d_{22} = \{\text{'F'}\}$. $Dom(Season) = d_{31} \cup$

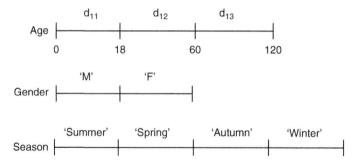

**Fig. 2.5** Decomposition of attribute domains

**Fig. 2.6** Coding of a
fragmentation schema

| Age | 1 | 2 | 3 | |
|---|---|---|---|---|
| Gender | 1 | 2 | | |
| Season | 1 | 2 | 3 | 4 |

$d_{32} \cup d_{33} \cup d_{34}$, where $d_{31} = \{$"Summer"$\}$, $d_{32} = \{$"Spring"$\}$, $d_{33} = \{$"Autumn"$\}$, and $d_{34} = \{$"Winter"$\}$. Sub domains of all three fragmentation attributes are represented in Fig. 2.5.

Domain partitioning of different fragmentation attributes may be represented by multidimensional arrays, where each array represents the domain partitioning of a fragmentation attribute. The value of each cell of a given array representing an attribute belongs to $(1..n_i)$, where $n_i$ represents the number of sub domain of the attribute (see Fig. 2.6). Based on this representation, fragmentation schema of each dimension table $D_j$ is generated as follows:

- All cells of a fragmentation attribute of $D_j$ have different values: this means that all sub domains will be used to partition $D_j$. For instance, the cells of each fragmentation attribute in Fig. 2.5 are different. Therefore, they all participate in fragmenting their corresponding tables (*Customer* and *Time*). The final fragmentation schema will generate 24 fragments of the fact table.
- All cells of a fragmentation attribute have the same value: this means that it will not participate in the fragmentation process. Table 2.1 gives an example of a fragmentation schema, where all sub domains of Season (of dimension table Time) have the same value; consequently, it will not participate in fragmenting the warehouse schema.
- Some cells has the same value: their corresponding sub domains will be merged into one. In Table 2.1, the first ([0, 18]) and the second (]18, 60[) sub domains of Age will be merged to form only one sub domain which is the union of the merged sub domains ([0, 60[). The final fragmentation attributes are: Gender and Age of dimension table *Customer*.

**Table 2.1** Example of a fragmentation schema

| Table | Attribute | SubDomain | SubDomain | SubDomain | SubDomain |
|-------|-----------|-----------|-----------|-----------|-----------|
| *Customer* | Age | 1 | 1 | 2 | |
| *Customer* | Gender | 1 | 2 | | |
| Time | Season | 1 | 1 | 1 | 1 |

4. Partition the fact table using referential partitioning based on the fragmentation schemes generated by our algorithm.

To illustrate this step, let us consider an example. Suppose that our algorithm generates a fragmentation schema represented by a coding described by Table 2.1. Based on this coding, DBA can easily generate scripts using DDL to create a partitioned warehouse. This may be done as follows: *Customer* is partitioned using the composite mode (*Range* on attribute *Age* and *List* on attribute *Gender*) and the fact table using the native referential partitioning mode.

```
CREATE TABLE Customer
(CID NUMBER, Name Varchar2(20), Gender CHAR, Age Number)
PARTITION BY RANGE (Age)
SUBPARTITION BY LIST (Gender)
SUBPARTITION TEMPLATE (SUBPARTITION Female VALUES ('F'),
SUBPARTITION Male VALUES ('M'))
(PARTITION Cust_0_60 VALUES LESS THAN (61),
PARTITION Cust_60_120 VALUES LESS THAN (MAXVALUE));
```

The fact table is also partitioned into 4 fragments as follows:

```
CREATE TABLE Sales (customer_id NUMBER NOT NULL,
product_id NUMBER NOT NULL,
time_id Number NOT NULL, price NUMBER,
quantity NUMBER,
constraint Sales_customer_fk foreign key(customer_id)
references CUSTOMER(CID))
PARTITION BY REFERENCE (Sales_customer_fk);
```

## 2.4 Experimental Results

We have conducted an intensive experimental study to evaluate our proposed strategies for choosing dimension tables to referential partition a fact table. Since, the decision of choosing the horizontal partitioning technique is done before populating the data warehouse; the development of a mathematical model estimating the cost of processing a set of queries is required. In this study, we develop a cost model estimating the number of inputs outputs required for executing this set of queries. It takes into account size of generated fragments of the fact tables ($\| F_i \|$, $1 \leq i \leq N$); the number of occupied pages of each fragment $F_i$, denoted by $|F_i| \left( = \frac{\| F_i \| \times TL}{PS} \right)$, where $PS$ and $TL$ represent respectively, the page size of disk and length of each

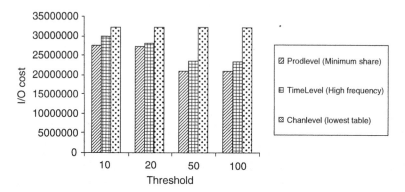

**Fig. 2.7** Theoretical results

instance of the fact table), buffer size, etc. The obtained results are then validated on Oracle11G DBMS in order to evaluate the effectiveness of our cost model.

**Dataset:** We use the dataset from the APB1 benchmark [21]. The star schema of this benchmark has one fact table *Actvars* (33 323 400 tuples) and four dimension tables: *Prodlevel* (9,900 tuples), *Custlevel* (990 tuples), *Timelevel* (24 tuples) and *Chanlevel* (10 tuples).

**Workload:** We have considered a workload of 55 single block queries with 55 selection predicates defined on 10 different attributes (Class_Level, Group_Level, Family_Level, Line_Level, Division_Level, Year_Level, Month_Level, Quarter_Level, Retailer_Level, All_Level). The domains of these attributes are split into: 4, 2, 5, 2, 4, 2, 12, 4, 4, 5 sub domains, respectively. We did not consider update queries (update and delete operations). Note that each selection predicate has a selectivity factor computed using SQL queries executed on the data set of APB1 benchmark. Our algorithms have been implemented using Visual C++ performed under Intel Pentium 4 with a memory of 3 Gb.

### 2.4.1  Theoretical Evaluation

The used cost model computes the inputs and outputs required for executing the set of 38 queries[1]. It takes into account the size of intermediate results of joins and the size of buffer. The cost final of 55 queries is computed as the sum of the cost of each query (for more details see [6]).

We use a hill climbing algorithm to select fragmentation schema of the APB1 warehouse due to its simplicity and less computation time [3]. Other algorithms may

---

[1]These queries are described in Appendix Section at: http://www.lisi.ensma.fr/ftp/pub/documents/papers/2009/2009-DOLAP-Bellatreche.pdf

be used such as simulated annealing and genetic [1]. Our hill climbing algorithm consists of the following two steps:

1. Find an initial solution and iteratively improve this solution by using hill climbing heuristics until no further reduction in total query processing cost. The initial solution is represented by multidimensional array, whose cells are filled in uniform way [3]. Each cell $i$ of fragmentation attribute $(A_k)$ is filled using the following formula:

$$Array[i]_k = \left\lfloor \frac{i}{n} \right\rfloor \qquad (2.3)$$

where $n$ is an integer $(1 \leq n \leq max_{1 \leq j \leq C}(n_j))$, where $C$ is the number of fragmentation attributes. To illustrate this distribution, let us consider three fragmentation attributes: *Gender*, *Season* and *Age*, where the numbers of sub domains of each attributes are respectively, 2, 4 and 3 (see Fig. 2.8). The generated fragments of partitioning schema corresponding to $n = 2, n = 3$ and $n = 4$ are 12, 4 and 2 respectively. If DBA wants an initial fragmentation schema with a large number of fragments, she/he considers n with a low value. In this case, all initial sub domains (proposed by DBA) have the same probability to participate on the fragmentation process.

2. The previous initial solution can be improved by introducing two specialized operators, namely Merge and Split, which allow us to further reduce the total query processing cost due to evaluating queries in $Q$. Let us now focus on the formal definitions of these operators. Given a fragmentation attribute $A_k$ of a dimension table $D_j$ having $FS(D_j)$ as fragmentation scheme, Merge operation takes as input two domain partitions of $A_k \in FS(D_j)$, namely and returns as output a new fragmentation scheme for $D_j$, denoted by $FS'(D_j)$, where these two domains are merged into a singleton domain partition of $A_k$. *Merge* reduces the number of fragments generated by means of the fragmentation scheme

| Gender | 0 | 1 |   |   |
|--------|---|---|---|---|
| Season | 0 | 1 | 1 | 2 |
| Age    | 0 | 1 | 2 |   |

n = 2

| Gender | 0 | 0 |   |   |
|--------|---|---|---|---|
| Season | 0 | 0 | 1 | 1 |
| Age    | 0 | 0 | 1 |   |

n = 3

| Gender | 0 | 0 |   |   |
|--------|---|---|---|---|
| Season | 0 | 0 | 0 | 1 |
| Age    | 0 | 0 | 0 |   |

n = 4

**Fig. 2.8** Different coding of an initial solution

$FS(D_j)$ of dimension table $D_j$, hence it is used when the number of generated fragments does not satisfy the maintenance constraint $W$.

Given a fragmentation attribute $A_k$ of a dimension table $D_j$ having $FS(D_j)$ as fragmentation scheme, *Split* takes as input a domain partition of $Ak$ in $FS(D_j)$ and returns as output a new fragmentation scheme for $D_j$, denoted by $FS(D_j)$, where that domain is split into two distinct domain partitions. Split increases the number of fragments generated by means of the fragmentation scheme $FS(D_j)$ of $D_j$.

On the basis of these operators running on fragmentation schemes of dimensional tables, the hill climbing heuristic still finds the final solution, while the total query processing cost can be reduced and the maintenance constraint $W$ can be satisfied.

The first experiments evaluate the quality of each criterion: *minimum share*, *high frequency*, and *largest size*. The size and frequency of each candidate dimension to partition the fact table are easily obtained from the data set of APB1 benchmark and the 38 queries. The share criterion is more complicated to estimate, since it requires more advanced techniques such histograms [11]. For our case, we calculate the *share* using SQL queries executed on the data set of APB1 benchmark created on Oracle11G. For instance, the *share* of the dimension table *CustLevel* is computed by the following query:

```
Select avg(Number) as AvgShare
    FROM (Select distinct customer_level, count(*) as Number
          From actvars
          Group By Customer_level);
```

For each criterion, we run our hill climbing algorithm with different values of the threshold (representing the number of generated fragments fixed by DBA) set to 10, 20, 50 and 100. For each obtained fragmentation schema, we estimate the cost of evaluating the 38 queries. At the end, 12 fragmentation schemes are obtained and evaluated. Figure 2.7 summarizes the obtained results. The main lessons behind these results are: (1) the minimum share criterion outperforms other criteria (frequency, maximum share), (2) minimum share and largest dimension table criteria give the same performance. This is because, in our real data warehouse, the table having the *minimum share* is the largest one and (3) the threshold has a great impact on query performance. Note that the fact of increasing the threshold does not mean getting a better performance, since when it is equal to 50 and 100; we got practically the same results. Having a large number of fragments increases the number of union operations which can be costly, especially when the size of manipulating partition instances is huge.

## 2.4.2 Validations on ORACLE 11G

To validate the theoretical results, we conduct experiments using Oracle11G. We choose this DBMS because it supports referential horizontal partitioning. The data set of ABP1 benchmark is created and populated using generator programs offered by APB1 [21]. During this validation, we figure out that *Composite partitioning* with more than two modes is not directly supported by Oracle11G DDL[2]. To deal with this problem, *we propose the use of virtual partitioning column proposed by Oracle11G*. A virtual column is an expression based on one or more existing columns in the table. The virtual column is only stored as meta-data and does not consume physical space. To illustrate the use of this column, let $D_i$ be a partitioned table in $N_i$ fragments. Each instance of this table belongs to a particular fragment. The identification of the relevant fragment of a given instance is done by matching partitioning key values with instance values. To illustrate this mechanism, suppose that dimension table *ProdLevel* is partitioned into eight partitions by the hill climbing algorithm using three attributes. A *virtual column PROD_COL* is added into *ProdLevel* in order to facilitate its partitioning using the *List* mode. mode:

```
CREATE TABLE Prodlevel(
CODE_LEVEL CHAR(12) NOT NULL, CLASS_LEVEL CHAR(12) NOT NULL, GROUP_LEVEL CHAR(12) NOT NULL,
FAMILY_LEVEL CHAR(12) NOT NULL, LINE_LEVEL CHAR(12) NOT NULL, DIVISION_LEVEL CHAR(12) NOT NULL,
PROD_COL NUMBER(5) generated always as (case when Class_Level
IN('ADX8MBFPWVIV','OC2WOOC8Q
6','LB2RKO0ZQCJD')
and  Group_Level='VTL9DOE3RSWQ' and Family_Level
in('JIHR5NBAZWGU','OX3BXTCVRRKU','M32G5M3AC4T5', 'Y45VKMTJDNYR') then 0
when Class_Level IN ('ADX8MBFPWVIV','OC2WOOC8QIJ6','LB2RKO0ZQCJD') and
Group_Level='VTL9DOE3RSWQ' and Family_Level  not in
('JIHR5NBAZWGU','OX3BXTCVRRKU','M32G5M3AC4T5', 'Y45VKMTJDNYR')
then 1 when Class_Level IN
('ADX8MBFPWVIV','OC2WOOC8QIJ6','LB2RKO0ZQCJD') and  Group_Level
not in ('VTL9DOE3RSWQ') and Family_Level in
('JIHR5NBAZWGU','OX3BXTCVRRKU','M32G5M3AC4T5','Y45VKMTJDNYR')
 then 2 when Class_Level IN ('ADX8MBFPWVIV','OC2WOOC8QIJ6','LB2RKO0ZQCJD')
 and Group_Level not in ('VTL9DOE3RSWQ') and
Family_Level   not in('JIHR5NBAZWGU','OX3BXTCVRRKU','M32G5M3AC4T5','Y45VKMTJDNYR')
then 3  when Class_Level NOT IN ('ADX8MBFPWVIV','OC2WOOC8QIJ6','LB2RKO0ZQCJD')
and  Group_Level='VTL9DOE3RSWQ' and Family_Level  in
('JIHR5NBAZWGU','OX3BXTCVRRKU','M32G5M3AC4T5','Y45VKMTJDNYR')
then 4 when Class_Level NOT IN ('ADX8MBFPWVIV','OC2WOOC8QIJ6','LB2RKO0ZQCJD')
and  Group_Level='VTL9DOE3RSWQ' and Family_Level  not in
('JIHR5NBAZWGU','OX3BXTCVRRKU','M32G5M3AC4T5','Y45VKMTJDNYR') then 5
Else  7  end), PRIMARY KEY (CODE_LEVEL))
PARTITION BY LIST(PROD_COL)
(PARTITION PT1 VALUES(0) TABLESPACE HCTB,PARTITION PT2 VALUES(1) TABLESPACE HCTB,
PARTITION PT3 VALUES(2) TABLESPACE HCTB, PARTITION PT4 VALUES(3) TABLESPACE HCTB,
PARTITION PT5 VALUES(4) TABLESPACE HCTB, PARTITION PT6 VALUES(5) TABLESPACE HCTB,
PARTITION PT7 VALUES(6) TABLESPACE HCTB, PARTITION PT8 VALUES(7) TABLESPACE HCTB);
```

No materialized views and advanced indexes are created, except indexes on primary and foreign keys.

Figure 2.9 compares the *native referential partitioning* and *user driven referential partitioning*. In both cases, the fact table is partitioned into eight partitions based on *ProdLevel* table. In user driven referential partitioning, the *ProdLevel* is horizontally partitioned using the *hash mode* on the primary key (Code_Level) and the fact table is horizontally partitioned using also the *hash mode* on the foreign key (the number of partitions using hash must be a power of 2). Both partitioning are compared to the non partitioning case. The result shows that native horizontal partitioning

---

[2]For example, we can partition a table using three fragmentation attributes.

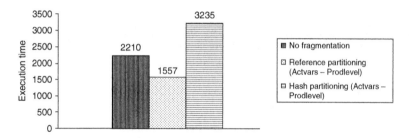

**Fig. 2.9** Native and user driven referential partitioning comparison

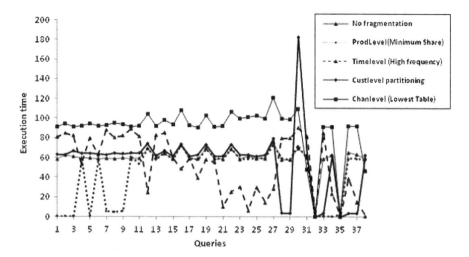

**Fig. 2.10** Individual evaluation of impact of criterion on query

outperforms largely the two other cases. *The user driven referential partitioning is not well adapted to optimize star join queries, where the non partitioning mode outperforms it.*

Figure 2.10 compares the quality of solution generated by each criteria: *Minimum Share, Highest frequency* and *Lowest table*. These criteria are also compared with the non partitioned case. To conduct this experiment, we execute three times our climbing algorithm with a threshold equal to 10. Each query is executed individually on each generated fragmentation schema (one schema per selection criterion) and its execution time (given in second) is computed using *Enterprise manager of Oracle 11G*. Based on the obtained results, we propose to DBA the following recommendations:

• To improve performance of queries containing only one join and selection predicates involving only one dimension table, DBA has to partition the fact table based on that dimension table; without worrying about its characteristic (minimum share, highest frequency, etc.). More clearly, queries 1, 2, 5 optimized with the *minimum share strategy* and queries 21, 22 and 23 with high frequency, etc. This case is very limited in real data warehouse applications, where queries involve all dimension tables.

- To speed up queries involving several dimension tables, but only one of them has selection predicates, DBA has to partition his/her fact table based on the dimension table involving selection predicates.
- To accelerate queries involving several dimension tables and each one contains selection predicates (this scenario is most representative in the real life), DBA has to partition his/her fact table based on the dimension table having a *minimum share*. *Query 34* of our query benchmark is an example of this scenario, where two dimension tables are used *ProdLevel* (with class_level = 'LB2RKO0ZQCJD) and *TimeLevel* (with two predicates Quarter_level = 'Q2' and Year_level = '1996'). Another example is about the *query 38 of our benchmark*, where two dimension tables are used, and each one has only one selection predicate. In this case, *TimeLevel* strategy (representing high frequency and also the table having a second minimum share) outperforms *ChanLevel* strategy representing the lowest table.

To complete our understanding, we use "*Set autotrace on*" utility to get execution plan of the query 34. First, we execute it on the non partitioned data warehouse and we generate its execution plan. We identify that Oracle optimizer starts by joining fact table with *ProdLevel* table (which has the minimum share criterion)[3]. Secondly, we execute the same query on a partitioned data warehouse obtained by fragmenting the fact table based on *ProdLevel*. The query optimizer starts by joining a partition of fact table with one of *ProdLevel* table and then does

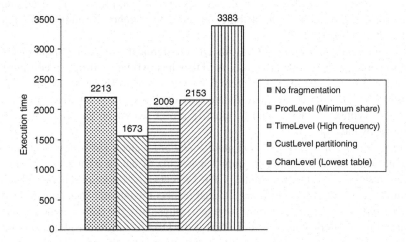

**Fig. 2.11** Evaluation of all strategies for overall queries

---

[3]The Oracle cost-based optimizer recognizes star queries.

other operation. Based on these two execution plans, we can conclude that query optimizer uses the same criteria (minimum share) for ordering join. Referential partitioning gives a partial order of star join queries problem. Figure 2.11 summarizes the performance of each criterion for all queries.

## 2.5   Advisor Tool

To facilitate the use of our partitioning strategies of a given relational data warehouse, we develop a tool, called, *ParAdmin* offering nice graphical interfaces to DBA. It supports primary and referential partitioning. It may be connected to any DBMS. For our study, we choose Oracle, since it supports referential partitioning. Based on a workload of queries $Q$ and the maintenance constraint representing the number of final fragments, it offers DBA several choices to simulate the partitioning. If they are satisfied, they can generate all the scripts necessary to partition the data warehouse to DBA. It is developed with under Visual C++. ParAdmin consists of a set of five modules: (1) *meta-base querying module*, (2) *managing queries module*, (3) *horizontal partitioning selection module (HPSM)* and (4) *horizontal partitioning module* and (5) *query rewriting module*.

*The meta-base querying module* is a very important module which allows the tool to work with any type of DBMS. From a type of DBMS, user name and password the module allows to connect to that account and to collect some information from the meta-base. These information concerns logical and physical levels of the data warehouse. Information of logical level includes tables and attributes in these tables. Information of physical level includes optimization techniques used and a set of statistics on tables and attributes of the data warehouse (number of tuples, cardinality, etc.) (Fig. 2.12).

*The managing queries module* allows DBA to define the workload of queries ($Q$) used to partition the data warehouse. An DBA has two possibilities to edit her/his queries: manually or by importing them from external files. This module offers DBA the possibility to add, delete, and update queries. It integrates a parser in order to check the syntax of queries and to identify errors (based on the used tables and attributes by each query).

*Horizontal partitioning selection module (HPSM)* selects partitioning schema of the data warehouse. It requires as inputs a schema of data warehouse, a workload and a threshold $W$. Using these data; HPSM selects a partitioning schema (PS) that minimizes the cost of the workload and that respect the maintenance constraint $W$.

*Horizontal partitioning module* fragments physically the data warehouse using partitioning schema obtained from HPSM if the DBA is satisfied with the proposed solution. It partitions dimension table(s) and propagates their partitioning to fact table.

*Query rewriting module* rewrites each global query on the generated fragmenta- tion schema. It starts with identifying valid fragments for each query; rewrites the

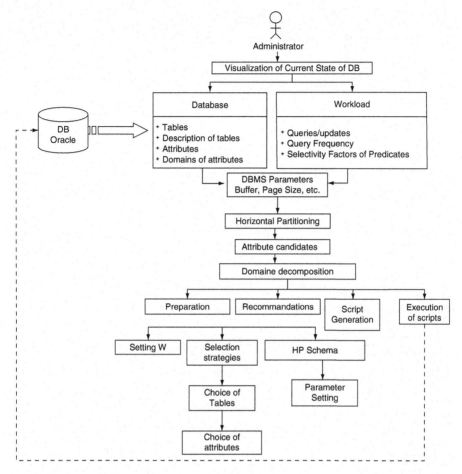

**Fig. 2.12** Steps of our tool

global query on each relevant fragment. Finally, it performs union of the obtained results.

The main functionalities of ParAdmin are described in Fig. 2.12. They were identified incrementally by our Ph.D. students of our laboratory (LISI) following my advanced databases course.

1. The first functionality is to display the current state of the database (the schema, attributes, size of each table, definition of each attribute, etc.) and the workload (description of queries, number of selection operations, selection predicates, etc.). Figure 2.13 shows an example of visualisation proposed by our tool.
2. The second functionality consists in preparing the partitioning process, by offering different choices to select dimension table(s) (minimum share, largest table, most frequently table) to be used to partition the fact table, fragmentation attributes of the chosen table(s), number of final fragments, etc. (Fig. 2.14).

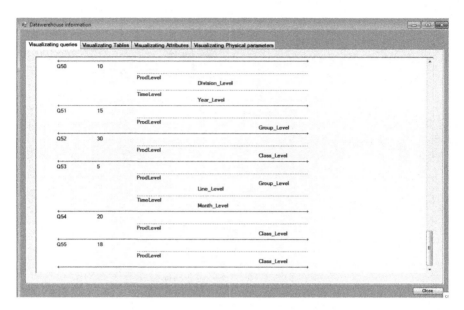

**Fig. 2.13** Visualization of current state of the database

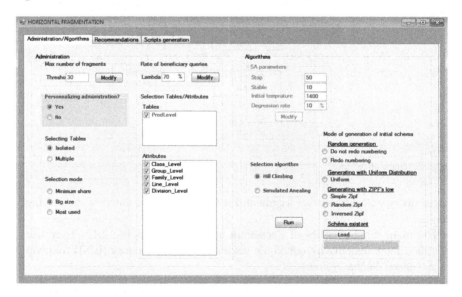

**Fig. 2.14** Choice of dimension table(s)

3. Once the preparation phase done, DBA may run the partitioning algorithm that proposes recommendations. DBA has the possibility to observe the cost of query before and after partitioning. As consequence, she/he can decide on considering or not the proposed partitioning. Her/his decision is based on the gain offered by horizontal partitioning for each query established using a threshold (Fig. 2.15).

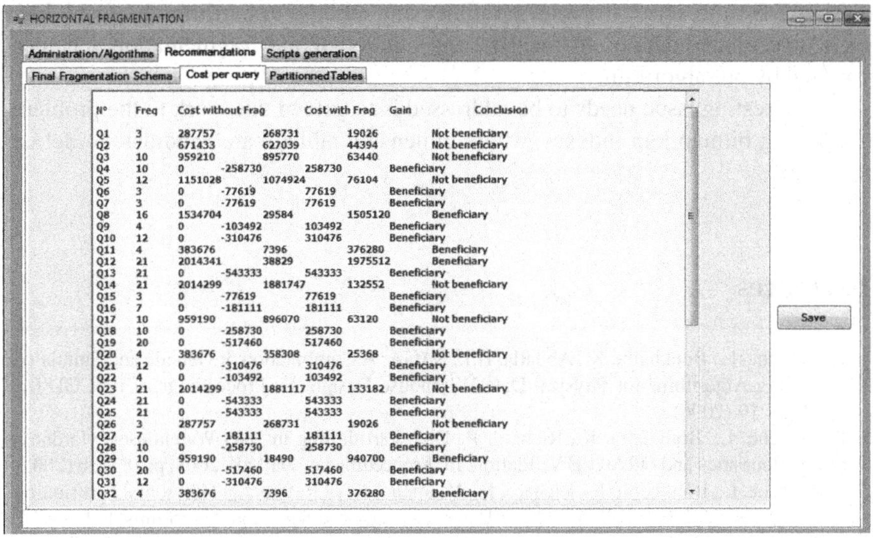

**Fig. 2.15**  Recommendations given to DBA

4. Finally, our tool proposes a functionality that generates scripts for primary and referential partitioning (if DBA is satisfied) corresponding to the target DBMS. They can be directly executed to really partition the warehouse.

## 2.6  Conclusion

Horizontal data partitioning is one of important aspects of physical design of advanced database systems. Two horizontal partitioning types are identified: *mono table partitioning* and *dependent-table partitioning*. The first one mainly used in traditional databases to optimize selections. The second one is well adapted for business intelligence applications in order to optimize selections and joins defined in mega queries involving a large number of tables. We propose a comprehensive procedure for using referential partitioning. It first selects relevant dimension table(s) to partition the fact table and discards some; even they are associated with selection predicates. To perform this selection, we propose three strategies based on three factors: *share*, *frequency* and *cardinality* of each dimension table. The share criterion reduces the size of intermediate results of joins; therefore, it may be used in cascade over dimension tables. Two types of experimental studies were conducted to validate our findings: one using a mathematical cost model and another using *Oracle11G* with the APB1 benchmark. The obtained results are encouraging and recommendations are given to assist administrators on using the referential partitioning. Our procedure can be easily incorporated in any DBMS supporting referential partitioning such as *Oracle11g*. A tool, called *ParAdmin* is presented to assist DBAs to select the relevant dimension table(s) to partition the fact table. It gives recommendations to measure the quality of the fragmentation schemes proposed by our algorithm.

An interesting issue needs to be addressed is to extend this work to the problem of selecting bitmap join indexes, where dimension table(s) are candidate to define each index.

# References

1. Bellatreche, L., Boukhalfa, K., Abdalla, H.I.: SAGA: A Combination of Genetic and Simulated Annealing Algorithms for Physical Data Warehouse Design. In: Proceedings of BNCOD'06, pp. 212–219 (2006)
2. Bellatreche, L., Boukhalfa, K., Richard, P.: Data Partitioning in Data Warehouses: Hardness Study, Heuristics and ORACLE Validation. In: Proceedings of DaWaK'2008, pp. 87–96 (2008)
3. Bellatreche, L., Boukhalfa, K., Richard, P., Woameno, K.Y.: Referential Horizontal Partitioning Selection Problem in Data Warehouses: Hardness Study and Selection Algorithms. In IJDWM. 5(4), 1–23 (2009)
4. Bellatreche, L., Karlapalem, K., Simonet. A.: Algorithms and Support for Horizontal Class Partitioning in Object-Oriented Databases. In the Distributed and Parallel Databases Journal, 8(2), 155–179 (2000)
5. Bellatreche, L., Woameno, K.Y.: Dimension Table Driven Approach to Referential Partition Relational Data Warehouses. In: ACM 12th International Workshop on Data Warehousing and OLAP (DOLAP), pp. 9–16 (2009)
6. Boukhalfa, K.: De la Conception Physique aux Outils d'Administration et de Tuning des Entrepts de Donnes. Poitiers University, France, PhD. Thesis. (2009)
7. Ceri, S., Negri, M., Pelagatti, G.: Horizontal Data Partitioning in Database Design. In: Proceedings of the ACM SIGMOD International Conference on Management of Data. SIGPLAN Notices, pp. 128–136 (1982)
8. Cho, W.S., Park, C.M., Whang, K.Y., So, S.H.: A New Method for estimating the number of objects satisfying an object-oriented query involving partial participation of classes. Inf. Syst. 21(3), 253–267 (1996)
9. Eadon, G., Chong, E.I., Shankar, S., Raghavan, A., Srinivasan, J., Das, S.: Supporting Table Partitioning By Reference in Oracle. In: Proceedings of SIGMOD'08, pp. 1111–1122 (2008)
10. Furtado, P.: Experimental evidence on partitioning in parallel data warehouses. In: Proceedings Of DOLAP, pp. 23–30 (2004)
11. Gibbons, P.B., Matias, Y., Poosala, V.: Fast incremental maintenance of approximate histograms. ACM Trans. Database Syst. 27(3), 261–298 (2002)
12. Gray, J., Slutz, D.: Data Mining the SDSS SkyServer Database. Microsoft Research, Technical Report MSR-TR-2002-01 (2002)
13. Karlapalem, K., Navathe, S.B., Ammar, M.: Optimal Redesign Policies to Support Dynamic Processing of Applications on a Distributed Database System. Information Systems, 21(4), 353–367 (1996)
14. Lei, H., Ross, K.A.: Faster Joins, Self-Joins and Multi-Way Joins Using Join Indices. In Data and Knowledge Engineering, 28(3), 277–298 (1998)
15. Legler, T., Lehner, W., Ross, A.: Query Optimization For Data Warehouse System With Different Data Distribution Strategies, In BTW, pp. 502–513 (2007)
16. Mahboubi, H., Darmont, J.: Data mining-based fragmentation of XML data warehouses. In: Proceedings DOLAP'08, pp. 9–16 (2008)
17. Munneke, D., Wahlstrom, K., Mohania, M.K.: Fragmentation of Multidimensional Databases. In: Proceedings of ADC'99 pp. 153–164 (1999)

18. Neumann, T.: Query simplification: graceful degradation for join-order optimization. In: Proceedings of SIGMOD'09, pp. 403–414 (2009)
19. Noaman, A.Y., Barker, K.: A Horizontal Fragmentation Algorithm for the Fact Relation in a Distributed Data Warehouse. In: Proceedings of CIKM'99, pp. 154–161 (1999)
20. Oracle Data Sheet: Oracle Partitioning (2007) White Paper: http://www.oracle.com/technology/products/bi/db/11g
21. OLAP Council: APB-1 OLAP Benchmark, Release II. http://www.olapcouncil.org/research/bmarkly.htm (1998)
22. Özsu, M.T., Valduriez, P.: Principles of Distributed Database Systems, Second Ed. Prentice Hall (1999)
23. Sanjay, A., Narasayya, V.R., Yang, B.: Integrating Vertical and Horizontal Partitioning Into Automated Physical Database Design. In: Proceedings of SIGMOD'04, pp. 359–370 (2004)
24. Simon, E.: Reality check: a case study of an EII research prototype encountering customer needs. In Proceedings of EDBT'08, pp. 1 (2008)
25. Stöhr, T., Märtens, H., Rahm, E.: Multi-Dimensional Database Allocation for Parallel Data Warehouses. In: Proceedings of VLDB2000, pp. 273–284 (2000)
26. Steinbrunn, M., Moerkotte, G., Kemper, A.: Heuristic and Randomized Optimization for the Join Ordering Problem. In VLDB Journal. **6**(3), 191–208 (1997)
27. Swami, A.N., Schiefer, K.B.: On the Estimation of Join Result Sizes. In: Proceedings of EDBT'04, pp. 287–300 (1994)
28. Sybase: Sybase Adaptive Server Enterprise 15 Data Partitioning. White paper (2005)

# Chapter 3
# Observer-Based Adaptive Fuzzy Control of Nonlinear Time-Delay Systems

**Shabnam Pourdehi, Dena Karimipour, Navid Noroozi, and Faridoon Shabani**

## 3.1 Introduction

The control of uncertain nonlinear systems is of practical importance since many real world systems exhibit nonlinear dynamic behavior. To make the issue even more difficult, in many practical cases, the mathematical model is poorly known or uncertain. Therefore, in modeling and analysis of such systems, one needs to handle unknown nonlinearities and/or uncertain parameters. One of the most useful techniques applied to uncertain nonlinear systems are adaptive control methods which are able to compensate the lack of precise knowledge of the system [1, 7, 13, 26, 31].

Most adaptive control schemes designed for the nonlinear systems assume all the states to be available and their controllers and adaptive laws are consisting of information from the states [7, 30, 31]. But in many practical systems the state vector is not measurable, so it's important to apply a state observer which is capable of estimating the unavailable states. The observer-based controllers utilize the estimation of the states instead of the states.

The phenomenon of time delay is frequently a source of instability and exists in various engineering systems. Time delay usually leads to unsatisfactory performances. Therefore, the problem of stabilization of time delay systems has received considerable attention over the past years [5, 6, 33, 39]. To overcome this difficulty, the Lyapunov–Krasovskii functionals are used for stability analysis and synthesis [6, 33].

The variable structure control with sliding mode which was introduced to control by [32], is an effective tool to control nonlinear systems. Sliding mode controllers are not only known to be robust to model uncertainties and to parameter variations, but also they show good disturbance rejection properties. There has been a wide

S. Pourdehi (✉) · D. Karimipour · N. Noroozi · F. Shabani
School of Electrical and Computer Engineering, Shiraz University, Shiraz, Iran
e-mail: Shabnamp_el83@yahoo.com

variety of applications of SMC in areas such as robotics [16,27], power control [40], aerospace [9,24], and chaos control [15,22,43].

On the other hand, since the work of [17] was proposed in 1974, fuzzy control has been an active research topic. If there exist uncertainties in nonlinear terms, or the nonlinear terms are completely unknown, the fuzzy control algorithms can be useful. Based on the universal approximation theorem in fuzzy logic control, several stable adaptive fuzzy control schemes [13,26] have been developed to overcome the difficulty of extracting linguistic control rules from experts and to cope with the system parameter changes [1, 6, 13, 26, 39]. Combining adaptive fuzzy control and sliding mode approach provides robust control schemes for nonlinear systems with uncertainty [1, 6, 18].

The implementation of control inputs of practical systems is frequently subject to nonlinearity as a result of physical limitations, for example, saturation, backlash, dead-zone, and so on. The presence of nonlinearities in control input may cause serious degradation of system performance, a reduced rate of response. Therefore, it is clear that the effects of input nonlinearity must be taken into account in control design and realization of nonlinear systems [1, 26, 36, 39].

Chaos is a very interesting nonlinear phenomenon and has been studied extensively over the past years. It has the random-like behavior usually seen in stochastic systems although it is associated with deterministic dynamics. The control and synchronization of chaotic systems have received increased research attention, since the classical work on chaos control first presented by [20] and [21] introduced the pioneering work on the synchronization of chaotic systems. Synchronization of chaotic systems has been used in a variety of fields such as chemical reactions, biological systems and secure communities. In particular, chaos synchronization has been widely investigated for applications in secure communication [2,8,28,38]. The idea is that chaotic signal can be used as a carrier and transmitted together with an information signal to a receiver. Due to these applications, various control methods have been proposed to synchronize chaotic systems such as sliding mode design [15, 43], adaptive control [29, 34], fuzzy control [11, 14]. In this paper, it is shown that the proposed controller is applicable for the synchronization of two different chaotic systems.

Most of the work cited before either consider a small group of nonlinear time-delayed systems or ignore the nonlinear effects in the control actuators. Furthermore, in many articles published on controlling time-delay systems, delay is assumed to be known. [33] proposed an adaptive fuzzy output tracking controller for a special class of perturbed time-delayed strict-feedback nonlinear systems, using the backstepping technique and Lyapunov–Krasovskii functionals. [6] has investigated the problem of variable structure adaptive fuzzy control for a class of nonlinear time-delay systems based on the Lyapunov–Krasovskii method. In both articles, the control inputs are assumed to be linear and the time-delay is known to the controller designer. In [39], a fuzzy sliding mode controller is employed to stabilize time-delayed systems subjected to input nonlinearities, where the system model is not a general class of nonlinear systems. To the best of our knowledge, the problem of control of a broad class of uncertain time-delayed nonlinear systems subjected to unknown delay, input nonlinearities and external disturbance has not been fully investigated yet.

Motivated by the aforementioned reasons, the purpose of this paper lies in the development of adaptive controllers to control a class of nonlinear systems with unknown nonlinear terms, unknown time-delay and disturbances, when dead-zone nonlinearity is present in the control input. An adaptive fuzzy sliding mode controller (SMC) and an adaptive observer-based controller are proposed to control this class of nonlinear system. Then, the Synchronization of two chaotic systems is studied which have the same form as the mentioned nonlinear system. The unknown nonlinear functions are approximated with fuzzy logic systems. The controllers have been designed to achieve the asymptotical stabilization even with the uncertainties and nonlinear input. These considerations make our design schemes more compatible with practical applications.

The rest of the paper is divided into eight sections. In Sect. 3.2, some properties of the uncertain nonlinear time-delayed system with nonlinearities are reviewed. Section 3.3 presents a brief description of fuzzy logic systems. Section 3.4 describes the design approaches of the sliding surface and the reaching law. Section 3.5 presents the observer-based adaptive controller designed to stabilize the system. In Sect. 3.6, the synchronization of two time-delayed chaotic systems is given as an application. Simulation examples are given to demonstrate the effectiveness of method proposed in this paper in Sect. 3.7. Finally, the conclusion is given in Sect. 3.8.

## 3.2  Preliminaries

A general description of uncertain time-delayed nonlinear dynamical systems is of the following form:

$$\dot{x} = (A + \Delta A(t))x + (A_d + \Delta A_d(t))x(t - d) + B(\phi(u) + F(x, x(t - d), \sigma(t))),$$

$$x(t) = \vartheta(t), \qquad t \in [-\bar{d} \ \ 0] \quad y = Cx, \tag{3.1}$$

where $x \in R^n$ and $u, y \in R$ are states, control input and output of the nonlinear system, respectively. $A \in R^{n \times n}$ is the state matrix and $A_d$ is the delay term matrix with appropriate dimension. $\Delta A(t)$ and $\Delta A_d(t)$ are time-varying uncertainties matrices of $A$ and $A_d$. $F(x, x(t - d), \sigma(t))$ is an unknown function representing uncertainties containing the current state, delayed state and time-varying disturbance parameter $\sigma(t)$. $\varphi(u) : R \rightarrow R$ is a continuous non-smooth function of $u$ which represents the input nonlinearity. $\vartheta(t)$ is a continuous vector-valued initial function. $d$ is the unknown delay parameter which assumed to be bounded by $\bar{d}$, $0 \leq d \leq \bar{d}$. The state of the system in (3.1) can be decomposed into two parts, $x_1 \in R^{n-1}, x_2 \in R$ and the matrix B is chosen as $B = [0, \dots, 0, 1]^T$.

**Assumption 1.** *Uncertainties* $\Delta A(t), \Delta A_d(t)$ *of the system in* (3.1) *satisfy the matching conditions*

$$\Delta A(t) = BE_1(t) \qquad \|E_1(t)\| \leq \delta_1$$
$$\Delta A_d(t) = BE_2(t) \qquad \|E_2(t)\| \leq \delta_2, \qquad (3.2)$$

where $\delta_1$ and $\delta_2$ are unknown positive scalars. These conditions are so-called matching conditions.

So $\|\Delta A\| \leq \gamma_1$, $\|\Delta A_d\| \leq \gamma_2$, where positive scalars $\gamma_1$ and $\gamma_2$ are not required to be known.

Thus system in (3.1) is rewritten in the following form:

$$\dot{x}_1 = A_{11}x_1 + A_{d11}x_1(t-d) + A_{12}x_2 + A_{d12}x_2(t-d),$$
$$x_1(t) = \bar{\vartheta}_1(t), \qquad t \in [-\bar{d} \quad 0], \qquad (3.3)$$

$$\dot{x}_2 = (A_{21} + \Delta A_{21})x_1 + (A_{d21} + \Delta A_{d21}(t))x_1(t-d) + (A_{22} + \Delta A_{22})x_2$$
$$+ (A_{d22} + \Delta A_{d22})x_2(t-d) + \phi(u) + F(x, x(t-d), \sigma(t)),$$
$$x_2(t) = \bar{\vartheta}_2(t), \qquad t \in [-\bar{d} \quad 0], \qquad (3.4)$$

where $A_{11}, \Delta A_{21}, A_{d11}, \Delta A_{d21}(t), \dots$ are the decomposed parts of $A$, $A_d$, $\Delta A(t)$ and $\Delta A_d(t)$ with appropriate dimensions. Then, the following assumptions are introduced.

**Assumption 2.** $F(x, x(t-d), \sigma(t))$ *satisfies the following decomposition:*

$$F(x, x(t-d, \sigma(t))) = F_1(x) + F_2(x(t-d)) + \Delta F(x(t), x(t-d), \sigma(t)), \quad (3.5)$$

where $F_1(x)$ and $F_2(x(t-d))$ are unknown continuous functions and $\Delta F(x(t), x(t-d), \sigma(t))$ is an uncertain function which satisfies

$$|\Delta F(x(t), x(t-d), \sigma(t))| \leq \alpha_0^* + \alpha_1^* \|x(t)\| + \alpha_2^* \|x(t-d)\|, \qquad (3.6)$$

where $\alpha_0^*, \alpha_1^*$ and $\alpha_2^*$ are unknown positive scalars.

**Assumption 3.** *The nonlinear input function* $\varphi(u)$ *satisfies dead-zone nonlinearity which is described as* [42]

$$\phi(u) = \begin{cases} m(u - b_1) & \text{for } u \geq b_1, \\ 0 & \text{for } -b_2 < u < b_1, \\ m(u + b_2) & \text{for } u \leq -b_2, \end{cases} \qquad (3.7)$$

where $m$ is a known positive scalar, but $b_1$ and $b_2$ are unknown positive scalars.

The function in (3.7) can be represented as

$$\phi(u) = mu + w(u) . \qquad (3.8)$$

It can be seen that it is composed of a line with the slope $m$, together with a term $w(u)$. It is clear that $w(u)$ is bounded, and satisfies $|w(u)| \le w^*$.

Before proceeding with the main results of this paper, in the next section, some preliminaries are stated about the approximation ability of the fuzzy logic systems.

## 3.3    Function Approximation with Fuzzy Logic Systems

The fuzzy system consists of the fuzzy rule base, fuzzy inference engine, fuzzifier and defuzzifier. A fuzzy rule base consists of the following fuzzy IF–THEN rules:

$$R^{(j)} : IF \ x_1 \ is \ A_1^j \& \ldots \& x_n is \ A_n^j, THEN \ y \ is \ B^j . \qquad (3.9)$$

The fuzzy logic systems with singleton fuzzifier, product inference engine, center-average defuzzifier are in the following form:

$$y(x) = \frac{\sum_{j=1}^{l} y^j \left( \prod_{i=1}^{n} \mu_{A_i^j}(x_i) \right)}{\sum_{j=1}^{l} \prod_{i=1}^{n} \mu_{A_i^j}(x_i)}, \qquad (3.10)$$

where $x = [x_1, x_2, \ldots, x_n] \in R^n$, $\mu_{A_j^i}(x_i)$ is the membership function of $A_j^i$, and $y^j$ represents a crisp value at which the membership function $\mu_{B^j}(y)$ for output fuzzy set reaches its maximum.

The universal fuzzy system with input vector $x \in \Omega_x$ for some compact set $\Omega_x \subset R^n$ is proposed here to approximate the uncertain term $y(x)$ where $\theta$ is a vector containing the tunable approximation parameters. From (3.10), the fuzzy system $y(x)$ can be expressed as

$$y(x) = \theta^T \xi(x), \qquad (3.11)$$

where $\theta = (\theta_1, \theta_2, \ldots, \theta_l)^T, \xi(x) = (\xi_1(x), \xi_2(x), \ldots, \xi_l(x))^T$ is a regressive vector with the regressor $\xi_j(x)$ defined as

$$\xi_j(x) = \frac{\prod_{i=1}^{n} \mu_{A_i^j}(x_i)}{\sum_{j=1}^{l} \prod_{i=1}^{n} \mu_{A_i^j}(x_i)}. \qquad (3.12)$$

The above fuzzy logic system is capable of uniformly approximating any continuous nonlinear function over a compact set $\Omega_x$ with any degree of accuracy. This property is shown by the following lemma.

**Lemma 1.** *For any given real continuous function $f(x)$ on a compact set $\Omega_x \subset R^n$ and a scalar $\varepsilon > 0$, there exists a fuzzy logic system $y(x)$ in the form of (3.10) such that*

$$\sup_{x \in \Omega_x} |f(x) - y(x)| \leq \varepsilon. \tag{3.13}$$

In the next sections, the unknown nonlinear terms are replaced with their fuzzy approximation to any desired degree in controller synthesis for system in (3.1).

## 3.4  Proposed Adaptive Fuzzy Sliding Mode Controller

This section proposes an adaptive variable structure control law to control the mentioned class of nonlinear systems using Lyapunov–Krasovskii stability theorem. Sliding mode control (SMC) is known to be an efficient control technique applicable to systems with profound nonlinearity and modeling uncertainty and it has attractive features such as fast response and good transient response.

**Theorem 1.** *Taking the sliding surface as follows*

$$S = \tilde{C}x = [-C \quad 1][x_1^T x_2]^T, \tag{3.14}$$

where $C = -\varepsilon^{-1}(PA_{12})^T$, if assumption 1 is met and also the following inequality is satisfied:

$$P(A_{11} + \gamma I) + (A_{11} + \gamma I)^T P + \varepsilon^{-1} PD_1 D_1^T P + \varepsilon^{-1} e^{2\gamma \bar{d}} PA_{d12} A_{d12}^T P$$

$$+ e^{2\gamma \bar{d}} PA_{d11} X A_{d11}^T P + X^{-1} + \varepsilon^{-1} e^{2\gamma \bar{d}} PD_2 D_2^T P$$

$$- \varepsilon^{-1}(PA_{12}) \times (PA_{12})^T < 0, \tag{3.15}$$

where P and X are positive matrices, $\varepsilon$ is a positive scalar and $d$ is the time delay, then the sliding motion is exponentially stable depending on the bound of time-delay with a given decay rate $\gamma$.

In this paper we assume $d$ to be unknown and just its upper bound is known. So, by some simple algebraic operations and using the definition of positive matrix, it can be shown that the inequality (3.15) is true even if replacing $d$ with its upper bound $\bar{d}$.

*Proof.* For system in (3.3), the following transformation is considered

$$x_1(t) = e^{-\gamma t} z(t). \tag{3.16}$$

We choose the Lyapunov–Krasovskii functional as

$$V = z^T Pz + \int_{t-d}^{t} \varepsilon z^T(v)[X^{-1} + C^T C]z(v)dv. \tag{3.17}$$

One can get

$$
\begin{aligned}
\dot{V} &= 2z^T P(A_{11} + A_{12}C + \gamma I)z + 2e^{\gamma d} z^T P(A_{d11} + A_{d12}C)z(t-d) \\
&\quad + \varepsilon z^T(t)[X^{-1} + C^T C]z(t) - \varepsilon z^T(t-d)[X^{-1} + C^T C]z(t-d) \\
&\leq 2z^T P(A_{11} + A_{12}C + \gamma I)z + \varepsilon^{-1} e^{2\gamma d} z^T PA_{d12} A_{d12}^T Pz \\
&\quad + \varepsilon z(t-d)^T C^T Cz(t-d) + \varepsilon^{-1} e^{2\gamma d} z^T PA_{d11} XA_{d11}^T Pz \\
&\quad + \varepsilon z(t-d)^T X^{-1} z(t-d) + \varepsilon z^T X^{-1} z + \varepsilon z^T C^T Cz - \varepsilon z(t-d)^T X^{-1} z(t-d) \\
&\quad - \varepsilon z(t-d)^T C^T Cz(t-d).
\end{aligned} \tag{3.18}
$$

Letting $C = -\varepsilon^{-1}(PA_{12})^T$, then

$$
\begin{aligned}
\dot{V} &= z^T[P(A_{11} + \gamma I) + (A_{11} + \gamma I)^T P]z + \varepsilon^{-1} e^{2\gamma d} z^T PA_{d12} A_{d12}^T Pz \\
&\quad + \varepsilon^{-1} e^{2\gamma d} z^T PA_{d11} XA_{d11}^T Pz + \varepsilon z^T X^{-1} z - \varepsilon^{-1} z^T PA_{12} A_{12}^T Pz. \tag{3.19}
\end{aligned}
$$

If the inequality (3.15) is satisfied, then the derivative of the Lyapunov–Krasovskii functional is negative definite and theorem 1 is proved.   □

In the rest of this section, the adaptive controller which can drive the trajectory of system in (3.1) onto the sliding surface is designed. For system in (3.1), consider the following controller:

$$
u = u_1 + u_2, \tag{3.20}
$$

where

$$
u_1 = -\frac{1}{m} sgn(\tilde{C}x) |\tilde{C}x|, \tag{3.21}
$$

$$
\begin{aligned}
u_2 &= -\frac{1}{m}\left[ \frac{1+g}{2\tilde{C}x} \|x\|^2 + e_0 sgn(\tilde{C}x) + e_1 sgn(\tilde{C}x)\|x\| \right. \\
&\quad \left. + \frac{1}{2} sgn(\tilde{C}x) |\tilde{C}x| \|\tilde{C}A_d\|^2 + \theta_1^T \xi_1(x) + \theta_2^T \xi_2(x) \right] - \beta \tilde{C}x, \tag{3.22}
\end{aligned}
$$

where $\beta$ is a positive scalar and $\phi$, $g$, $e_0$, $e_1$, $\theta_1$ and $\theta_2$ are adaptive parameters, whose adaptive laws are as follows

$$
\dot{e}_0 = k_0 |\tilde{C}x|, \dot{e}_1 = k_1 |\tilde{C}x| \|x\|, \dot{\theta}_1 = \Lambda_1 \tilde{C}x\xi_1(x), \dot{\theta}_2 = \Lambda_1 \tilde{C}x\xi_1(x), \dot{g} = k_2 \|x\|^2, \tag{3.23}
$$

where $\Lambda_1$ and $\Lambda_2$ are positive matrices with proper dimensions, $k_1$, $k_2$ and $k_3$ are positive scalars. In the above controller, $u_2$ contains two fuzzy logic systems used to approximate the unknown nonlinear functions.

**Theorem 2.** *By using the control law in (3.20)–(3.22) and the adaptive laws in (3.23), the states of system in (3.1) converge to the previously defined sliding surface in (3.14).*

*Proof.* The candidate for the Lyapunov–Krasovskii functional for system in (3.1) is considered as follows:

$$V = \frac{1}{2m}S^2 + \frac{1}{2m}\int_{t-d}^{t}|f_2(x(\tau))|^2\,d\tau + \frac{(\beta_2^{*2}+1)}{2m}\int_{t-d}^{t}\|x(\tau)\|^2\,d\tau$$

$$+ \frac{1}{2m}\left(k_0^{-1}\bar{e}_0^2 + k_1^{-1}\bar{e}_1^2 + \bar{\theta}_1^T\Lambda_1^{-1}\bar{\theta}_1 + \bar{\theta}_2^T\Lambda_2^{-1}\bar{\theta}_2 + \frac{1}{2}k_2^{-1}\bar{g}^2\right), \quad (3.24)$$

where $\bar{e}_0 = e_0^* - e_0, \bar{e}_1 = e_1^* - e_1, \bar{\theta}_1 = \theta_1^* - \theta_1, \bar{\theta}_2 = \theta_2^* - \theta_2, \bar{g} = g^* - g$.
The time derivative along the state trajectories of the system in (3.1) is

$$\dot{V} = \frac{1}{m}S\dot{S} + \frac{1}{2m}[|f_2(x(t))|^2 - |f_2(x(t-d))|^2] + \frac{(\beta_2^{*2}+1)}{2m}[\|x(t)\|^2$$

$$- \|x(t-d)\|^2] + \frac{1}{m}\left(k_0^{-1}\bar{e}_0\dot{\bar{e}}_0 + k_1^{-1}\bar{e}_1\dot{\bar{e}}_1 + \bar{\theta}_1^T\Lambda_1^{-1}\dot{\bar{\theta}} + \bar{\theta}_2^T\Lambda_2^{-1}\dot{\bar{\theta}}_2\right.$$

$$\left. + \frac{1}{2}k_2^{-1}\bar{g}\dot{\bar{g}}\right). \quad (3.25)$$

According to assumption 1, we get

$$\|\tilde{C}\Delta A\| \le \gamma_1^*,$$
$$\|\tilde{C}\Delta A_d\| \le \gamma_2^*. \quad (3.26)$$

Based on (3.8) and (3.26), we have

$$\frac{1}{m}S\dot{S} = \frac{1}{m}|\tilde{C}x|\left[\|\tilde{C}\Delta A\|\|x\| + \|\tilde{C}\Delta A_d\|\|x(t-d)\|\right]$$

$$+ \frac{1}{m}\tilde{C}x[\tilde{C}Ax + \tilde{C}A_d x(t-d) + F_1(x(t)) + F_2(x(t-d))]$$

$$+ \frac{1}{m}|\tilde{C}x||\Delta F| + \frac{1}{m}\tilde{C}x\phi(u)$$

$$\le \frac{1}{m}|\tilde{C}x|\left[\gamma_1^*\|x\| + \gamma_2^*\|x(t-d)\|\right] + \frac{1}{m}\tilde{C}x[\tilde{C}Ax + F_1(x(t))]$$

$$+ \frac{1}{m}\left[\frac{1}{2}|\tilde{C}x|^2\|\tilde{C}A_d\|^2 + \frac{1}{2}\|x(t-d)\|^2\right]$$

$$+ \frac{1}{m}\left[\frac{1}{2}|\tilde{C}x|^2 + \frac{1}{2}|F_2(x(t-d))|^2\right]$$

$$+ \frac{1}{m}|\tilde{C}x|[|\Delta F| + |w|] + \frac{1}{m}\tilde{C}x(mu). \quad (3.27)$$

Using (3.6), one can obtain

$$|\Delta F| + |w| \leq (\alpha_0^* + \omega^*) + \alpha_1^* \|x\| + \alpha_2^* \|x(t - d)\|. \tag{3.28}$$

Substituting (3.26)–(3.28) into (3.25), one can obtain

$$
\begin{aligned}
\dot{V}(t) \leq \; & \frac{1}{m}\tilde{C}x[(\alpha_0^* + \omega^*) + (\alpha_1^* + \gamma_1^*)\|x\| + (\alpha_2^* + \gamma_2^*)\|x(t - d)\|] \\
& + \frac{1}{m}\tilde{C}x[\tilde{C}Ax + F_1(x(t))] + \frac{1}{m}|\tilde{C}x|^2\|\tilde{C}A_d\|^2 + \frac{1}{2m}\|x(t-d)\|^2 + \frac{1}{2m}|\tilde{C}x|^2 \\
& + \frac{1}{2m}|F_2(x(t - d))|^2 + \tilde{C}xu + \frac{1}{2m}|F_2(x(t))|^2 - \frac{1}{2m}|F_2(x(t - d))|^2 \\
& + \frac{(\beta_2^{*2} + 1)}{2m}\|x\|^2 - \frac{(\beta_2^{*2} + 1)}{2m}\|x(t - d)\|^2 \\
& + \frac{1}{m}\left(k_0^{-1}\bar{e}_0\dot{\bar{e}}_0 + k_1^{-1}\bar{e}_1\dot{\bar{e}}_1 + \bar{\theta}_1^T \Lambda_1^{-1}\dot{\bar{\theta}}_1 + \bar{\theta}_2^T \Lambda_2^{-1}\dot{\bar{\theta}}_2 + \frac{1}{2}k_2^{-1}\bar{g}\dot{\bar{g}}\right), \tag{3.29}
\end{aligned}
$$

$\beta_0^*$, $\beta_1^*$ and $\beta_2^*$ are considered as

$$\alpha_0^* + \omega^* = \beta_0^*, \quad \alpha_1^* + \gamma_1^* = \beta_1^*, \quad \alpha_2^* + \gamma_2^* = \beta_1^*. \tag{3.30}$$

The unknown nonlinear continuous functions are defined as the following

$$
\begin{aligned}
\tilde{C}Ax + F_1(x(t)) &= \tilde{F}_1(x(t)), \\
\frac{1}{2\tilde{C}x}|F_2(x(t))|^2 &= \tilde{F}_2(x(t)). \tag{3.31}
\end{aligned}
$$

From the knowledge of Sect. 3.3, two fuzzy logic systems $\hat{F}_1(x|\theta_1)$ and $\hat{F}_2(x|\theta_2)$ can be used to approximate $\tilde{F}_1(x(t))$ and $\tilde{F}_2(x(t))$.

$$\hat{F}_1(x|\theta_1) = \theta_1^T \xi_1(x), \; \hat{F}_2(x|\theta_2) = \theta_2^T \xi_2(x). \tag{3.32}$$

According to lemma1, there exist optimal approximation parameters $\theta_1^*$ and $\theta_2^*$, such that $\theta_1^{*T} \xi_1(x)$ and $\theta_2^{*T} \xi_2(x)$ can approximate $\tilde{F}_1(x(t))$ and $\tilde{F}_2(x(t))$ to any desired degree. We assume that the minimum approximation error satisfies the following assumption.

**Assumption 4.** *The approximation error between* $\tilde{F}_1(x(t))$, $\tilde{F}_2(x(t))$, $\theta_1^{*T} \xi_1(x)$ *and* $\theta_2^{*T} \xi_2(x)$ *satisfies the following inequality*

$$[\tilde{F}_1 + \tilde{F}_2 - \theta_1^{*T} \xi_1(x) - \theta_2^{*T} \xi_2(x)] \leq h_0^* + h_1^* \|x\|, \tag{3.33}$$

where $h_0^*$ and $h_1^*$ are unknown scalars. These two parameters $\theta_1^*$ and $\theta_2^*$ will be learned through the adaptive algorithms in (3.23).

Substituting controller in (3.21), (3.30)–(3.32) into the above inequality, we can obtain

$$
\dot{V}(t) \leq \frac{1}{m}\left|\tilde{C}x\right|\left[\beta_0^* + \beta_1^*\|x\|\right] + \frac{1}{m}\tilde{C}x[\tilde{F}_1(x(t)) + \tilde{F}_2(x(t))] + \frac{1}{m}\left[\frac{1}{2}\left|\tilde{C}x\right|^2\right.
$$
$$
\left. + \frac{\beta_2^{*2}}{2}\|x(t-d)\|^2\right] + \frac{1}{2m}\left|\tilde{C}x\right|^2\left\|\tilde{C}\Delta A_d\right\|^2 + \frac{1}{2m}\left|\tilde{C}x\right|^2
$$
$$
+ \tilde{C}x\left[-\frac{1}{m}sgn(\tilde{C}x)\left|\tilde{C}x\right|\right] + \frac{1}{2m}\|x(t-d)\|^2 + \tilde{C}xu_2
$$
$$
+ \frac{(\beta_{2*}^2 + 1)}{2m}\|x\|^2 - \frac{(\beta_{2*}^2 + 1)}{2m}\|x(t-d)\|^2
$$
$$
+ \frac{1}{m}\left(k_0^{-1}\bar{e}_0\dot{\bar{e}}_0 + k_1^{-1}\bar{e}_1\dot{\bar{e}}_1 + \bar{\theta}_1^T\Lambda_1^{-1}\dot{\bar{\theta}}_1 + \bar{\theta}_2^T\Lambda_2^{-1}\dot{\bar{\theta}}_2 + \frac{1}{2}k_2^{-1}\bar{g}\dot{\bar{g}}\right). \quad (3.34)
$$

By doing some algebraic operations and replacing $\beta_2^{*2}$ with $g*$ and using controller in (3.22), we get

$$
\dot{V}(t) \leq \frac{1}{m}\tilde{C}x[\tilde{F}_1(x(t)) + \tilde{F}_2(x(t)) - \theta_{1*}^T\xi_1(x) - \theta_{2*}^T\xi_2(x)]
$$
$$
+ \frac{1}{m}\tilde{C}x[\bar{\theta}_{1T}\xi_1(x) + \bar{\theta}_{2T}\xi_2(x)] + \frac{1}{m}\left|\tilde{C}x\right|[\beta_{0*} + \beta_{1*}\|x\|] - \frac{e_0}{m}\left|\tilde{C}x\right|
$$
$$
- \frac{e_1}{m}\left|\tilde{C}x\right|\|x\| - \frac{1+g}{2m}\|x\|^2 + \frac{(1+g*)}{2m}\|x\|^2 - \beta(\tilde{C}x)^2
$$
$$
+ \frac{1}{m}\left(k_0^{-1}\bar{e}_0\dot{\bar{e}}_0 + k_1^{-1}\bar{e}_1\dot{\bar{e}}_1 + \bar{\theta}_{1T}\Lambda_1^{-1}\dot{\bar{\theta}}_1 + \bar{\theta}_{2T}\Lambda_2^{-1}\dot{\bar{\theta}}_2 + \frac{1}{2}k_2^{-1}\bar{g}\dot{\bar{g}}\right). \quad (3.35)
$$

Letting $\beta_0^* + h_0^* = e_0^*$, $\beta_1^* + h_1^* = e_1^*$ and using assumption 4, we have

$$
\dot{V}(t) \leq \frac{1}{m}\tilde{C}x[\bar{\theta}_1^T\xi_1(x) + \bar{\theta}_2^T\xi_2(x)] + \frac{1}{m}\left|\tilde{C}x\right|[(e_0^* - e_0) + (e_1^* - e_1)\|x\|]
$$
$$
+ \frac{1}{2m}[(1+g*) - (1+g)]\|x\|^2 - \beta(\tilde{C}x)^2 + \frac{1}{m}(k_0^{-1}\bar{e}_0\dot{\bar{e}}_0 + k_1^{-1}\bar{e}_1\dot{\bar{e}}_1
$$
$$
+ \bar{\theta}_1^T\Lambda_1^{-1}\dot{\bar{\theta}}_1 + \bar{\theta}_2^T\Lambda_2^{-1}\dot{\bar{\theta}}_2 + \frac{1}{2}k_2^{-1}\bar{g}\dot{\bar{g}}). \quad (3.36)
$$

Substituting adaptive laws in (3.23) into (3.36), we get the following inequality

$$
\dot{V}(t) \leq -\beta(\tilde{C}x)^2. \quad (3.37)
$$

According to the corollary of Barbalat's lemma in Appendix 1, from (3.37) it is easy to see that the dynamics of system in (3.1) will enter into the desired sliding mode. So based on Lyapunov–Krasovskii stability theorem in Appendix 1, the asymptotical stability of the resulting closed-loop system is proved.                □

*Remark 1.* A dead-zone input function is only a special case of nonlinear functions in this form. Therefore, it is straightforward to ensure that the developed control law of SMC can be applied to control the systems with other bounded nonlinearities in this form e.g. backlash hysteresis.

*Remark 2.* In the control law in (3.20), we can use high slope saturation function $sat(\frac{S}{\delta})$ instead of signum function, $sgn(S)$ where $\delta$ is a small positive constant. In addition, $sgn(S)$ can be replaced by a fuzzy logic system in the same idea with [18] and [23].

Controlling large-scale systems in engineering applications, due to their widely applicability to many practical systems such as power systems and spacecrafts has become one of the most interesting topics [35, 37, 41]. The proposed controller in this section can be applied to large-scale systems.

Consider a class of uncertain interconnected large-scale system with state delays composed of N subsystems:

$$\dot{x}_i = (A_i + \Delta A_i(t))x_i + (A_{di} + \Delta A_{di}(t))x_i(t - d_i) + B_i(\phi_i(u_i))$$

$$+F_i(x_i, x_i(t - d_i), \sigma_i(t))) + \sum_{i \neq j, j=1}^{N} B_j H_{ij}(x_j(t - \tau_{ij}), t) \,,$$

$$x_i(t) = \vartheta_i(t), \qquad t \in [-\bar{d}_i \quad 0] \,, \qquad for \ i = 1, \ldots, N \,.$$

$$y_i = C_i x_i \,, \qquad for \ i = 1, \ldots, N \tag{3.38}$$

$H_{ij}$ are the interconnected matrices of the jth subsystem to the ith subsystem. $d_i$ and $\tau_{ij}$ are unknown fixed delays.

**Assumption 5.** *There exist some unknown positive constants $\varsigma_{ij}$, such that*

$$\left\| H_{ij}(x_j(t - \tau_{ij}), t) \right\| \leq \varsigma_{ij} \tag{3.39}$$

**Corollary 1.** *For the uncertain system given in (3.38) which satisfies assumptions 1–5, if the control input is given by (3.20)–(3.22) with the adaptive laws by (3.23), then the closed-loop system is asymptotically stable.*

*Proof.* Satisfying assumption 5, the controller can be implemented without using any coupling state. It is really decentralized and the controller can be used separately for any subsystem. The proof is omitted.                □

## 3.5  Observer Design

In many practical systems, only partial information of the system states is available, so it is important to design a state observer and propose a control scheme which uses only the observed states. In this section, an adaptive observer is designed for system in (3.1), which utilizes only the observed states. Based on Lyapunov–Krasovskii stability theorem, stability of observed states and the states of system in (3.1) is proved.

For system in (3.1) the following observer is considered

$$\dot{\hat{x}} = A\hat{x} + Bmu(t) + B\hat{\theta}_1(y - C\hat{x}),  \tag{3.40}$$

$\hat{x} \in R^n$ is the vector of observed states and $\hat{\theta}_1$ is an adaptive parameter.

Since the time-delay is not assumed to be known, the delayed states are not used in the dynamics of the proposed observer. Furthermore, the uncertainties, disturbances and unknown functions are not applied in the observer dynamics. To prove the stability of the system with observer the following assumptions are considered.

**Assumption 6.** *There exist matrix $L$ and positive matrix $P$ satisfying*

$$P(A - LC) + (A - LC)^T P = -Q_1,$$
$$B^T P = C,  \tag{3.41}$$

where $Q_1$ is also a positive definite matrix.

**Theorem 3.** *If assumption 1 and the following inequality are satisfied*

$$P[(A_{11} + \gamma I) + (A_{11} + \gamma I)^T - \varepsilon^{-1} A_{12} A_{12}^T]P < 0  \tag{3.42}$$

then by choosing the sliding surface as the following

$$S = \left[ \frac{-\varepsilon^{-1}}{2} A_{12}^T P \quad 1 \right] \hat{x} = 0,  \tag{3.43}$$

where P and X are positive matrices and $\varepsilon$ is a positive scalar, the sliding motion will be exponentially stable with a given decay rate $\gamma$.

To rewrite the sliding surface we have

$$S = \tilde{D}\hat{x} = [-D \quad I][\hat{x}_1 \quad \hat{x}_2^T]^T = -D\hat{x}_1 + \hat{x}_2 = 0,  \tag{3.44}$$

$$\dot{\hat{x}}_1 = A_{11}\hat{x}_1 + A_{12}D\hat{x}_1.  \tag{3.45}$$

*Proof.* For system in (3.43), the following transformation is considered

$$\hat{x}_1 = e^{-\gamma t}\hat{z}.  \tag{3.46}$$

We choose the Lyapunov candidate as

$$V = \hat{z}^T P \hat{z}. \tag{3.47}$$

Taking the time derivative of $V$, one can get

$$\dot{V} = 2\hat{z}^T P \dot{\hat{z}} = 2\hat{z}^T P(A_{11} + A_{12}D + \gamma I)\hat{z} = \hat{z}^T [P(A_{11} + \gamma I) + (A_{11} + \gamma I)^T P]\hat{z}$$
$$+ 2\hat{z}^T PA_{12}D\hat{z}. \tag{3.48}$$

If $D = \frac{-\varepsilon^{-1}}{2}A_{12}^T P$ and the inequality (3.42) is satisfied, then the derivative of the Lyapunov function is negative definite and based on Lyapunov stability theorem in Appendix 1, theorem 3 is proved. □

In the rest of this section, we design a controller based on the observer in (3.40), which makes the closed loop system stable. For systems in (3.1) and (3.40), the following controller is considered

$$u = -\frac{1}{m}(\hat{\theta}_1(y - C\hat{x}) + \hat{\theta}_2^T \xi_2(\hat{x}) + \hat{\theta}_3^T \xi_3(e) + \hat{\theta}_4) - \beta S, \tag{3.49}$$

where $\beta$ is a positive scalar and $\theta_1$, $\theta_4 \in \mathfrak{R}$, $\theta_2$, $\theta_3 \in \mathfrak{R}^n$ are adaptive parameters, whose adaptive laws are as follows

$$\dot{\hat{\theta}}_1 = k_1 \| e^T PB \|,$$
$$\dot{\hat{\theta}}_2 = S \Lambda_2 \xi_2(\hat{x}),$$
$$\dot{\hat{\theta}}_3 = S \Lambda_3 \xi_3(e),$$
$$\dot{\hat{\theta}}_4 = k_4 S. \tag{3.50}$$

$\Lambda_2$, $\Lambda_3 \in \mathfrak{R}^{n \times n}$ are positive definite matrices and $k_1$, $k_4$ are positive scalars.

**Theorem 4.** *Consider system in (3.1) and the adaptive observer in (3.40), subjected to assumption 6. If the control input in (3.49) and adaptive laws in (3.50) are selected, then the states $\hat{x}$ of the adaptive observer and the error vector $e$ converge to zero asymptotically.*

*Proof.* From (3.1) and (3.32) the following error dynamics can be obtained:

$$\dot{e}(t) = Ae(t) + \Delta A(t)x(t) + A_d x(t-d) + \Delta A_d(t)x(t-d) + B(F(x, x(t-d), \sigma(t)))$$
$$-\hat{\theta}_1(y - C\hat{x}) + w(u)). \tag{3.51}$$

To prove the asymptotically stability of system in (3.1), it must be shown that $\hat{x}$ and $e$ converge to zero. For this purpose the following Lyapunov–Krasovskii functional is selected.

$$V = \frac{1}{2}S^2 + e^T Pe + \frac{1}{\rho_5} \int_{t-d}^{t} |F_2(x(\tau))|^2 \, d\tau + \beta_1 \int_{t-d}^{t} (\|\hat{x}(\tau)\|^2 + \|e(\tau)\|^2) d\tau$$

$$+ \frac{1}{2}(2k_1^{-1}\bar{\theta}_1^2 + \bar{\theta}_2^T \Lambda_2^{-1}\bar{\theta}_2 + \bar{\theta}_3^T \Lambda_3^{-1}\bar{\theta}_3 + k_4^{-1}\bar{\theta}_4^2), \tag{3.52}$$

where $\bar{\theta}_i = \theta_i^* - \hat{\theta}_i$ for $i = 1, \ldots, 4$.

The time derivative along the trajectories of system states and the observer will be

$$\dot{V} = S\dot{S} + 2e^T P\dot{e} + \frac{1}{\rho_5}(|F_2(x(t))|^2 - |F_2(x(t-d))|^2) + \beta_1[(\|\hat{x}(t)\|^2 + \|e(t)\|^2)$$

$$- (\|\hat{x}(t-d)\|^2 + \|e(t-d)\|^2)] + k_1^{-1}\bar{\theta}_1\dot{\bar{\theta}}_1 + \bar{\theta}_2^T \Lambda_2^{-1}\dot{\bar{\theta}}_2 + \bar{\theta}_3^T \Lambda_3^{-1}\dot{\bar{\theta}}_3$$

$$+ k_4^{-1}\bar{\theta}_4\dot{\bar{\theta}}_4. \tag{3.53}$$

Based on (3.6) and assumption 6, one can get

$$S\dot{S} + 2e^T P\dot{e} = \tilde{D}x(\tilde{D}A\hat{x} + mu(t) + \hat{\theta}_l(y - C\hat{x}))$$

$$+ e^T[P(A - LC) + (A - LC)^T P]e + 2e^T PLCe$$

$$+ 2e^T P(\Delta A(t)x(t) + A_d x(t - d)$$

$$+ \Delta A_d(t)x(t - d)) + 2e^T PB(F_1(x) + F_2(x(t - d))$$

$$+ \Delta F(x(t), x(t - d), \sigma(t)))$$

$$- 2e^T PB\hat{\theta}_l(y - C\hat{x}) + 2e^T PBw(u) \le \tilde{D}x(\tilde{D}A\hat{x} + mu(t)$$

$$+ \hat{\theta}_l(y - C\hat{x})) - e^T Q_1 e$$

$$+ \|B^T Pe\|^2 \left(\rho_1 + \rho_2 + \rho_4 + \rho_5 + \frac{1}{\mu} + \frac{2}{\gamma}(\alpha_1^{*2} + \alpha_2^{*2})\right)$$

$$+ (\|e\|^2 + \|\hat{x}\|^2)\left(\gamma + \frac{\delta_1^2}{\rho_1}\right) + (\|e(t - d)\|^2$$

$$+ \|\hat{x}(t - d)\|^2)\left(\gamma + \frac{\delta_2^2}{\rho_2} + \frac{1}{\rho_3}\right)$$

$$+ \rho_3 \|e^T PA_d\|^2 + \mu e^T PLL^T Pe + \frac{1}{\rho_4}|F_1(x)|^2$$

$$+ \frac{1}{\rho_5}|F_2(x(t - d))|^2 - 2e^T PB\hat{\theta}_l(y - C\hat{x})$$

$$+ 2e^T PBw(u) + 2\alpha_0^* \|e^T PB\|. \tag{3.54}$$

If we let

$$\rho_1 + \rho_2 + \rho_4 + \rho_5 + \frac{1}{\mu} + \frac{2}{\gamma}(\alpha_1^2 + \alpha_2^2) = 2\theta_1^*, \tag{3.55}$$

$$\gamma + \frac{\delta_2^2}{\rho_2} + \frac{1}{\rho_3} = \beta_1, \tag{3.56}$$

$$\gamma + \frac{\delta_1^2}{\rho_1} + \beta_1 = \beta_2, \tag{3.57}$$

we will have

$$\dot{V} \leq \tilde{D}x(\tilde{D}A\hat{x} + mu(t) + \hat{\theta}_I(y - C\hat{x})) - e^T(Q_1 - \rho_3 e^T PA_d A_d^T Pe$$
$$- \mu e^T PLL^T Pe - \beta_2 I)e + \beta_2 \|\hat{x}\|^2 + \frac{1}{\rho_4}|F_1(x)|^2 + \frac{1}{\rho_5}|F_2(x)|^2$$
$$+ 2\|e^T PB\|(w^* + \alpha_0^*) + \bar{\theta}_2^T \Lambda_2^{-1}\dot{\hat{\theta}}_2 + \bar{\theta}_3^T \Lambda_3^{-1}\dot{\hat{\theta}}_3 + k_4^{-1}\bar{\theta}_4\dot{\hat{\theta}}_4. \tag{3.58}$$

The following assumptions are considered on the nonlinear functions $F_1(x(t))$ and $F_2(x(t))$

$$|F_1(x(t))| = |F_1(\hat{x}(t) + e(t))| \leq \sqrt{\eta_1}G_1(\hat{x}) + \sqrt{\eta_2}G_2(e),$$
$$|F_2(x(t))| = |F_2(\hat{x}(t) + e(t))| \leq \sqrt{\eta_3}H_1(\hat{x}) + \sqrt{\eta_4}H_2(e). \tag{3.59}$$

Two fuzzy logic systems $\hat{F}_1(x|\hat{\theta}_2)$ and $\hat{F}_2(x|\hat{\theta}_3)$ are used to approximate the unknown nonlinear terms appeared in the inequality (3.58).

$$\hat{F}_1(x|\hat{\theta}_2) = \hat{\theta}_2^T \xi_2(\hat{x}), \quad \hat{F}_2(x|\hat{\theta}_3) = \hat{\theta}_3^T \xi_3(e). \tag{3.60}$$

According to lemma1, there exist optimal approximation parameters $\theta_2^*$ and $\theta_3^*$ such that $\theta_1^{*T}\xi_1(x)$ and $\theta_2^{*T}\xi_2(x)$ can approximate the following nonlinear functions.

$$\frac{1}{\tilde{D}x}(\beta_2\|\hat{x}\|^2 + \frac{2\eta_1}{\rho_4}G_1^2(\hat{x}) + \frac{2\eta_3}{\rho_5}H_1^2(\hat{x})) + \tilde{D}A\hat{x} = \theta_2^{*T}\xi_2(\hat{x}) + \omega_2,$$

$$\frac{1}{\tilde{D}x}\left(\frac{2\eta_2}{\rho_4}G_2^2(e) + \frac{2\eta_4}{\rho_5}H_2^2(e) + 2\|e^T PB\|(w^* + \alpha_0^*)\right) = \theta_3^{*T}\xi_3(e) + \omega_3,$$

$$\tag{3.61}$$

where $\omega_2$ and $\omega_3$ are the fuzzy approximation errors. Considering

$$Q_1 - \rho_3 e^T PA_d A_d^T Pe - \mu e^T PLL^T Pe - \beta_2 I = Q, \tag{3.62}$$

we get

$$\dot{V} \leq -e^T Q e + \tilde{D} x (m u(t) + \hat{\theta}_1 (y - C\hat{x}) + {\theta_2^*}^T \xi_2(\hat{x}) + \omega_2$$
$$+ {\theta_3^*}^T \xi_3(e) + \omega_3) \bar{\theta}_{2T} \Lambda_2^{-1} \dot{\hat{\theta}}_2 + \bar{\theta}_{3T} \Lambda_3^{-1} \dot{\hat{\theta}}_3 + k_4^{-1} \bar{\theta}_4 \dot{\hat{\theta}}_4. \quad (3.63)$$

If we let $\omega_2 + \omega_3 = \theta_4^*$, we can obtain

$$\dot{V} \leq -e^T Q' e - \beta S^2. \quad (3.64)$$

According to corollary of Barbalat's lemma in Appendix 1, it can be shown from (3.64) that error and the sliding surface converge to zero, so the states of system in (3.1) are stable.                                                                              □

## 3.6  Application of the Proposed Controller to Chaos Synchronization

In the most synchronization approaches used for the chaotic systems, master–slave or drive–response formulism is employed. In essence, the chaos synchronization problem entails controlling the dynamic behavior of a "slave" system by means of a control input, such that its oscillation, following a period of transition, mimics that of the "master" system. A large number of developed methods for chaos synchronization assume the chaotic systems involved are identical, with known or unknown parameters. Nevertheless, some often in practice, in cases such as laser array, biological systems and cognitive processes, it is rarely assumed that the structure of drive and response systems are exactly identical. Therefore, it is essential to investigate synchronization of two different chaotic systems.

In this section the parameter differences between the master system and the slave system are considered as the system uncertainties. They are assumed to be time-varying but norm-bounded. In what follows, the results in Theorems 1 and 2 are invoked to synchronize two non-identical chaotic systems.

Consider the master system described by

$$\dot{y} = A y(t) + A_d y(t - d) + B F_1(y(t)) + B F_2(y(t - d)). \quad (3.65)$$

The slave or response system is given by

$$\dot{x} = (A + \Delta A) x(t) + (A_d + \Delta A_d) x(t - d) + B F_1(x(t)) + B F_2(x(t - d))$$
$$+ B \Delta F(x(t), x(t - d), \sigma(t)) + B \Phi(u). \quad (3.66)$$

**Assumption 7.** *The nonlinear functions in system (3.1) and (3.2) satisfy the following inequalities:*

$$|F_1(x(t)) - F_1(y(t))| \leq \vartheta_1 f_1(e(t)),$$
$$|F_2(x(t-d)) - F_2(y(t-d))| \leq \vartheta_2 f_2(e(t-d)), \tag{3.67}$$

where $\vartheta_1$ and $\vartheta_2$ are unknown positive scalars and $f_1(.)$ and $f_2(.)$ are unknown nonlinear functions.

**Assumption 8.** $\Delta F(x(t), x(t-d), \sigma(t))$ *is an uncertain function which is unknown, but bounded.*

$$|\Delta F(x(t), x(t-d), \sigma(t))| \leq \alpha_0^*, \tag{3.68}$$

where $\alpha_0^*$ is an unknown positive scalars.

It is noted that the condition $\|y(t)\| \leq l_1^* < \infty$ can be easily satisfied since one of the properties of chaotic system is that the chaotic trajectory is limitary where $l_1^*$ is an unknown and sufficiently constant.

Subtracting (3.66) from (3.65), the following error system equation is formed:

$$\dot{e} = Ae(t) + \Delta Ax(t) + A_d e(t-d) + \Delta A_d x(t-d) + BF_1(x(t))$$
$$- BF_1(y(t)) + BF_2(x(t-d)) - BF_2(y(t-d))$$
$$+ B\Delta F(x(t), x(t-d), \sigma(t)) + B\Phi(u). \tag{3.69}$$

The main objective is to design a suitable controller to synchronize the drive and response systems such that

$$\|x - y\| \to 0 \text{ as } t \to \infty \Rightarrow \|e\| \to 0 \text{ as } t \to \infty. \tag{3.70}$$

By applying the designed controller in theorems 2 and 3, the above objective can be reached.

**Corollary 2.** *Consider the sliding surface is taken as follows*

$$S = \tilde{C}e = [-C \ 1][e_1^T e_2]^T, \tag{3.71}$$

where $C = -\varepsilon^{-1}(PA_{12})^T$. If assumption 1 is met and the following inequality is satisfied:

$$P(A_{11} + \gamma I) + (A_{11} + \gamma I)^T P + \varepsilon^{-1} PD_1 D_1^T P$$
$$+ \varepsilon^{-1} e^{2\gamma \bar{d}} PA_{d12} A_{d12}^T P + e^{2\gamma \bar{d}} PA_{d11} XA_{d11}^T P + X^{-1}$$
$$+ \varepsilon^{-1} e^{2\gamma \bar{d}} PD_2 D_2^T P - \varepsilon^{-1}(PA_{12}) \times (PA_{12})^T < 0, \tag{3.72}$$

where P and X are positive matrices, $\varepsilon$ is a positive scalar and $\bar{d}$ is bound of the time delay, then the sliding motion will be exponentially stable depend on the bound of time-delay with a given decay rate $\gamma$.

*Proof.* Proof of this theorem is similar to the proof of theorem 1.                    □

In the rest of this section, we will design the controller which can drive the trajectory of system in (3.69) onto the sliding surface. For system in (3.69), consider the following controller:

$$u = u_1 + u_2, \tag{3.73}$$

where

$$u_1 = -\frac{1}{m}sgn(\tilde{C}e)\left|\tilde{C}e\right|, \tag{3.74}$$

$$u_2 = -\frac{1}{m}\left[\frac{1+g}{2\tilde{C}e}\|e\|^2 + e_0sgn(\tilde{C}e) + e_1sgn(\tilde{C}e)\|e\| \right.$$
$$\left. + \frac{1}{2}sgn(\tilde{C}e)\left|\tilde{C}e\right|\left\|\tilde{C}A_d\right\|^2 + \theta_1^T\xi_1(e) + \theta_2^T\xi_2(e)\right] - \beta\tilde{C}e, \tag{3.75}$$

and $\beta$ is a positive scalar. $\phi$, $g$, $e_0$, $e_1$, $\theta_1$ and $\theta_2$ are adaptive parameters, whose adaptive laws are as follows

$$\dot{e}_0 = k_0\left|\tilde{C}e\right|, \dot{e}_1 = k_1\left|\tilde{C}e\right|\|e\|, \dot{\theta}_1 = \Lambda_1\tilde{C}e\xi_1(e), \dot{\theta}_2 = \Lambda_1\tilde{C}e\xi_1(e), \dot{g} = k_2\|e\|^2,$$
$$\tag{3.76}$$

where $\Lambda_1$ and $\Lambda_2$ are positive matrices with proper dimensions and $k_0$, $k_1$, $k_2$ and $k_3$ are positive scalars.

**Corollary 3.** *By using the above control law in (3.71), $\|e\| \to 0$ as $t \to \infty$, which means that synchronization occurs between system in (3.65) and system in (3.66).*

*Proof.* Given in Appendix 2.                                                           □

*Remark 3.* One important application of chaos synchronization is secure communication. The proposed method for chaos synchronization can be used in this application. Chaotic system in (3.65) is used as transmitter and the information signal is mixed at the transmitter end to generate a chaotic transmit signal, which is then transmitted to the receiver end. The receiver is considered to be the chaotic dynamic system in (3.66). Once synchronization between these two systems is achieved, the information signal is recovered by the receiver.

## 3.7  Simulation Results

In this section three examples are presented to show the effectiveness of the proposed methods. In the First example, the proposed controller in Sect. 3.4 is used to stabilize a large-scale power system. The observer-based adaptive control scheme is applied to a nonlinear time-delay system in the second example. The third example investigates the synchronization of two different Chua's circuits to verify the method introduced in Sect. 3.6.

## 3.7.1 Proposed Adaptive Fuzzy Sliding Mode Controller for Stabilization of the Large-Scale System

In this sub-section, the performance of the proposed control scheme in Sect. 3.4 is evaluated by applying the method to a model of three machines operating onto an infinite-bus system [19]. This system is a large-scale power system which is made up of three subsystems. This large-scale system is described as follows: Subsystem 1:

$$
\begin{bmatrix} \dot{\delta}_1 \\ \dot{x}_1 \end{bmatrix} = \begin{bmatrix} 0 & 1 \\ 0 & -\frac{D_1}{M_1} \end{bmatrix} \begin{bmatrix} \delta_1 \\ x_1 \end{bmatrix} + \begin{bmatrix} 0 \\ 1 \end{bmatrix} \begin{bmatrix} -\frac{y_1 E^2}{M_1} \sin(\delta_1 + 2.83) + \frac{Pm_1}{M_1} \end{bmatrix}
$$

$$
+ \begin{bmatrix} 0 \\ 1 \end{bmatrix} \begin{bmatrix} -\frac{y_2 E^2}{M_1} \sin(\delta_1 - \delta_2 + 3.34) \end{bmatrix}, \tag{3.77}
$$

Subsystem 2:

$$
\begin{bmatrix} \dot{\delta}_2 \\ \dot{x}_2 \end{bmatrix} = \begin{bmatrix} 0 & 1 \\ 0 & -\frac{D_2}{M_2} \end{bmatrix} \begin{bmatrix} \delta_2 \\ x_2 \end{bmatrix} + \begin{bmatrix} 0 \\ 1 \end{bmatrix} \begin{bmatrix} \frac{Pm_2}{M_2} \end{bmatrix} + \begin{bmatrix} 0 \\ 1 \end{bmatrix} \begin{bmatrix} -\frac{y_2 E^2}{M_2} \sin(\delta_2 - \delta_1 - 2.34) \end{bmatrix}
$$

$$
-\frac{y_3 E^2}{M_2} \sin(\delta_2 - \delta_3 - 3.04) \end{bmatrix}, \tag{3.78}
$$

Subsystem 3:

$$
\begin{bmatrix} \dot{\delta}_3 \\ \dot{x}_3 \end{bmatrix} = \begin{bmatrix} 0 & 1 \\ 0 & -\frac{D_3}{M_3} \end{bmatrix} \begin{bmatrix} \delta_3 \\ x_3 \end{bmatrix} + \begin{bmatrix} 0 \\ 1 \end{bmatrix} \begin{bmatrix} \frac{Pm_3}{M_3} \end{bmatrix} + \begin{bmatrix} 0 \\ 1 \end{bmatrix} \begin{bmatrix} -\frac{y_3 E^2}{M_3} \sin(\delta_3 - \delta_2 + 3.04) \end{bmatrix}, \tag{3.79}
$$

with the parameters $\frac{M_1}{y_3 E^2} = \frac{M_2}{y_3 E^2} = \frac{M_3}{y_3 E^2} = 0.5$, $\frac{D_1}{y_3 E^2} = \frac{D_2}{y_3 E^2} = \frac{D_3}{y_3 E^2} = 0.005$, $\frac{Pm_1}{y_3 E^2} = \frac{Pm_2}{y_3 E^2} = \frac{Pm_3}{y_3 E^2} = 0.1$ and $\frac{y_1}{y_3} = \frac{y_2}{y_3} = 1$. The initial conditions and parameters of the dead-zone function are given as follows:

$$
x_i(0) = \delta_i(0) = 0.8, \ e_{0i}(0) = 0.2, \ e_{1i}(0) = 0.2, \ g_i(0) = 0.2,
$$

$$
\theta_{1i}(0) = [0.2 \ 0.2 \ 0.2 \ 0.2 \ 0.2]^T,
$$

$$
m_i = 2, \ b_{1i} = -3, \ b_{2i} = 2, \quad for \quad i = 1, 2, 3. \tag{3.80}
$$

The sliding surfaces are $S_i = \delta_i(t) - 2x_i(t)$ for $i = 1, 2, 3$.

By properly choosing the positive matrices $P_i$, $X_i$ and positive scalar $\varepsilon_i$ for $i = 1, 2, 3$, these sliding surfaces satisfy the condition in theorem 1.

Figures 3.1–3.3 display the state trajectories of subsystems in (3.77)–(3.79) with the control law in (3.20). In these figures, one can see that the asymptotical stabilization of the mentioned system is achieved. Moreover, Figs. 3.4– 3.6 display the dynamics of the sliding surfaces of the subsystems and Figs. 3.7– 3.9 show the

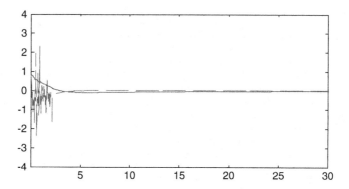

**Fig. 3.1** The trajectories of the states of subsystem 1, $\delta_1$ (*solid line*) and $x_1$ (*dash-dotted*)

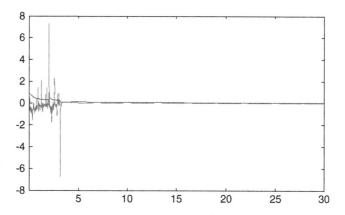

**Fig. 3.2** The trajectories of the states of subsystem 2, $\delta_2$ (*solid line*) and $x_2$ (*dash-dotted*)

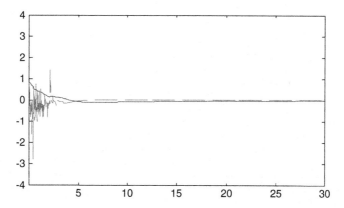

**Fig. 3.3** The trajectories of the states of subsystem 3, $\delta_3$ (*solid line*) and $x_3$ (*dash-dotted*)

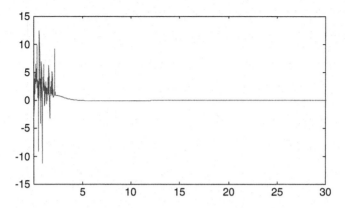

**Fig. 3.4** The dynamics of the sliding surface of subsystem 1

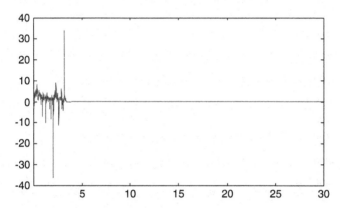

**Fig. 3.5** The dynamics of the sliding surface of subsystem 2

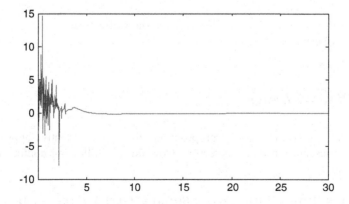

**Fig. 3.6** The dynamics of the sliding surface of subsystem 3

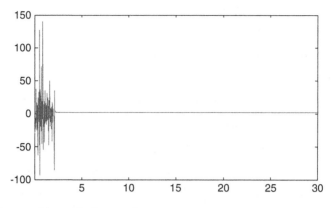

**Fig. 3.7** The control input of subsystem 1

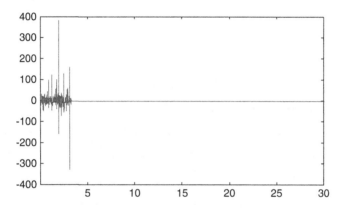

**Fig. 3.8** The control input of subsystem 2

behavior of their control signals. These figures show effectiveness of our proposed methodology performance.

### 3.7.2  Observer Design

To show the effectiveness of the proposed observer-based adaptive approach an example is presented in this sub-section. Consider the following nonlinear time-delay system

$$\dot{x} = \bar{A}(t)x + \bar{A}_d(t)x(t-d) + B(\phi(u) + F_1(x) + F_2(x(t-d)))$$
$$+\Delta F(x(t), x(t-d), \sigma(t)), \tag{3.81}$$

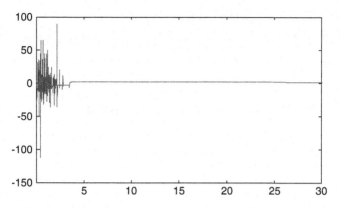

**Fig. 3.9** The control input of subsystem 3

where $x = [x_{11}\ x_{12}\ x_2]^T$. $\bar{A}(t)$ and $\bar{A}_d(t)$ are defined as follows:

$$\bar{A}(t) = \begin{bmatrix} -10 & 1 & 1 \\ 1 & -8 & 1 \\ 5 + \cos(t) & 4 + 2\sin(t) & 2 + \cos(t) \end{bmatrix},$$

$$\bar{A}_d(t) = \begin{bmatrix} 1 & 0 & 1 \\ 0 & 1 & 1 \\ 3 + \sin(t) & 4 + \cos(t) & 2 + \sin(t) \end{bmatrix}. \tag{3.82}$$

The time delay is considered to be $0.2\,s$ and $B = [0\ 0\ 1]^T$ and $F_1(x)$, $F_2(x(t - d))$, $\Delta F(x(t), x(t-d), \sigma(t)))$ are unknown. For simulation we choose them as the following

$$F_1(x) = \frac{1}{2}\frac{\|x\|^2}{1 + \|x\|^2}, \quad F_2(x(t - d)) = \frac{\|x_d\|^2}{3 + \|x_d\|^2},$$

$$\Delta F(x(t), x(t - d), \sigma(t))) = \sin(t)\|x\|^2. \tag{3.83}$$

If we choose $\gamma = 1$, $\varepsilon = 1$ and $P = X = I$, inequality (3.42) is satisfied and the sliding surface is $S(\hat{x}) = \hat{x}_{11} + \hat{x}_{12} + \hat{x}_2 = 0$. The membership functions are chosen as follows

$$\mu_{F_i^1}(x_i) = \exp(-(x_i + 5)^2), \quad \mu_{F_i^1}(x_i) = \exp(-(x_i + 2.5)^2),$$

$$\mu_{F_i^1}(x_i) = \exp(-(x_i)^2), \quad \mu_{F_i^1}(x_i) = \exp(-(x_i - 5)^2),$$

$$\mu_{F_i^1}(x_i) = \exp(-(x_i - 2.5)^2). \tag{3.84}$$

The parameters of the dead-zone function and the initial conditions are selected to be

$$m_1 = 2, \quad b_1 = -3, \quad b_2 = 2,$$

$$\theta_1(0) = 0, \quad \theta_2(0) = \theta_3(0) = [0.1 \ \ 0.1 \ \ 0.1 \ \ 0.1 \ \ 0.1]^T, \quad \theta_4(0) = 0,$$

$$x(t) = [-.2 \ \ 0.9 \ \ -0.6]^T \, for \quad t \in [-0.2 \ 0]. \tag{3.85}$$

The trajectories of the system states and the observed states are shown in Fig. 3.10. Figure 3.11 presents the errors between these two.

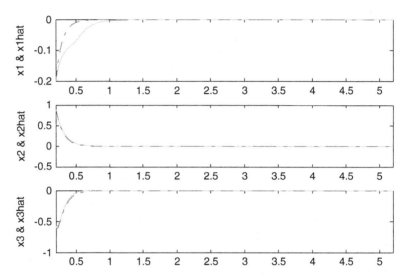

**Fig. 3.10** The trajectories of the states (*solid line*) and state observations (*dash-dotted*)

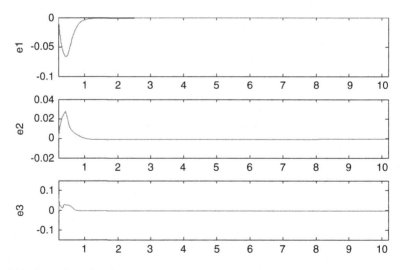

**Fig. 3.11** The trajectories of errors between states and their observations

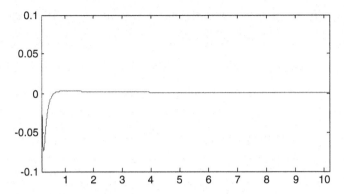

**Fig. 3.12** The dynamics of the sliding surface

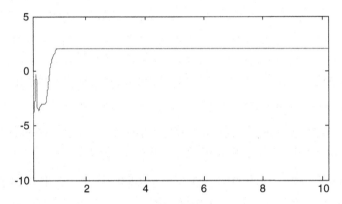

**Fig. 3.13** The control input

As it's obvious from these figures, the states and their observations converge to each other and both tend to zero. Figure 3.12 shows the dynamics of the sliding surface which was designed to control the system. The control input is illustrated in Fig. 3.13.

### 3.7.3  Chaos Synchronization

In this sub-section the proposed controller in (3.73) is used to synchronize two chaotic Chua's circuits. Consider a time-delay Chua's circuit as the master system, described by

$$
\begin{bmatrix} \dot{y}_1 \\ \dot{y}_2 \\ \dot{y}_3 \end{bmatrix} = \begin{bmatrix} 0 & -b & 0 \\ 1 & -1 & 1 \\ 0 & a & -m_1 a \end{bmatrix} \begin{bmatrix} y_1(t) \\ y_2(t) \\ y_3(t) \end{bmatrix} + \begin{bmatrix} -c & 0 & 2c \\ 0 & 0 & -c \\ 0 & 0 & -c \end{bmatrix} \begin{bmatrix} y_1(t-d) \\ y_2(t-d) \\ y_3(t-d) \end{bmatrix}
$$
$$
+ B(F_1(y(t)) + F_2(y(t-d))), \tag{3.86}
$$

with $a = 9$, $b = 14.28$, $c = 0.01$, $m_0 = -\frac{1}{7}$, $m_1 = \frac{2}{7}$, $d = 0.2$ and the nonlinear functions as the following

$$F_1(y(t)) = -\frac{a}{2}(m_0 - m_1)(|y_3(t) + 1| - |y_3(t) - 1|),$$

$$F_2(y(t - d)) = -2c\sin(0.1y_3(t - d)). \tag{3.87}$$

Also the system in (3.65) with the uncertainties is chosen as the slave system. The uncertainties in (3.66) are given as

$$A(t) = \begin{bmatrix} 0 & 0 & 0 \\ 0 & 0 & 0 \\ 0.5\cos(t) & 0.3\sin(2t) & 0.8\sin(2t) \end{bmatrix},$$

$$\Delta A_d(t) = \begin{bmatrix} 0 & 0 & 0 \\ 0 & 0 & 0 \\ \cos(3t) & 0.4 & 1 + \sin(3t) \end{bmatrix},$$

$$\Delta F(x(t), x(t - d), \sigma(t)) = 0.5(|0.1x_3(t) + 0.1| - |0.1x_3(t) - 0.1|). \tag{3.88}$$

Let us define the initial conditions for these two Chua's circuits and the parameters of the dead-zone function as follows:

$$y(t) = (0.1, 0.1, 0.1)^T, \quad x(t) = (0.5, 0.1, 0.3)^T \text{ for } t \in [-0.2, 0]$$

$$e_0(0) = 0.2, \quad e_1(0) = 0.2, \quad g(0) = 0.2,$$

$$\theta_1(0) = \theta_2(0) = [0.2\ 0.2\ 0.2\ 0.2\ 0.2]^T,$$

$$m = 2, b_1 = -3, b_2 = 2. \tag{3.89}$$

The sliding surface is $S = -e_1(t) + 2e_2(t) + e_3(t)$. By properly choosing the positive matrices $P$, $X$ and positive scalar $\varepsilon$, this sliding surface satisfies the condition in Corollary 2.

The problem of the synchronization between two chaotic systems is now transformed into another problem on how to choose the control law to force the error vector to converge to zero, which it is preferred to be done as fast as possible. Figs. 3.14–3.16 display the synchronization between these two chaotic circuits with control law in (3.73). Obviously, the synchronization errors converge asymptotically to the origin as shown in Fig. 3.17. The dynamic of sliding surface are depicted in Fig. 3.18. The control signal $u$ due to our proposed method is shown in Fig. 3.19.

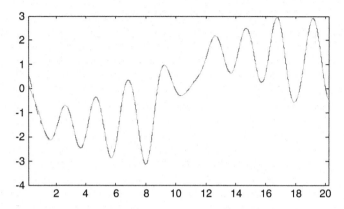

**Fig. 3.14** The trajectories of $y_1$(*solid line*) and $x_1$ (*dash-dotted*)

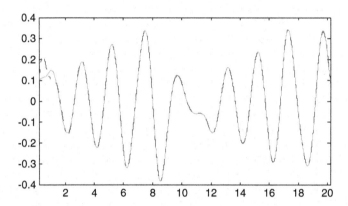

**Fig. 3.15** The trajectories of $y_2$(*solid line*) and $x_2$ (*dash-dotted*)

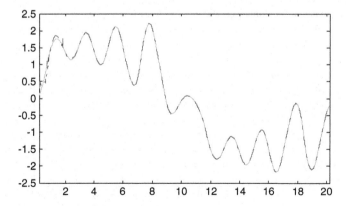

**Fig. 3.16** The trajectories of $y_3$(*solid line*) and $x_3$ (*dash-dotted*)

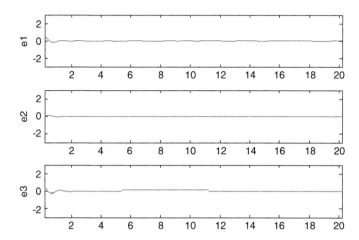

**Fig. 3.17** Synchronization errors

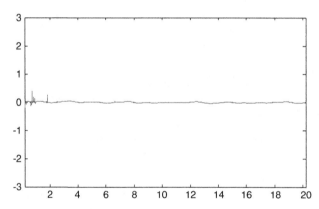

**Fig. 3.18** The dynamics of sliding surface

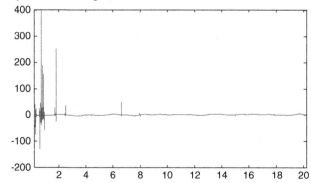

**Fig. 3.19** The control input

## 3.8   Conclusion

In this paper, a class of nonlinear systems with unknown time delay and input nonlinearity is addressed. Non-ideal effects such as unknown parameters, uncertainties, input nonlinearities and unknown time-delay are considered from a practical point of view. By appropriately choosing Lyapunov–Krasovskii functionals, an adaptive fuzzy control scheme is presented for a class of nonlinear systems which guarantees the closed-loop system asymptotically stable. In addition, a stabilizing adaptive fuzzy observer-based control scheme is designed for the same system. We investigate variable structure controllers for a nonlinear system with input nonlinearity and unknown time-delay only requiring the time-delay to be bounded. Simulation example is included to show the effectiveness of the obtained theoretic results. The proposed control laws are also employed for synchronization of two non-identical time-delayed chaotic systems. To verify the effectiveness of the proposed methods, the schemes were applied to two time-delay chua's circuits.

## Appendix 1

In this appendix, some lemmas are reviewed which are needed to derive the main result.

1. Young's inequality [4], Theorem 156: For any two vectors $x$ and $y$, the following inequality holds:

$$x^T y \leq \varepsilon x^T x + \frac{1}{\varepsilon} y^T y, \tag{3.90}$$

   in which $\varepsilon$ is a positive scalar.
2. Barbalat's lemma [10]: Let $f: \Re \rightarrow \Re$ be a uniformly continuous function on $[0, \infty)$. Suppose that $\lim_{t \to \infty} \int_0^t f(\tau) d\tau$ exists and is finite. Then, $f(t) \rightarrow 0$ as $t \rightarrow \infty$.
   Corollary of Barbalat's lemma [25]: If $g, \dot{g} \in L_\infty$ and $g \in L_p$, for some $p \in [1, \infty)$, then $g(t) \rightarrow 0$ as $t \rightarrow \infty$.
3. Lyapunov–Krasovskii Stability Theorem [3, 12] Consider the delayed differential equation

$$\dot{x}(t) = f(t, x_t). \tag{3.91}$$

Suppose that $f$ is continuous and $f : \Re \times C \rightarrow \Re^n$ takes $\Re \times ($ bounded sets of $C)$ into bounded sets of $\Re^n$, and $u, v, w : \Re^+ \rightarrow \Re^+$ are continuous and strictly monotonically nondecreasing functions, $u(s), v(s), w(s)$ are positive for $s > 0$ with $u(0) = v(0) = 0$. If there exists a continuous functional $V : \Re \times C \rightarrow \Re$ such that

$$u(\|x(t)\|) \leq V(t, x_t) \leq v(\|x(t)\|),$$
$$\dot{V}(t, x_t) \leq -w(\|x(t)\|), \tag{3.92}$$

where $\dot{V}$ is the derivative of $V$ along the solution of the above delayed differential equation, then the solution $x = 0$ of this equation is uniformly asymptotical stable.

4. Lyapunov Stability Theorem [10]: Consider the following autonomous differential equation

$$\dot{x}(t) = f(t, x). \tag{3.93}$$

where $f : D \to \Re^n$ is a locally Lipschitz map from a domain $D \subset \Re^n$ into $\Re^n$. Let $x = 0$ be an equilibrium point for (A.A.4) and $D \subset \Re^n$ be a domain containing $x = 0$. Let $V : D \to \Re$ be a continuously differentiable function such that

$$V(0) = 0, and \, V(x) > 0, in \, D - \{0\} \tag{3.94}$$

$$\dot{V}(x) \leq 0, in \, D \tag{3.95}$$

Then, $x = 0$ is stable. Moreover, if

$$\dot{V}(x) < 0, in \, D - \{0\} \tag{3.96}$$

Then $x = 0$ is asymptotically stable.

## Appendix 2: Proof of Corollary 3

Let us define a Lyapunov function as:

$$V = \frac{1}{2m}S^2 + \frac{\vartheta_2^2}{2m}\int_{t-d}^{t}|f_2(e(\tau))|^2\,d\tau + \frac{(\beta_2^{*2}+1)}{2m}\int_{t-d}^{t}\|e(\tau)\|^2\,d\tau$$
$$+ \frac{1}{2m}\left(k_0^{-1}\bar{e}_0^2 + k_1^{-1}\bar{e}_1^2 + \bar{\theta}_1^T \Lambda_1^{-1}\bar{\theta}_1 + \bar{\theta}_2^T \Lambda_2^{-1}\bar{\theta}_2 + \frac{1}{2}k_2^{-1}\bar{g}^2\right), \tag{3.97}$$

$\bar{e}_0 = e_0^* - e_0, \bar{e}_1 = e_1^* - e_1, \bar{\theta}_1 = \theta_1^* - \theta_1, \bar{\theta}_2 = \theta_2^* - \theta_2, \bar{g} = g^* - g.$

The time derivative of $V$ along the trajectories of the dynamic model (3.69) is

$$\dot{V} = \frac{1}{m}S\dot{S} + \frac{\varphi_2^2}{2m}\left[|f_2(e(t-d))|^2\right] + \frac{(\beta_2^{*2}+1)}{2m}\left[\|e(t)\|^2 - \|e(t-d)\|^2\right]$$
$$+ \frac{1}{m}(k_0^{-1}\bar{e}_0\dot{e}_0 + k_1^{-1}\bar{e}_1\dot{e}_1 + \bar{\theta}_1^T \Lambda_1^{-1}\dot{\bar{\theta}} + \bar{\theta}_2^T \Lambda_2^{-1}\dot{\bar{\theta}}_2 + \frac{1}{2}k_2^{-1}\bar{g}\dot{g}). \tag{3.98}$$

Regarding assumptions 1 and 7, one can obtain

$$\frac{1}{m}S\dot{S} = \frac{1}{m}|\tilde{C}e|\left[\|\tilde{C}\varDelta A\|\,\|x\| + \|\tilde{C}\varDelta A_d\|\,\|x(t-d)\|\right]$$

$$+\frac{1}{m}\tilde{C}x\Big[\tilde{C}Ae + \tilde{C}A_d e(t-d) + F_1(x(t)) - F_1(y(t)) + F_2(x(t-d))$$

$$-F_2(y(t-d))\Big] + \frac{1}{m}|\tilde{C}e|\,|\varDelta F| + \frac{1}{m}\tilde{C}x\phi(u)$$

$$\leq \frac{1}{m}|\tilde{C}e|\left[\delta_1 l_1^* + \delta_2 l_1^* + \delta_1\,\|e\| + \delta_2\,\|e(t-d)\|\right]$$

$$+\frac{1}{m}\tilde{C}e[\tilde{C}Ae + \vartheta_1 sgn(\tilde{C}e)f_1(e(t))] + \frac{1}{m}\left[\frac{1}{2}|\tilde{C}e|^2\,\|\tilde{C}A_d\|^2\right.$$

$$+\frac{1}{2}\,\|e(t-d)\|^2\bigg] + \frac{1}{m}\left[\frac{1}{2}|\tilde{C}e|^2 + \frac{\vartheta^2}{2}|f_2(e(t-d))|^2\right]$$

$$+\frac{1}{m}|\tilde{C}e|\,[|\varDelta F| + |w|] + \frac{1}{m}\tilde{C}e(mu). \tag{3.99}$$

The unknown nonlinear continuous functions are defined as follows

$$\tilde{C}Ae + \vartheta_1 sgn(\tilde{C}e)f_1(e(t)) = \tilde{F}_1(e(t)),$$

$$\frac{\vartheta_2^2}{2\tilde{C}e}|f_2(e(t))|^2 = \tilde{F}_2(e(t)). \tag{3.100}$$

We use two fuzzy logic systems $\hat{F}_1(e|\theta_1)$ and $\hat{F}_2(e|\theta_2)$ to approximate $\tilde{F}_1(e(t))$ and $\tilde{F}_2(e(t))$.

$$\hat{F}_1(e|\theta_1) = \theta_1^T\xi_1(e),$$

$$\hat{F}_2(e|\theta_2) = \theta_2^T\xi_2(e). \tag{3.101}$$

Substituting (3.100) into (3.98), one can derive the following inequality

$$\dot{V}(t) \leq \frac{1}{m}|\tilde{C}e|\left[(\alpha_0^* + \omega^* + \delta_1 l_1^* + \delta_2 l_1^*) + \delta_1\,\|e\|\right] + \frac{1}{2m}|\tilde{C}e|^2$$

$$+\frac{\delta_2^2 + 1}{2m}\,\|e(t-d)\|^2 + \frac{1}{m}\tilde{C}e[\tilde{F}_1(e(t))] + \frac{1}{2m}|\tilde{C}e|^2\,\|\tilde{C}A_d\|^2$$

$$+\frac{1}{m}\left[\frac{1}{2}|\tilde{C}e|^2 + \frac{\vartheta^2}{2}|f_2(e(t-d))|^2\right] + \tilde{C}eu$$

$$+\frac{\tilde{C}e}{m}\left(\frac{\vartheta^2}{2\tilde{C}e}|f_2(e(t))|^2\right) - \frac{\vartheta^2}{2m}|f_2(e(t-d))|^2$$

$$+\frac{(\beta_2^{*2} + 1)}{2m}[\|e(t)\|^2 - \|e(t-d)\|^2]$$

$$+ \frac{1}{m} \left( k_0^{-1} \bar{e}_0 \dot{e}_0 + k_1^{-1} \bar{e}_1 \dot{e}_1 + \bar{\theta}_1^T \Lambda_1^{-1} \dot{\bar{\theta}}_1 + \bar{\theta}_2^T \Lambda_2^{-1} \dot{\bar{\theta}}_2 + \frac{1}{2} k_2^{-1} \bar{g} \dot{\bar{g}} \right).$$

(3.102)

$\beta_0^*$, $\beta_1^*$ and $\beta_2^*$ are considered as

$$\alpha_0^* + \omega^* + \delta_1 l_1^* + \delta_2 l_1^* = \beta_0^*, \ \delta_1 = \beta_1^*, \ \delta_2 = \beta_2^*.$$

(3.103)

Replacing the above equality in (3.102), we get the inequality which is the same as (3.34). So by some operation similar to the proof of theorem 2, we get

$$\dot{V} \leq -\beta(\tilde{C}e)^2.$$

(3.104)

Using the corollary of Barbalat's lemma in Appendix 1, it can be shown from (3.104) that synchronization error and the sliding surface converge to zero, so the slave system synchronizes with the master system.

# References

1. Chiang, C.C., Yang, C.C.: Robust Adaptive Fuzzy Sliding Mode Control for a Class of Uncertain Nonlinear Systems with Unknown Dead-Zone. In: 2006 IEEE International Conference on Fuzzy Systems, pp. 492—497. IEEE Press, Vancouver (2006)
2. Grzybowski, J.M.V., Rafikov, M., Balthazar, J.M.: Synchronization of the unified chaotic system and application in secure communication, Commun. Nonlinear Sci Numer. Simul. **14**, 2793–2806 (2009)
3. Hale, J.K., Lunel, S.M.V.: Introduction to functional DiPerential equations. Springer, New York (1993)
4. Hardy, G., Littlewood, J.E., Polya, G.: Inequalities. Cambridge University Press, Cambridge, U.K (1989)
5. Hua, C., Feng, G., Guan, X.: Robust controller design of a class of nonlinear time delay systems via backstepping method. Automatica **44**, 567–573 (2008)
6. Hua, C., Guan, X., Duan, G.: Variable structure adaptive fuzzy control for a class of nonlinear time-delay systems. Fuzzy Sets Syst. **148**, 453–468 (2004)
7. Hua, C., Li, F., Guan, X.: Observer-based adaptive control for uncertain time-delay systems. Inform. Sci. **176**, 201–214 (2006)
8. Hyun, C.H., Park, C.W., Kim, J.H., Park, M.: Synchronization and secure communication of chaotic systems via robust adaptive high-gain fuzzy observer. Chaos Solitons Fractals **40**, 2200–2209 (2009)
9. Jafarov, E.M., Tasaltin, R.: Robust sliding-mode control for the uncertain MIMO aircraft model F-18. IEEE Trans. Aerosp. Electron. Syst. **36**, 1127–1141 (2000)
10. Khalil, H.K.: Nonlinear systems. Prentice Hall, Englewood Cliffs, NJ (2003)
11. Kim, J.H., Park, Ch.W., Kim, E., Park, M.: Fuzzy adaptive synchronization of uncertain chaotic systems. Phys. Lett. **A 334**, 295–305 (2005)
12. Kolmanovskii, V.B., Myshkis, A.D.: Introduction to the theory and applications of functional Differential equations. Kluwer Academic Publishers, Dordrecht (1999)
13. Labiod, S., Boucherit, M.S., Guerra, T.M.: Adaptive fuzzy control of a class of MIMO nonlinear systems. Fuzzy Sets Syst. **151**, 59–77 (2005)

14. Lee, W.K., Hyun, Ch.H., Lee, H., Kim, E., Park, M.: Model reference adaptive synchronization of T–S fuzzy discrete chaotic systems using output tracking control. Chaos Solitons Fractals **34**, 1590–1598 (2007)
15. Lin, J.Sh., Yan, J.J., Liao, T.L.: Chaotic synchronization via adaptive sliding mode observers subject to input nonlinearity. Chaos Solitons Fractals **24**, 371–381 (2005)
16. Liu, T.S., Lee, W.S.: A repetitive learning method based on sliding mode for robot control. J. Dyn. Syst. Meas. Contr. **122**, 40–48 (2000)
17. Mamdani, E.H.: Applications of fuzzy algorithms for simple dynamic plants. Proc. IEE **121**(12), 1585–1588 (1974)
18. Noroozi, N., Roopaei, M., Zolghadri, J.M.: Adaptive fuzzy sliding mode control schemes for uncertain systems. Commun. Nonlinear Sci. Numer. Simul. **14**, 3978–3992 (2009)
19. Okuno, H., Takeshita, M., Kanari, Y.: OGY Control by Asymptotically Transition Method in Power System. In: 2002 Society of the Instrument and Control Engineers (SICE) Annual Conference, vol. 5, pp. 3163–3168, Osaka (2002)
20. Ott, E., Grebogi, C., Yorke, J.A.: Controlling chaos. Phys. Rev. Lett. **64**, 999–1196 (1990)
21. Pecora, L.M., Carroll, T.L.: Synchronization in chaotic systems. Phys. Rev. Lett. **64**, 821–824 (1990)
22. Roopaei, M., Zolghadri, J.M.: Synchronization of a class of chaotic systems with fully unknown parameters using adaptive sliding mode approach. Chaos **18**, 043112 (2008)
23. Roopaei, M., Zolghadri, J.M.: Chattering-free fuzzy sliding mode control in MIMO uncertain systems. Nonlinear Anal. Theor. Methods Appl. **71**, 4430–4437 (2009)
24. Salamci, M., Ozgoren, M.K., Banks, S.P.: Sliding mode control with optimal sliding surfaces for missile autopilot design. J. Guid. Contr. Dynam. **23**, 719–727 (2000)
25. Sastry, Sh., Bodson, M.: Adaptive Control. Prentice Hall, Englewood Cliffs, NJ (1989)
26. Shahnazi, R., Pariz, N., Vahidian, K.A.: Adaptive fuzzy output feedback control for a class of uncertain nonlinear systems with unknown backlash-like hysteresis. Commun. Nonlinear Sci. Numer. Simul. **15**, 2206–2221 (2010)
27. Slotine, J.E., Sastry, S.S.: Tracking control of non-linear systems using sliding surfaces, with application to a robot arm. Int. J. Contr. **38**, 465–492 (1983)
28. Sun, Y., Cao, J., Feng, G.: An adaptive chaotic secure communication scheme with channel noises. Phys. Lett. **A 372**, 5442–5447 (2008)
29. Sun, M., Tian, L., Jia, Q.: Adaptive control and synchronization of a four-dimensional energy resources system with unknown parameters. Chaos Solitons Fractals **39**, 1943–1949 (2009)
30. Ting, C.S.: An observer-based approach to controlling time-delay chaotic systems via Takagi–Sugeno fuzzy model. Inform. Sci. **177**, 4314–4328 (2007)
31. Tong, S.H., Li, H.X., Wang, W.: Observer-based adaptive fuzzy control for SISO nonlinear systems. Fuzzy Sets Syst. **148**, 355–376 (2004)
32. Utkin, V.I.: Variable structure systems with sliding modes. IEEE Trans. Autom. Contr. **AC-22**, 212–222 (1977)
33. Wang, M., Chen, B., Liu, X., Shi, P.: Adaptive fuzzy tracking control for a class of perturbed strict-feedback nonlinear time-delay systems. Fuzzy Sets Syst. **159**, 949–967 (2008)
34. Xiao, J.W., Yi, Y.: Coupled-adaptive synchronization for Chen chaotic systems with different parameters. Chaos Solitons Fractals **33**, 908–913 (2007)
35. Xu, J., Chen, S.: Decentralized Adaptive Controller Design for Uncertain Large-Scale Time-Delay Systems. In: 6th International Conference on Intelligent Systems Design and Applications (ISDA'06), vol. 2, pp. 173–177, China (2006)
36. Yan, J.J.: Design of robust controllers for uncertain chaotic systems with nonlinear inputs. Chaos Solitons Fractals **19**, 541–547 (2004)
37. Yan, J.J., Chang, W.D., Hung, M.L.: An adaptive decentralized synchronization of master–slave large-scale systems with unknown signal propagation delays. Chaos Solitons Fractals **29**, 506–513 (2006)
38. Yeh, J.P., Wu, K.L.: A simple method to synchronize chaotic systems and its application to secure communications. Math. Comput. Modell. **47**, 894–902 (2008)
39. Yu, F.M., Chung, H.Y., Chen, Sh.Y.: Fuzzy sliding mode controller design for uncertain time-delayed systems with nonlinear input. Fuzzy Sets Syst. **140**, 359–374 (2003)

40. Zhang, Y., Changxi, J., Utkin, V.I.: Sensorless sliding mode control of induction motors. IEEE Trans. Ind. Electron. **47**, 1286–1297 (2000)
41. Zhou, J.: Decentralized adaptive control for large-scale time-delay systems with dead-zone input. Automatica **44**, 1790–1799 (2008)
42. Zhou, J., Wen, C.: Adaptive Backstepping Control of Uncertain Systems. In: Thoma, M., Morari, M. (eds.) LNCIS, vol. 372, pp. 83–96, Springer, Berlin (2008)
43. Zribi, M., Smaoui, N., Salim, H.: Synchronization of the unified chaotic systems using a sliding mode controller. Chaos Solitons Fractals **42**, 3197–3209 (2009)

# Chapter 4
# Evolution Management of Data Integration Systems by the Means of Ontological Continuity Principle

**Ladjel Bellatreche, Guy Pierra, and Eric Sardet**

## 4.1 Introduction

Digital repositories of information are springing up everywhere and interconnectivity between computers around the world is being established which creates a mine of information. Many studies and integration systems were proposed to exploit this information by providing end users and decision makers a unified view of this information. The spectrum ranges from early multi-database systems [16] over data warehouse systems [4, 5], mediator systems [27, 32], to peer to peer systems [13, 19, 28]. Note that any integration system shall take into consideration two main aspects: (1) resolution of syntactic , schematic and semantic conflicts and (2) evolution managements.

Concerning conflict resolution, the more difficult task to integrate heterogeneous data sources is the identification of the equivalent concepts (and properties) used by these sources. To do so, different categories of semantic conflicts should be solved. Goh et al. [10] suggest the following taxonomy: naming conflicts, scaling conflicts, confounding conflicts and representation conflicts.

- Naming conflicts: occur when naming schemes of concepts differ significantly. The most frequently case is the presence of synonyms and homonyms. For instance, the status of a person means her familial status or her employment status.
- Scaling conflicts: occur when different reference systems are used to measure a value (for example price of a product can be given in dollar or in euro).

L. Bellatreche (✉) · G. Pierra
LISI/ENSMA – Poitiers University, 1, avenue Clément Ader, 86960 Futuroscope, France
e-mail: bellatreche@ensma.fr; pierra@ensma.fr

E. Sardet
CRITT Informatique CRCFAO, 86960 Futuroscope – France
e-mail: sardet@ensma.fr

T. Özyer et al. (eds.), *Recent Trends in Information Reuse and Integration*,
DOI 10.1007/978-3-7091-0738-6_4, © Springer-Verlag/Wien 2012

– Confounding conflicts: occur when concepts seem to have the same meaning, but differ in reality due to different evaluation contexts. For example, the weight of a person depends on the date where it was measured.
– Representation conflicts : arise when two source schemas describe the same concept in different ways. For example, in a source, student's name is represented by two elements *FirstName* and *LastName* and in another one it is represented by only one element Name.

To resolve these different conflicts, more and more integration systems following either mediator [10], materialized [5] or peer to peer architectures [19] use ontologies [11]. The use of ontology provides a shared vocabulary that labels concepts of a domain. Ontologies are then used to describe the semantic of sources and to make their contents more explicit. The instances of any source or integration system referencing an ontology are called ontology-based instances [1].

The need for evolution management results from the fact that most of the sources operate in an asynchronous manner. They may modify their structures or remove some data without any prior "public" notification. Consequently, the relation between an integrated system and its sources is often *slightly coupled*. This situation may generate maintenance anomalies [7, 34]. Nevertheless, the problem of source evolution has received a little attention in the literature, especially in the context of ontology-based integration systems (OBISs) compared to the first issue (conflict resolution).

Our work is concentrated on engineering domains, where several standard domain ontologies have been developed or are currently under development. Some of them are associated with maintenance agencies allowing updating continuously these ontologies. Examples of already standardized domain ontologies include: Electronic Components (IEC 61360-4), Laboratory Measuring Instruments (ISO 13584-501), Machining Tools (ISO 13399), Mechanical Fasteners (ISO 13584-511). Examples of domain ontologies under development include: Optics and Optronic (ISO 23584), Bearing (ISO 23768).

In fact, when such shared ontologies exist in a domain, developing OBIS may provide three following capabilities:

1. To integrate automatically a large set of data sources.
2. To manage evolution of both ontologies and ontology instances.
3. To support an asynchronous evolution between sources.

In [5], we already showed that when a shared (e.g., standardized) domain ontology exists, and when each local source a priori references the shared ontology, an automatic integration becomes possible. More precisely, an automatic and a reliable integration of data sources is possible if the source owners a priori agree on a common (but partial) shared vocabulary. Indeed, the articulation between the local ontologies and the shared one allows an automatic resolution of the different conflicts. In order to make this automation possible, three major assumptions were defined: (1) each data source participating in the integration process shall contain its own ontology. We call such a source an *ontology-based database* (OBDB) [3]. (2) Each local source a priori references a shared ontology by subsumption relationships "as much as possible" (i.e., (a) each local class must reference its smallest subsuming class in the shared ontology and (b) only a property whose meaning does not exist in the shared ontology may be defined in a local ontology, otherwise, it should be imported though the subsumption relationship (*OntoSub*, see 3.1). This pair of requirements are called smallest subsuming class reference requirements (S2CR2)) [4]. (3) A local ontology may restrict and extend the shared ontology as much as needed. The next challenge is to leave as much autonomy as possible to each OBDB, and in particular to support asynchronous changes and evolution.

In traditional databases, changes fall into two categories [26]: (1) schema changes (add/modify /drop attributes or tables) and (2) content changes (insert/update/delete instances). To tackle the problem of schema changes, two different approaches were proposed: *schema evolution* and *schema versioning*. The first approach consists in updating a schema and transforming data from an old schema into a new one (only the current version of a schema is present). In contrast, the second approach keeps track of the history of all versions of a schema. This approach is suitable for data warehousing environment, where decision makers may need historical data [18]. Content changes, like insert or update, are not considered as an issue, since it is a basic functionality of a database. In OBISs, these two categories of changes exist and may be addressed by adapting the existing solutions, but a third category of changes that concerns ontology evolution (the shared and the local ones) needs to be supported.

When the shared ontology evolves over the time and none global clock exits enforcing all the sources and the integration system to evolve at the same time, various sources to be integrated may reference the same shared ontology as it was at various points in time. Therefore, the problem of integration turns to an *asynchronous evolution problem*. Although the evolution was largely studied [21], to the best of our knowledge, none of these systems considered the problem of asynchronous evolution of ontologies.

To manage asynchronous evolution, the following issues should be addressed: (1) the management of evolution of ontologies (local and shared) in order to maintain consistency between ontologies and the data originating from various sources [14], (2) the management of the life cycle of ontology instances (periods where an instance was alive) and (3) the capability to interpret each ontology instance of the integration system, even if it is described using a different version of the shared ontology than the one that is referenced by the integration system.

The contribution and outline of our paper are as follows: Section 2 positions our work with a respect to the existing and similar ones. Section 3 presents our integration approach based on a priori articulation between a shared ontology and local ontologies. Section 4 proposes an approach for managing evolution of schemas and ontology instances. Section 5 proposes a framework for managing ontology changes and suggests distinguishing between ontology evolution and ontology revolution. Section 6 proposes a floating version model to support asynchronous evolution of ontologies. Section 7 presents an implementation of our approach using the Express language and ECCO environment. Section 8 concludes the paper by summarizing the main results and suggesting future work.

## 4.2   Related Work

Many works have been done on evolution of ontologies, schemas and instances. Schema and instance evolutions have been largely studied in the context of object-oriented databases [8, 26], where the key concern was the capability to migrate instances from a previous schema to a new schema. Thus, evolution has been distinguished from versioning as two different policies for managing schema and instance evolution [26]. Database evolution means schema evolution with instance migration. Only the current schema exists at each time and all instances conform to this schema. Database versioning means recording snapshots of each particular schema state together with the corresponding instance data.

Most of research works on ontology evolution also recognized the importance of some levels of instance data compatibility through ontology evolution. [21] classify the different kinds of ontology changes and discuss their effects on instance data. But they do not define any policy for managing evolving ontologies and their related instance data. [9] also investigated the various kinds of ontology changes and they proposed a classification of these changes. They also did not defined global policies for integrating data corresponding to different versions within a same ontology. Other works have been proposed to automatically identify the differences between several ontology versions, like PROMPTDIFF [21]. This system uses heuristics to compare different versions of ontologies and outline their differences. But integrating these different ontologies and their corresponding instance data was not considered in these works. Contrariwise a number of ontology-based integration systems have been proposed. We can cite for example, Picsel [6, 25] and Observer [17]. But none of these systems deals with the problem of ontology evolution.

Our approach is data-integration oriented. Like in [21], we investigated the various kinds of changes that may occur in an ontology, but our goal was to identify how these changes may be taken into account by an integration system (in a warehouse architecture). Like the work on databases, we have distinguished two policies. (a) Ontology evolutions follow the ontological continuity principle for which the floating version mechanism allows to build a single ontology providing access to all instances data. (b) In ontology revolutions, deprecated parts of ontologies are identified and documented. This leaves free the integration system administrator either to leave in the deprecated classes their existing instances; this could be viewed as the versioning strategy in database evolution; or to modify and align instances with the new ontology if he/she considers that such a transformation is meaningful. To the best of our knowledge, our proposal is the first that addresses ontology evolution in ontology-based integration systems.

## 4.3   An a Priori Integration Approach

In this section, we propose a formal model of ontologies and of OBDB in order to facilitate the presentation of our proposed integration process.

### 4.3.1   Formal Model for Ontologies

Formally, the ontology $O$ can be defined as a 4-tuple $< C, P, Applic, Sub >$, where:

- $C$ is the set of the classes used to describe the concepts of a given domain, each class is associated with a version number.
- $P$ is the set of all properties used to describe the instances of the classes of $C$.
- $Applic$ is a function defined as $Applic : C \rightarrow 2^P$. It associates to each class of the ontology, the properties that are applicable for each instance of this class and that may be used, in the database, for describing its instances. Note that for each $c_i \in C$, only a subset of $Applic(c_i)$ may be used in any particular database, for describing $c_i$ instances.
- $Sub$ is the subsumption relationship defined as $Sub : C \rightarrow 2^C$,[1] where for a class $c_i$ of the ontology, it associates its direct subsumed classes.[2] $Sub$ defines a partial order over $C$. In our model, there exists two kinds of subsumption relationships: $Sub = OOSub \cup OntoSub$, where:

  • $OOSub$ is the usual object-oriented subsumption with inheritance relationship. Through $OOSub$, applicable properties are inherited. $OOSub$ must define a single hierarchy.
  • $OntoSub$ is a subsumption relationship without the inheritance. Through $OntoSub$ (also called *case-of* in the PLIB ontology model [23]), whole or part of applicable properties of a subsuming class may be explicitly imported by a subsumed class,

    $OntoSub$ is in particular used as an *articulation operator* allowing to connect local ontologies to a shared ontology while insuring a large independence for local ontologies. Through this relationship, a local class may import or map all or any of the properties that are defined in referenced class(es) when the reference was established. The referencing class may also define additional properties. Updating the referenced class would not affect automatically the referencing class.

---

[1] $2^C$ denotes the power set of $C$.
[2] $C_1$ subsumes $C_2$ iff $\forall x \in C_2, x \in C_1$.

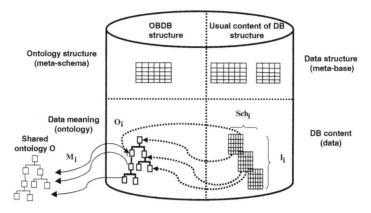

**Fig. 4.1** The structure of an ontology-based database

### 4.3.2 Formal Model for OBDB

Let $S = \{S, \ldots, S_n\}$ be a set of sources participating in the integration process. We assume that each source $S_i$ has a local ontology $O_i$ that references/extends the shared ontology $O$. We can define formally each source $S_i$ as a 5-tuple: $< O_i, Sch_i, I_i, Pop_i, M_i >$ (Fig. 4.1), where:

- $O_i$ is an ontology ($O_i :< C_i, P_i, Applic_i, Sub_i >$).
- $Sch_i : C_i \rightarrow 2^{P_i}$ associates to each ontology class $c_{i,j}$ of $C_i$ the properties which are effectively used to describe the instances of the class $c_{i,j}$. This set may be any subset of $Appli(c_{ij})$ (as the role of an ontology is to conceptualize a domain, the role of a database schema is to select only those properties that are relevant for its target application).
- $I_i$ is the set of instances of the source $S_i$.
- $Pop_i : C_i \rightarrow 2^{I_i}$ is the set of instances of each class.
- The mapping $M_i$ represents the *articulation* between the shared ontology $O$ and the local ontology $O_i$. It is defined as a function: $M_i : C \rightarrow 2^{C_i}$, that defines the subsumption relationships without inheritance holding between $C$ and $C_i$.[3]

### 4.3.3 Formal Model of Integration Process

Several automatic integration operators may be defined in the above context [4]. For simplicity reason, we just outline below the *ExtendOnto* integration operators, where the contents of the various sources are duplicated within a data warehouse

---

[3]Note that $\forall i \colon C_i \cap C = \phi$, each source populates only local classes.

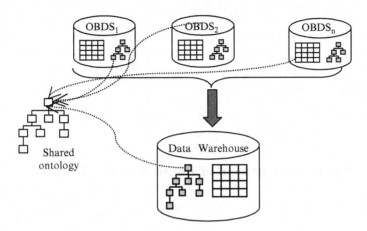

**Fig. 4.2** The structure of our data warehouse

(materialized integration approach [4]) and where the ontology of data warehouse consists of the shared ontology extended by the local extensions of this ontology done in the local ontologies of all the sources that have been added in the warehouse. Thanks to the articulation mappings $(M_i)$,[4] we note that all warehouse data that may be interpreted by the shared ontology (i.e., of which the class is subsumed by a shared ontology class) may be accessed through this ontology, whatever source they came from.[5]

The ontology-based data warehouse $DW$ has the same source structure (Fig. 4.2): $DW :< O_{DW}, Sch_{DW}, I_{DW}, Pop_{DW}, \phi >$, where:

1. $O_{DW}$ is the warehouse ontology. It is computed by integrating local ontologies into the shared one. Its components are computed as follows:

   - $C_{DW} = C \cup (\cup_{1 \leq i \leq n} C_i)$.
   - $P_{DW} = P \cup (\cup_{1 \leq i \leq n} P_i)$.
   - $Applic_{DW}(c) = \begin{cases} Applic(c), \text{ if } c \in C \\ Applic_i(c), \text{ if } c \in C_i \end{cases}$
   - $Sub_{DW}(c) = \begin{cases} Sub(c) \cup_{1 \leq i \leq n} M_i(c), \text{ if } c \in C \\ Sub_i(c), \text{ if } c \in C_i \end{cases}$

2. $I_{DW} = \cup_{1 \leq i \leq n} I_i$.
3. The instances are stored in tables as in their sources.

---

[4] All articulation mapping ensure the S2CR3 requirements.

[5] Another integration scenario, called *ProjOnto*, assumes that source instances are extracted after a projection operation on the shared ontology.

- $\forall c_i \in C_{DW} \wedge \forall c_i \in C_i (1 \le i \le n)$:

    (a) $Sch_{DW}(c_i) = Sch_i(c_i)$
    (b) $Pop_{DW}(c_i) = Pop_i(c_i)$

- $\forall c \in C$

    (a) $Sch_{DW}(c) = Applic(c) \cap (\cup_{c_j \in Sub_{DW}(c)} Sch(c_j))$.
    (b) $Pop_{DW}(c) = \cup_{c_j \in Sub_{DW}(c)} Pop(c_j)$

## 4.4   Evolution Management of Contents and Schemas

In this section, we present a mechanism to identify classes, properties and instances and the life cycle of instances.

To identify classes and properties, we assume that universal identifiers (UI) are defined in the ontology [4]. In our work, we use the UI defined in ISO 29002-5: *identification scheme*. Moreover, each ontology concept is associated with an ontology version number (OntologyVersion). In order to recognize instances of the data warehouse over time and over data sources, any ontology class (either shared or local) is associated with a *semantic key*. It is composed by the representation (in a string form) of values of one or several applicable properties of this class. This key identifies unambiguously the same instance over time and data sources even if some property values are added or removed.

### 4.4.1   The Life Cycle of Ontology Instances

In some situations, it may be useful to know the existence of instances in the warehouse at any previous point in time. Two solutions are possible to satisfy this requirement.

- In the *explicit storage* approach [2,31], all the versions of each table are explicitly stored. This solution has two advantages: (1) it is easy to implement and allows an automation of the process of updating of data, and (2) query processing is straightforward in cases where we precise the version on which the search will be done. On the other hand, the query processing cost can be very important if the query needs an exploration of all versioned data of the warehouse. Another drawback is due to the storage of the replicated data.
- In the *implicit storage* approach [31]: only one version of each table $T$ is stored. This schema is obtained by making the *union* of all properties appearing in the various versions. For each data warehouse updating, one adds all existing instances to each source's tables. Instances are supplemented by null values. This solution avoids the exploration of several versions of a given table. The major drawbacks of this solution are: (1) the problem of replicated data is still present,

(2) the implementation is more difficult than the previous one concerning the automatic computation of the schema of stored tables (the names of columns may have changed in the sources); (3) the layout of the life cycle of data is difficult to compute (*valid time* [31]) and (4) null values cannot be easily implemented (resulting from the instance or from storage mechanism).

Our solution follows the second approach and solves the problems in the following way:

1. The problem of *replicated data* is solved thanks to the single semantic identification (value of the semantic key) of each instance of data.
2. The problem of the updating process of table schemata is solved through the use of universal identifiers (UI) for all the properties.
3. The problem of the representation of the instances life cycle is solved by a pair of properties: $(ContentVersion_{min}, ContentVersion_{max})$. It enables us to know the validation period of a given instance.[6]
4. The problem of the semantic ambiguity of the null values is handled by archiving the $Sch$ functions of the various versions of each class. This archive enables one to determine the true schema of the version of a table at any point in time, and thus the initial representation of each instance.

## 4.5 Ontology Evolution Management

In this section, we first propose taxonomy of ontology evolution into (normal) evolution and epistemological revolution.

### 4.5.1 An Ontology Evolution Framework

It is well known that different changes in ontologies have different effects both on ontology consistency and on ontology instances [20]. Thus managing ontology evolution needs first to discuss the various kinds of changes that may occur in ontology.

For us, the major characteristics of an ontology, that distinguishes it from e.g., conceptual model (in database design) or knowledge model (in artificial intelligence) is that on ontology is not only a specification of a conceptualization (definition that may apply also to a number of conceptual models and knowledge models), but also a specification of a shared conceptualization. This means, from logical point of view, that an ontology is (or should be) a logic theory of a part of the world, shared by a whole community, and allowing their members to understand

---

[6]Or $(t_{min}, t_{max})$ if timestamps are used.

each others. An ontology should be like a set theory (for mathematicians), mechanics (for mechanical engineers) or analytical accounting (for accountants). For this kind of theories, two kinds of changes may be identified: *normal evolution*, and *epistemological revolution*. Normal evolution of a theory is its deepening. New truths and more detailed truths are added to the old truths. But, along with this kind of evolution, what was true yesterday remains true today. For instance, more detailed classes are added, new properties are defined, new phenomena are discovered and value ranges are extended.

But it sometimes also arrives that some axioms of a theory are challenged or become false. In this case, it is not any more an evolution. It is a *revolution*, where two different logical systems will coexist or be opposed. It was the case for instance when quantic mechanics theory was proposed as an alternative to rational mechanics, or when non-Euclidean geometry was proposed an alternative to Euclidean geometry. In such case no complete formal mapping may be defined between both theories. The rationale of the changes can often be only informally described, and each member of the community across which the previous ontology was shared, must understand what should be done with legacy information.

The ontology evolution framework that we propose that we have implemented in the PLIB ontology model [24] is based on the above distinction:

– On ontology evolution only extends the logical theory defined by the ontology.
– An ontology revolution introduces an inconsistency in the current logical theory. This inconsistency is clearly identified and shall be available to allow a user to process the ontology inconsistency.

More precisely, an ontology evolution must fulfil the ontological continuity principle defined as follow:


Principle of ontological continuity:

*if we consider an ontology as a set of axioms, then ontology evolution must ensure that any true axiom for a certain version of an ontology will remain true for all its later versions.*


### 4.5.2   Constraints on Ontology Evolutions

In this section, we discuss the constraints for each kind of concept (classes, properties and instances) that result from the ontological continuity principle. Let $O^k :<$ $C^k, P^k, Sub^k, Applic^k >$ and $O^{k+1} :< C^{k+1}, P^{k+1}, Sub^{k+1}, Applic^{k+1} >$ be an ontology in version $k$ and an ontolgy in version $(k + 1)$ resulting from the evolution of $O^k$, respectivly.

We have already precised that each manipulated ontological concept (e.g., class and property) is associated with an *universal identifier* (UI). These identifiers have two parts: a code (unique) identifying each concept and a version, identifying version of that concept. UI is then the concatenation of code and version ($IU :=$ $code||version$). The code of the same concept is invariant. Each change concerning its definition changes only its version. We denote by $code(c)$ and $version(c)$ the functions associating to each concept $c$ of the ontology (class, property or datatype) their corresponding code and version. These two functions are extended to deal with sets. Let $C = \{c_i, i = 1, n\}$ be a set of classes; $code(C) = \{code(c_i)|c_i \in C\}$ and $version(C) = \{version(c_i)|c_i \in C\}$.

Each reference between elements referring this UI is itself implicitly versionned by the versions of its extremities. Finally, each class or property definition contains, in particular, the date from which this version is valid. Note that this characteristic is supported by PLIB ontology model that we manipulate [22]. It is also supported by OWL via the *versionInfo* label. An *owl:versionInfo* statement generally has as its object a string giving information about this version.

The following constraints shall be fulfilled to ensure the ontological continuity principle.

**Permanence of the classes** Existence of a class is an axiom that could not be denied across evolution. Moreover, the definition of a class could be refined, but this should not exclude any instance that was member of the class in the previous version. Thus, evolution of classes shall ensure (1) $\forall k, code(C^k) \subset code(C^{k+1})$, (2) $\forall c_i' \in C^k, c_i'' \in C^{k+1} s.t code(c_i') = code(c_i''), \forall x \in c_i' \Rightarrow x \in c_i''$.

*Example 1.* If an individual is a car, it will remain forever (not revolution) a car. But it may be defined as a "convertible", convertible being a new subclass of car. Thus, a class definition can be refined without restricting the set of the possible instances of that class. This means that: (1) the definition class may itself evolve, (2) each class definition will be associated with a new version number. This number is only required when the set of possible instances of the new class is different from (i.e., greater than) the set of possible instances. Else (e.g. typo correction), another mechanism may be used (e.g., a time stamp or a revision number interval to the class definition) and (3) definition (intentional) of each class will include definitions (intentional) of its earlier versions.

**Permanence of subsumption** Subsumption is also an ontological concept that cannot be denied. Let $Sub^* : C \rightarrow 2^C$ be the transitive closure of direct subsumption relationship $Sub$ and $Sub^{*k}$. The $Sub^*$ is the $Sub^*$ function in $O^k$. We have $\forall c \in C^k, code(Sub^{*k}(c)) \subset code(Sub^{*k+1}(c))$.

Note that this constraint obviously allows some evolution of the subsumption hierarchy, for example by intercalating intermediate classes between two other classes linked by a subsumption relation.

**Permanence of properties** Also: $\forall k, code(P^k) \subset code(P^{k+1})$.

Taking into account the ontological continuity principle, a property range and property domain only could be increased, but not decreased like in object oriented database evolution (in ORION system [8]).

**Description of instances** The fact that a property $p \in Applic(c)$ means that $p$ is rigid [12] for every instance of $c$. This is still an axiom that may not be reversed:

$\forall c \in C^k, code(Applic^k(c)) \subset code(Applic^{k+1}(c))$

Note that this does not require that the same properties are always used to describe the instances of the same class. As described in Sect. 3.3, schematic evolution does not depend only on ontology evolutions. It depends also, and mainly, on the organizational objectives of each particular database version.

### 4.5.3  Representation of Ontology Revolution and Corrections of Errors

Ontology revolution means that some previous ontology axioms become inconsistent with the new ontology version, and that it cannot be left as an axiom in the new ontology.

But, in a number of cases, the revolution means that the previous ontology already contained errors or drawbacks and that it should be corrected, else it would be in any case not sound.

When the ontology concepts which are erroneous have not yet been used for creating instances or when instances where created in a closed user environment, the erroneous instance descriptions can be corrected concurrently with the ontology.

When the shared ontology is distributed in an open user environment where all possible instances are not accessible to the ontology provider, instance corrections cannot be performed concurrently with the ontology. In this case, the need is to correct the ontology errors but also to provide a mechanism that allows ontology users to understand and process the error corrections in an asynchronous manner. For each data set that contains instances, processing errors means (1) recognizing which instance descriptions are in errors and (2) defining how erroneous descriptions should be corrected to be in line with the corrected ontology. In such an environment, we propose to introduce a mechanism called "deprecation". *Deprecation* means that:

- The ontology is corrected, this means that new elements with new codes are introduced in the ontology to replace erroneous ontology elements, but
- To ensure backward compatibility of the corrected ontology with preexisting instances, the erroneous ontology elements remain also in the corrected ontology
- All these erroneous elements are associated with an *is_deprecated* attribute with a true value, the meaning of which being: "this ontology class, property or value shall no longer be used for defining new instances", and
- A second attribute associated with each *is_deprecated attribute*, called *is_deprecated_interpretation*, is used to specify, how an instance that references deprecated elements should be changed to be in line with the up-dated ontology.

Note that the specification in *is_deprecated_interpretation* may be either informal, to explain to a ontology user how the corresponding data should be processed, or formal to direct a computer how to correct automatically the data.

*Example 2.* If in a class $C_1$ an applicable property $P_1$ whose value was supposed to be expressed in meters, is replaced by a property $P_2$ which has the same meaning but whose value shall be expressed in microns, (1) the $P_1$ is_deprecated attribute is set to true, and (2) its is_deprecated_interpretation attribute could be set to: "the value of this property shall now be expressed in microns and recorded in property $P_2$".

*Example 3.* In the above example, the value of the is_deprecated_interpretation attribute of $P_1$ could be represented, if this approach has been agreed by the community that uses the reference dictionary, as an expression using the syntax of the EXPRESS language, and representing the values of properties by the property identifiers. In this case the content could be set to: "$P_2 := P_1 \times 1000$", where $P_1$ and $P_2$ are property UIs.

## 4.6   Floating Version Model: A Global Access to Current Instances

Before presenting our floating version model, we indicate the updating scenario of our data warehouse in the engineering domain [22]. There may exist one or several shared ontologies. The process is the same for each ontology, thus we consider a signed shared one. The shared ontology evolves over the time and each class or property is associated with a version number. The current version of a shared ontology is available over the Internet and may be totally or partially downloaded through Web services. Moreover, at given moments, chosen by the data warehouse administrator, the current version of a source $S_i$ is loaded in the warehouse. This version includes its local ontology, that references "as much as possible" the shared one, the mapping $M_i$ between local ontology $O_i$ and the shared ontology $O$, and its current content (*certain instances eventually already exist in the warehouse, others are new, others are removed*). This scenario is common in the engineering domain [22], where an engineering data warehouse consolidates descriptions (i.e., electronic catalogues) of industrial components of a whole of suppliers. Therefore, in this scenario, the maintenance process is carried out each time that a new version of an electronic catalogue of a supplier is received.

Our floating version model is able to support two kinds of user services: (1) it allows to provide an access via a single ontology to the set of all instances that have been recorded in the data warehouse over the time and/possibly (2) it also allows to record the various versions of the ontologies (shared and local) and to trace the life cycle of instances (full multi-version management). In this section we discuss how these objectives will be achieved.

The principal difficulty due to source autonomy is that, in some situations, when two different sources are loaded, let's say $S_i$ and $S_j$, a same class $c$ of the shared ontology $O$ can be referred by an articulation mapping (i.e., subsumption) in different versions. For example, classes $c_i^n$ of $S_i$ and $c_j^p$ of $S_j$ may refer to $c^k$ (class $c$ with version $k$) and $c^{k+l}$ (class $c$ with version $k + l$), respectively. According to the principle of ontological continuity, it is advisable to note that:

1. All subsumed classes for $c^k$ are also subsumed for $c^{k+l}$.
2. All applicable properties in $c^k$ are also applicable in $c^{k+l}$.

Thus, the subsumption relation between $c^k$ and $c_i^n$ also holds between $c^{k+l}$ and $c_i^n$. Moreover, all the properties that were applicable from $c^k$ for example, querying $c_i^n$ instances are also applicable from $c^{k+j}$. Therefore, all accesses to $c_i^n$ data that could be done from class $c^k$ may also done from class $c^{k+l}$.

This remark leads us to propose a model, called the *floating version model*, which enables to reach all the data in the data warehouse via only one version of each class of the warehouse ontology. This set of versioned classes, called the *current version* of the warehouse ontology, is such that the current version of each class $c^f$ is higher or equal to the largest version of that class referenced by a subsumption relationship at the time of any maintenance. In practice, this condition is satisfied as follows:

– If an articulation $M_i$ references a class $c^f$ with a version lower than $f$, then $M_i$ is updated in order to reference $c^f$.
– If an articulation $M_i$ references a class $c^f$ with a version greater than $f$, then the warehouse connect itself to the shared ontology server, loads the last version of the shared ontology and migrates all references $M_i$ $(i = 1..n)$ to new current versions.

*Example 4.* During the maintenance process of a class $C_1$ (Fig. 4.3) that references the shared ontology class $C$ with version 2 **(1)**, the version of $C$ in warehouse ontology is 1 **(2)**. In this case, the warehouse downloads the current version of the shared ontology **(3)**. This one being 3, class $C_1$ is modified to reference version 3 **(4)**.

We described below the two automatic maintenance processes that our floating version model makes possible.

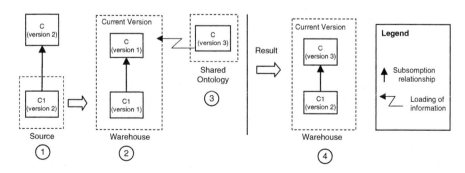

**Fig. 4.3** An example of floating versions

### 4.6.1  Simplified Version Management

If the only requirements of users is to be able to browse all current instances of the data warehouse in an uniform way (i.e., whatever the version of the shared ontology referenced by the various sources) then, at each maintenance step: (1) ontology description of the various classes of the data warehouse ontology are possibly replaced by newer versions, and (2) the table associated to each class coming from a local ontology in the data warehouse is simply replaced by the corresponding current table in the local source.

### 4.6.2  A Full Multi-version Management

Note that in the previous case (Sect. 5.5), the articulation between a local ontology class and a shared ontology class stored in the current version of the data warehouse may not be its original definition (see the Fig. 4.3). If the data warehouse user also wants to keep all the instances that existed over the time, and to browse instances through the ontological definitions that existed when these instances were loaded, it is necessary to archive also all the versions of the warehouse ontology. This scenario may be useful, for example, to know the exact domain of an enumeration-valued property when the instance was defined. By implementing this possibility, we get a *multi-version data warehouse* which archives also all versions of classes having existed in the data warehouse life, and all the relations in their original forms. Note that the principle of ontological continuity seems to make seldom necessary this complex archive.

The multi-version data warehouse has three parts (see Fig. 4.4):

1. *Current ontology*. It contains the "floating version" of the warehouse ontology. This version provides a generic data access interface to all instance data, whenever they were introduced in the warehouse.

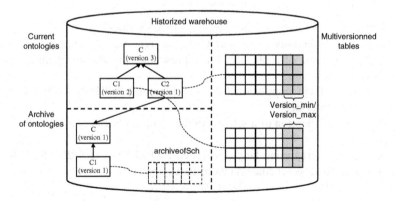

**Fig. 4.4** Structure of warehouse integrated system

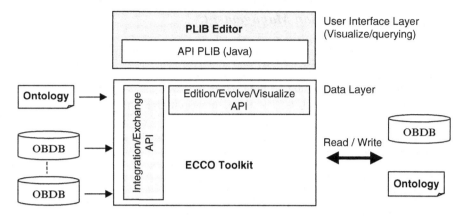

**Fig. 4.5** Architecture of our prototype

2. *Ontology archive*. It contains all versions of each class and property of the warehouse ontology. This part gives to users the true definitions of versions of each concept. Versions of table schema $T_i$ are also historized by archiving the function $Sch^k(c_i)$ of each version $k$ of $c_i$, where $T_i$ corresponds to the class $c_i$.
3. *Multi-versioned tables*. It contains all instances and their time_min and time_max.

## 4.7    Implementation of Our Approach

In order to validate our work, we have developed a prototype integrating several OBDSs [33, 34] (Fig. 4.5), where ontologies and instance data are described using the PLIB ontology model [23] specified by the Express language [29]. Such ontologies and instance data are exchangeable as instances of EXPRESS files ("physical file"). To process EXPRESS files, we used the ECCO Toolkit of PDTec which offers the following main functions [30]:

1. Edition, syntax and semantic checker of EXPRESS models.
2. Generation of functions (Java and C++) for reading, writing and checking integrity constraints of a physical file representing population of instances of an EXPRESS schema.
3. Manipulation of the population (physical file) of EXPRESS models using a graphical user interface.
4. Access to the description of a schema in the form of objects of a meta-model of EXPRESS.
5. Support of a programming language called EXPRESS-C. EXPRESS-C allows managing an Express schema and its instance objects.

An ontology and an OBDS may be created via an editor called PLIBEditor. It is used also to visualize, edit and update ontologies (both shared and local) and instance data. It uses a PLIB API developed under ECCO. PLIBEditor proposes a QBE-like graphical interface to query instance data from the ontologies. This interface relies on the OntoQL query language [15] to retrieve the result of interactively constructed queries.

We have developed a set of integration API allowing the automatic integration both in the simplified version management scenario, and in the full multi-version management scenario.

Figure 4.6 shows the view offered to users over the content of the data warehouse after integration. The data warehouse ontology (left side) consists of the shared ontology extended by all the subsumed classes of the integrated local ontology. It provides for hierarchical access and query over the data warehouse content:

- A query over a shared ontology class allows querying all the classes subsumed either through *OOSub* or *OntoSub* relationships (see Sect. 3.1). As consequence, instance data are integrated from all sources (see left side of Fig. 4.6).
- Hierarchical access allows also to go down until classes that came from any particular ontology (see right side of Fig. 4.6).

This system has been used as a pilot project for *Peugeot Citron Car Company* in France. This company has several part providers over the world. Our system allows integrating data into a data warehouse and ensures the evolution of source's providers.

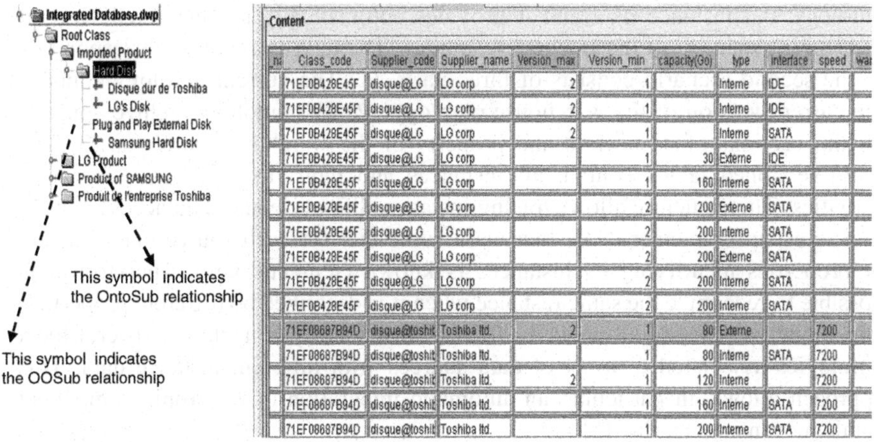

**Fig. 4.6** Integrated hierarchical access and integrated querying over the data warehouse

## 4.8 Conclusion

In this paper, we presented the asynchronous versioning problem, where autonomous ontology-based data sources are integrated in an ontology-based data warehouse. The sources that we considered are those containing local ontologies referencing in an a priori manner a shared one by subsumption relationships. These sources are autonomous and heterogeneous, and we assume that ontologies, schemas, and data may evolve over the time. Our integration process integrates first ontologies and then the data. The presence of ontologies allows an automation of the integration process, but it makes the management of autonomous sources more difficult. Concerning ontology changes, we proposed to distinguish ontology evolution and ontology revolution. Ontology evolution must respect the principle of ontological continuity that ensures that an axiom that was true for a particular version will remain true over all successive evolutions. This assumption allows the management of each old instance using a new version of the ontology. Ontology revolution results into a new logical system that cannot be formally mapped onto the previous one. For ontology revolution, we proposed to tag as deprecated all the classes and properties whose axiomatic definition is no longer consistent with the new ontology, and to document informally or possibly formally the inconsistency rationale.

For ontology evolution, we have proposed two scenarios for the automatic integration of autonomous ontology-based data sources into an ontology-based data warehouse. Both scenarios are based on the floating version model, where the integration process always maintains a single version of the data warehouse ontology. This version, called the current ontology, allows interpreting all the instance data in the data warehouse. The first scenario just updates the data warehouse ontology and instance data and it provides automatically an integrated view of the current state of all the integrated sources. The data warehouse corresponding to the second scenario consists of three parts. (1) The current ontology contains the current version of the warehouse ontology. (2) The ontology archive contains all the versions of each class and property of the warehouse ontology. (3) The multi-versioned tables contain all instances and their first and last version of activities. This structure allows tracing instances life cycle and data access is done in a transparent manner. Note that we assume that for each source class, the set of properties identifying its instances is known (semantic key). Therefore it is possible to recognize the same instance when it is described by different properties. Our model was validated under ECCO environment by considering several local ontologies associated to a set of data sources with different instance data. This approach allows, in particular, an automatic integration of electronic component catalogues in engineering [5].

Concerning the perspectives, we are currently investigating how the floating version model might be adapted to mediator architecture and to peer to peer environments.

# References

1. Alexaki, S., Christophides, V., karvounarakis, G., Plexousakis, D., Tolle, K.: The ics-forth rdfsuite: Managing voluminous rdf description bases. In: Proceedings of SemWeb'01 (2001)
2. Bebel, B., Eder, J., Koncilia, C., Morzy, T., Wrembel, R.: Creation and management of versions in multiversion data warehouse. In: Proceedings of the 2004 ACM symposium on Applied computing, pp. 717–723 (2004)
3. Bellatreche, L., Ait Ameur, Y., Chakroun, C.: A design methodology of ontology based database applications. Logic Journal of the IGPL, Oxford University Press (2010)
4. Bellatreche, L., Pierra, G., Nguyen Xuan, D., Dehainsala, H., Ait Ameur, Y.: An a priori approach for automatic integration of heterogeneous and autonomous databases. In: Proceedings of DEXA'04, pp. 475–485 (2004)
5. Bellatreche, L., Xuan, N.D., Pierra, G., Dehainsala, H.: Contribution of ontology-based data modeling to automatic integration of electronic catalogues within engineering databases. In Comput. Ind. J. **57**(8-9), 711–724 (2006)
6. Bressan, S., et al.: The context interchange mediator prototype. In: Proceedings of the ACM SIGMOD International Conference on Management of Data, pp. 525–527 (1997)
7. Chen, S., Liu, B., Rundensteiner, E.A.: Multiversion-based view maintenance over distributed data sources. ACM Trans. Database Syst. **4**(29), 675–709 (2004)
8. Clamen, S.M.: Type evolution and instance adaptation. Technical report, Carnegie Mellon University, CMU-CS-92-133R (1992)
9. Flouris, G., Plexousakis, D., Antoniou, G.: A classification of ontology change. In Semantic Web Applications and Perspectives (2006)
10. Goh, C.H., Bressan, S., Madnick, E., Siegel, M.D.: Context interchange: New features and formalisms for the intelligent integration of information. ACM Trans. Inform. Syst. **17**(3), 270–293 (1999)
11. Gruber, T.: A translation approach to portable ontology specification. Know. Acquis. **5**(2), 199–220 (1995)
12. Guarino, N., Welty, C.A.: Ontological analysis of taxonomic relationships. In: Proceedings of ER'2000, pp. 210–224 (2000)
13. Halevy, A.Y., Ives, Z.G., Mork, P., Tatarinov, I.: Piazza: data management infrastructure for semantic web applications. In: Proceedings of WWW, pp. 556–567 (2003)
14. Heflin, J., Hendler, J.: Dynamic ontologies on the web. American Association for Artificial Intelligence (2000)
15. Jean, S., et al.: Ontoql: an exploitation language for obdbs. In VLDB PhD Workshop, pp. 41–45 (2005)
16. Landers, T.A., Rosenberg, R.: An overview of multibase. In: Proceedings of the Second International Symposium on Distributed Data Bases, pp. 153–184 (1982)
17. Mena, E., Kashyap, V., Sheth, A.P., Illarramendi, A.: OBSERVER: An approach for query processing in global information systems based on interoperation across pre-existing ontologies. In: Conference on Cooperative Information Systems, pp. 14–25 (1996)
18. Morzy, T., Wrembel, R.: Modeling a multiversion data warehouse: A formal approach. In: Proceedings of ICEIS'03 (2003)
19. Nejdl, W., et al.: Edutella: a p2p networking infrastructure based on rdf. In WWW, pp. 604–615 (2002)
20. Noy, N.F., Klein, M.: Ontology evolution: Not the same as schema evolution. Knowl. and Inform. Syst. J. **6**(4), 428–440 (2004)
21. Noy, N.F., Musen, M.A.: Ontology versioning in an ontology management framework. IEEE Intell. Syst. **19**(4), 6–13 (2004)
22. Pierra, G.: Context representation in domain ontologies and its use for semantic integration of data. J. Data Semantics (JODS'08). **10**, 174–211 (2008)
23. Pierra, G., Potier, J.C., Sardet, E.: From digital libraries to electronic catalogues for engineering and manufacturing. Int. J. Comput. Appl. Tech. (IJCAT), **18**, 27–42 (2003)

24. Pierra, G., Sardet, E.: Methodology For Structuring Part Families. ISO-IS 13584-42.2. ISO Genève (2010)
25. Reynaud, C., Giraldo, G.: An application of the mediator approach to services over the web. Special track Data Integration in Engineering, Concurrent Engineering (CE'2003) – the vision for the Future Generation in Research and Applications, pp. 209–216 (2003)
26. Roddick, J.F.: A survey of schema versioning issues for database systems. Inform. and Software Tech. **37**(7), 383–393 (1995)
27. Roth, M.T., et al.: The garlic project. In: Proceedings of the ACM SIGMOD International Conference on Management of Data, pp. 557–557 (1996)
28. Rousset, M.-C., Adjiman, P., Chatalic, P., Goasdoué, F., Simon, L.: Somewhere: A scalable peer-to-peer infrastructure for querying distributed ontologies. In: OTM Conferences, pp. 698–703 (2006)
29. Schenk, D., Wilson, P.: Information Modelling The EXPRESS Way. Oxford University Press (1994)
30. Staub, G., Maier, M.: Ecco tool kit – an environnement for the evaluation of express models and the development of step based it applications. User Manual (1997)
31. Wei, H.-C., Elmasri, R.: Study and comparison of schema versioning and database conversion techniques for bi-temporal databases. Sixth International Workshop, TIME-99 Proceedings, pp. 88–98 (1999)
32. Wiederhold, G.: Mediators in the architecture of future information systems. IEEE Comput. **25**(3), 38–49 (1992)
33. Xuan, D.N.: Intégration de bases de données hétérogènes par articulation a priori d'ontologies: application aux catalogues de composants industriels. PhD thesis, Poitiers University, France (2006)
34. Xuan, D.N., Bellatreche, L., Pierra, G.: A versioning management model for ontology-based data warehouses. In: Proceedings of DaWaK'06, pp. 195–206 (2006)

# Chapter 5
# Automated Learning for Assessment of School Readiness: Effectiveness of Reusing Information from Child's Social Network

Iyad Suleiman, Shang Gao, Maha Arslan, Tamer Salman, Faruk Polat, Reda Alhajj, and Mick Ridley

## 5.1  Introduction

The transition from kindergarten to elementary school is considered one of the critical periods in the life of a child, during which he/she acquires fundamental skills and ways of critical learning and thinking. Not every student who attends kindergarten is necessarily ready to go to grade one; this is somehow true by considering the fact that kindergartens vary in the material they cover from almost nothing to advanced topics which could include a larger portion of the topics covered in grade one. For the child, this is a profound change and a transition from a small, intimate setting to a larger setting with more expectations and various demands made upon him/her [5, 16, 20]. The transition also requires the child to adjust to and develop emotional and social relationships in a new environment [14]. Generally,

I. Suleiman (✉) · M. Ridley
Dept of Computing, University of Bradford, Bradford, UK

S. Gao
Computer Science Department, University of Calgary, Calgary, Alberta, Canada

M. Arslan
Haifa University, Haifa, Israel

T. Salman
Dept of Computer Science, Technion, Haifa, Israel

F. Polat
Dept of Computer Engineering, Middle East Technical University, Ankara, Turkey

R. Alhajj
Computer Science Department, University of Calgary, Calgary, Alberta, Canada

Department of Computer Science, Global University, Beirut, Lebanon
e-mail: alhajj@ucalgary.ca

T. Özyer et al. (eds.), *Recent Trends in Information Reuse and Integration*,
DOI 10.1007/978-3-7091-0738-6_5, © Springer-Verlag/Wien 2012

the coordination is done by the kindergarten teacher, usually in cooperation with the parents, and is a critical component in the children's development. It may be more convenient to parents and more critical to the child to be correctly assessed as not ready to proceed to grade one. This may save the child from facing some emotional and educational shocks that may push him/her away from school. He/she may turn from a kid willing to go to school and eager to meet new friends into someone who will hate school and will try his/her level best to skip every school day. Here it is worth stressing the need for and the benefit of developing a standard curriculum for kindergarten students in order to avoid the transition shock and to produce more accurate and trustable results from the readiness test. It is not realistic to expect from a little child to know something he/she never heard of before. It is wiser to blame the system than blaming the kids and classifying them incorrectly. As a result, we assume that all kids to undertake a readiness test have been exposed to the least knowledge and background expected for the test. Once this is guaranteed, the outcome from the test can be trusted and hence can be used to draw conclusions and make decisions regarding whether a given child is ready to attend school or not.

An external advisory system takes part in addressing developmental delays through communications clinicians and educational psychologist as needed. In a normal environment where standards are applied, they conduct observations of kindergarten children, receive reports from the kindergarten teacher and give psychological tests to children who appear to have developmental delays. These tests may be accompanied by questionnaires and testing instruments.

The most common "diagnosis" utilized by psychologists regarding children with learning deficits, in cases when the intellectual measures are satisfactory, is "childish for his age" or "emotionally immature" [7]. This is the fundamental diagnosis which is the basis for the decision of whether to have the child spend an additional year in compulsory kindergarten or advance to first grade. When the intelligence test scores seem low, there is generally a referral to special education where the child may find a more appropriate environment that matches their capabilities and expectations. They may eventually excel in such environment. They may also start to build their own skills and capabilities quickly so that they may be moved to the normal education system in short time. Parent should understand the situation of their child carefully and thoroughly; they should avoid rush decisions that may only lead to have their kid suffering for the rest of his/her life. Hence, it is to the benefit of the child first and then to the benefit of the parents and the school system to run the test on time, understand the results. and proceed accordingly.

Today, it is possible to diagnose various forms of dyslexia, language and attention deficits, motor deficits and cognitive mathematical deficits. It is possible to begin an early intervention at a young age, as needed and acceptable [8, 11]. Thus, school readiness is more than just about children. School readiness, in the broader sense, involves children, families, early environments, schools, and communities. Children are not innately ready or not ready for school. Their skills and development are strongly influenced by their families and through their interactions with other people and environments before coming to school. Assessing school readiness is important to the education of young children. Assessment helps measuring the current state of children's development and knowledge and can be used to guide classroom and individual education programming [4, 8, 18, 20].

Realizing its importance, school readiness has attracted the attention of the research communities in Europe, North America, Australia, and New Zealand, e.g., [1, 15]. For instance, [9] studied the impact of school readiness on later achievements. [10] conducted a readiness study on British children in the 1970 cohort. [13], studied the impact of school readiness on the future of students. [12] discussed policy planning based on the school readiness test. [19] discussed the influence of parents on school readiness. [23] as well as [24] studied the influence of school readiness measures.

Realizing the importance of conducting the school readiness test and in our effort to contribute to the field, we want to develop effective automated assessment process. Explicitly speaking, we have integrated a modified Genetic Algorithm (GA) into a computerized assessment tool for school readiness. Our goal is to create a computerized assessment tool that can learn the user's skills and adjust the assessment tests accordingly. The user plays various sessions from various games, while the GA selects the upcoming session or group of sessions to be chosen for the user according to his skills and status. In this paper, we stress the importance of information reuse by describing the modified GA and the learning procedure. We integrate a penalizing system into the GA and a fitness heuristic for best choice selection. We present two methods for learning, a memory system and a no-memory system. Furthermore, we present several methods for the improvement of the speed of learning.

The rest of this paper is organized as follows. Section 5.2 covers the basic components of the proposed genetic algorithm based method for accessing child readiness to attend school. Section 5.3 discusses the constraints incorporated into the genetic algorithm process in order to avoid certain events. Section 5.4 presents the assessment learning process. The multi-sessions mode is discussed in Section 5.5. Section 5.6 includes conclusions and future enhancements.

## 5.2  Basic Components of the Proposed GA Based Method

Assessment of a person's abilities and skills is an important task for many organizations. Examples include evaluating a child's readiness for school or determining an employee's aptitude for a position. Assessment involves various parameters related to the test subject's aptitude, such as motor skills, linguistic development, or deductive capabilities, among others. Assessment is carried out by many bodies, such as public education systems, commercial testing companies, and recruiters. The major question to be answered is whether the test subject has the required capabilities and is ready to undertake the next step in his/her life.

To facilitate the assessment process, computer-based assessment systems are used. A recent innovation is an adaptive computer-based assessment tool that analyzes the subject's skill and dynamically adapts the assessment tests accordingly. A computer-based apparatus is used for evaluating school readiness of a child by having the child play a series of games comparing the child's performance with a database of performance results for a population. The apparatus includes a processor for processing the child's performance data and comparing it with the performance results of the population, and for applying a GA to determine the most appropriate next test for the child. The population for the comparison is chosen from the social network of the tested child by considering his/her friends, his/her brothers and sisters, his/her classmates in the previous/current grade, his/her neighborhood, etc. Incorporating the social network in the process is essential because it leads to a more reasonable population for comparison. For instance, a child interacts with his/her environment, including sisters and brothers, classmates, friends, who he/she meets almost daily. So, it is important to realize how the child performance compared to each of these groups individually and to the different combinations of these groups. A normal child is expected to perform at least at the average. We will first start by describing the employed GA process and then we will present the utilization of the social network model in the assessment process.

A GA is a type of optimization technique for finding an optimal solution to a problem of the kind that has a number of possible solutions [3, 6, 22, 25]. In other words, given a large search space, the GA is an effective process that tries to find an optimal or close to optimal solution from the search space. In a GA, a solution is represented as a chromosome interchangeably called individual and consists of genes. The GA applies genetic operators, such as crossover and mutation, to a population of chromosomes to evolve a new generation. Crossover produces new chromosomes by processing of the best existing chromosomes. One point and two point crossover are very common where one and two cut points are, respectively, specified in the parent chromosomes; then parts of the parents are exchanges/swapped to produce the children which are added to the population. However, uniform distribution based cross-over fits better to the application described in this paper and hence has been adapted in the GA. Mutation on the other hand, modifies certain gene(s) within the chromosome in order to guarantee better convergence. At the end of the crossover and mutation process, the size of the population almost doubles and hence some evaluation is needed to decide on chromosomes to leave and those to be moved to the next phase or generation; this is done by applying a fitness function on the chromosomes. The fitness of the chromosomes is evaluated to determine those to survive for the next generation. The process of generating new chromosomes and selecting the most fit is repeated until a threshold is satisfied or an upper bound is hit on the number of generations. At the end of the process, the best chromosome guides the assessment process. If the student is fast enough is taking the test then an intermediate result from the GA may be used to guide the assessment, with the GA process advancing towards the optimal solution. This will help in keeping the student busy instead of getting upset of waiting for the next step and may be end up not taking the test seriously enough.

A problem affecting current adaptive computer-based assessment tools is due to the fact that there are so many possible sets of solutions that great deal of the processing time and processor power are required to reach an optimal or near optimal solution. The presented approach provides an improved GA for adaptive computer-based assessment methods and systems. The improvement consists of applying constraints to the chromosomes so as to optimize the solution set leading to faster convergence. Two types of constraints are used: hard constraints, which filter out solutions that violate non-negotiable requirements and soft constraints, which carry a penalty, reducing a chromosome's fitness evaluation.

### 5.2.1 Closer Look at the Employed GA-Process

The GA based method is described by an example driven approach, with reference to the accompanying figures, in which components are designated by reference numerals.

Figure 5.1 illustrates an exemplary computer-based assessment tool, which can be adapted to include the presented method. Server 11 is a computing device, which can operate stand alone or on a network that can serve a plurality of users. Server 11 is provided with a core program 12 that comprises a GA component 13 and an assessment component 14.

The GA (13) receives from the assessment (14) the user's performance data at various stages of the assessment and evolves for each such stage an optimal or near optimal solution concerning the content to use for the next stage of the assessment. The assessment receives the solution from the GA, builds the assessment output using content in the database (15) according to the received solution, and sends the assessment output to an input/output unit (17).

The input/output unit can be one or more of various known computer peripherals for enabling a user to communicate with the computer program. For example, a display or speakers for output and a keyboard, a mouse, or a joystick for input. The input/output unit can be implemented in various ways, such as a dedicated terminal, an independent computer, or part of the same computer as server 11. If physically

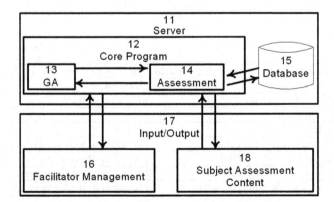

**Fig. 5.1** An adaptive computer-based assessment tool in accordance with a preferred embodiment of the presented method

separate from server 11, the input/output unit can be connected to server 11 via various communication links, such as direct serial or parallel connection or TCP/IP Ethernet connection. The input/output unit comprises subject assessment content unit 18, wherein assessment information is presented to a subject, for example on a display, and the user's response is received, for example via a keyboard and mouse. The input/output unit can also comprise facilitator management unit 18, whereby a facilitator can control system feed behavior. Finally, the tool can be adapted to serve other purposes such as job-candidate evaluation.

## 5.2.2  From User to GA

When the subject completes an assessment session, a record 22 (Fig. 5.2) of the result is passed to the GA (13), which is in charge of selecting the next session (or group of sessions) for that category of the assessment. The result record 22 includes results of the session and also information about the user (for example, age or experience level) and the difficulty level of the session. The result record 22 is produced from the information received by the assessment (14) from a subject interacting with subject assessment content (18). The result record 22 is used by the GA to create a chromosome that embeds the performance formula; in other words, the items in record 22 are used to construct the chromosomes for the GA.

The GA then creates a population based on that chromosome and it then evolves subsequent generations using crossover and mutation until some stopping criterion is met (typically that a chromosome meets a fitness evaluation threshold or maximum number of generations is reached). The fittest chromosome is translated into a record of new session information 20 (Fig. 5.2). The information is used by the assessment unit to select the content from the database (15) for the next session.

20                                       22

| Data Structure: New Session Information | Data Structure: Result Record |
|---|---|
| Level number | Level number |
| Maximal number of tries | Number of tries |
| Maximal time | Time |
| Minimal time | Time |
| Maximal number of help clicks | Number of help clicks |
| Maximal number of help clicks | Number of solve clicks |
| Number of needed right answers | Number of right answers |
| Number of possible wrong answers | Number of wrong answers |

**Fig. 5.2** A representative table of data structures for use with a genetic algorithm to generate new session content

An exemplary GA algorithm is presented in the sequel. There are two main cases: The first case is an assessment system with no memory, in which performance depends only on the most recent result. The second case is an assessment system with n-cell memory in which we evaluate the user's performance according to the last n-1 sessions that he/she participated in. In the latter case, we can draw information regarding the user's history and his/her learning curve. This allows for reuse of previous results achieved in evaluating the new achievements of the child involved in the assessment. Further, it leads to perception to tell how the child improved over time as opposed to the no memory version which does not reuse any past information.

### 5.2.3  Chromosome Generation and Selection

Each chromosome is a suggestion for a choice of content for the next session. The chromosome is a vector of genes that are properties of a session. For instance, the i-th element of the vector, which is the i-th gene in the chromosome, contains the maximal time that the session should last.

The chromosome population consists of a preset number of chromosomes. The speed of convergence of the iterating GA depends strongly on the initial population. The range that the initial population covers includes (or come close to include) the current skills of the user, under the assumption that the learning and skills curves are consistent and continuous. If they are not, a regression procedure can be added.

The current skills of the user are drawn from the last game played. Hence, an appropriate chromosome for that game is the base chromosome from which to create the initial population. There are two ways of creating the population of chromosomes: one is according to a uniform distribution around the base chromosome and the other is according to a normal distribution around the base chromosome with preset standard deviation parameter vector $\sigma = (\sigma_1, \sigma_2, \ldots, \sigma_l)$, where $\sigma_i$ is the standard deviation from the base for gene $i$.

Given large enough deviation values, the first generation of chromosomes will contain the user's current state and the appropriate offer with high probability for the next assessment exercise. The standard deviation vector is balanced to the best trade-off empirically. After the chromosome population is created, the GA starts to create new generations in which better chromosomes produce new chromosomes using crossover and mutation. A chromosome is better than another chromosome when its evaluation value yields better scoring from the fitness or evaluation function.

The evaluation function is a sum of two terms, the first is a difficulty term and the second is a penalty term. The difficulty term is a distance function between the chromosome information and the learning curve factor of the user: $\delta$. The target is the chromosome that best achieves minimal distance between the anticipated $\delta$ and the difference between the chromosome information and the given results. Furthermore, learning factor $\delta$ is adjusted to the user after every activation of the

GA, i.e., $\delta$ is a grade for the improvement in the GA anticipation for the user. Of course, $\delta$ is negative when the user is having difficulties with the current skill needs.

The fitness sub-function would make a decision whether the offered game is suitable for the user according to his result in the last game and will yield an appropriate improvement factor $\delta$. The improvement factor will be used to determine his fitness in the next game. For example, we start with an improvement factor $\delta_{initial} = 0$, or any other preset value that corresponds to the normal learning curve of the user. In the next round, the improvement factor will be a value that states whether the user's skills are increasing or decreasing and it will be the value $\delta$ which is appropriate to the chromosome that was chosen. If the user's result was insufficient, i.e., the game was too difficult for him/her, then the next game that would fit him/her should not be harder than the previous one (maybe even easier) and hence, the improvement factor will be increased to be zero or a small positive value, such that the game after it will have more or less the same difficulty to make sure the user's skills have really improved and he is stable in his/her current level. This is important in order to keep the child involved and eager to proceed with the test; otherwise he/she will give up early and the outcome may not be the anticipated one.

Assuming $\delta$ is given, the optimal chromosome $Ch$ will yield the closest improvement to $\delta$. Hence, the fitness evaluation function is set such that the difference between the optimal chromosome and the last result is as close as possible to $\delta$. If functions $f_i$ are positive metrics, then the optimal chromosome satisfies:

$$\min_{Ch \in population} \left\| \begin{array}{l} f_1(Ch.Level - Re.Level) + \\ f_2(Re.Time - Ch.MaxTime) + \\ f_3(Ch.MaxHelpClicks - Re.HelpClicks) + \\ f_4(-Re.Tries) + \\ f_5(-Re.Wrongs) + \\ f_6(-Re.SolveClicks) + \\ f_7(-Re.ExitClicks) + \\ -\delta \end{array} \right\|$$

where $Ch$ is the chromosome and $Re$ is the result.

The functions $f_i$ can be any loss functions; the $L_\infty$ norm metric functions seem to be satisfying; however, other metric functions can be experimentally tested, such as $L_2$ norm metrics.

An alternative method for integrating the improvement factor is an n-cell memory machine, where the last n-results are included in the fitness evaluation procedure and then the embedded $\delta$ factor is computed within the fitness evaluation. In this method, the input to the evaluation function is the chromosome as well as the results record that the user achieved by interacting with the software. Theoretically, there is no difference between the two methods, since they both use the same amount of information, hence have the same power of computation.

## 5.3   Empowering the GA Process with Constraints

Due to the specific setting of the problem, it is necessary to dictate the occurrence or prevention of several events. These dictates arise from the psychological and logical aspects of the games and the requirements of a reasonable order of the gaming procedure. Therefore, there is a set of conditions that the user must follow. The modification of the GA to include this set of conditions comprises the proposed method, which is described in more detail in this section. The conditions are divided into two groups, according to the level of the seriousness of each condition. Some conditions form merely advice to the tester, and can be treated as guidelines for testing the child. Other conditions form a necessity and must be adhered to; incorporating them does make difference.

The logical difference between the two kinds of conditions dictate different methods of implementation, hence they are divided into two sets of constraints: soft constraints (SC) that may be broken and hard constraints (HC) that must be satisfied.

### 5.3.1   The Soft Constraints

The penalty term is imposed on the GA due to requirements of the assessment tool. The design of the assessment tool contains several constraints. We divide the constraints into two categories, hard constraints (HC) and soft constraints (SC). The HCs are induced from "must" and "must not" statements in the design. The SCs are induced from "recommended" and "strongly recommended" statements in the design.

The solution we present for integrating the SCs into the system is a penalty term in the fitness evaluation function. We assign a penalty value for each constraint and sum it in the fitness evaluation function when the constraint is broken. A list of the SCs 30 can be seen in the table in Fig. 5.3.

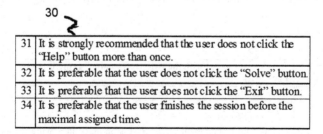

| 30 | |
|----|--|
| 31 | It is strongly recommended that the user does not click the "Help" button more than once. |
| 32 | It is preferable that the user does not click the "Solve" button. |
| 33 | It is preferable that the user does not click the "Exit" button. |
| 34 | It is preferable that the user finishes the session before the maximal assigned time. |

**Fig. 5.3**  A representative table of soft constraints in accordance with a preferred embodiment of the presented method

The penalty term in the fitness evaluation function is as follows:

$$\sum_{i=1}^{\#SCs} C(i) \cdot SCB(i, Chromosome)$$

where #SCs is the number of soft constraints, $C(i)$ is the penalty factor for constraint $i$ and SCB() is a Boolean function implemented as follows:

$$SCB(i, Chromosome) = \begin{cases} 0, & Chromosome\ satisfies\ SC(i) \\ 1, & otherwise \end{cases}$$

So, the fitness evaluation function for a chromosome is:

$$FitnessEvaluation(Chromosome) = Difficulty(Chromosome)$$
$$+ \sum_{i=1}^{\#SCs} C(i) \cdot SCB(i, Chromosome)$$

As formulated above, there is some redundancy because certain terms also appear in the SCs penalty term. This does not affect the quality of the solution, since it will only increase the penalty on these terms and hence has a uniform effect.

### 5.3.2  The Hard Constraints

The hard constraints (HCs) are induced from requirements that are presented by the use of "must" or "must not" statements. A representative list of HCs 40 is shown in the table in Fig. 5.4.

A scheme for implementing the hard constraints is shown in Fig. 5.5. An input chromosome 50 is checked 52 for satisfying HCs. If it is satisfied, the chromosome is output 58, if not; it is checked 54 to see whether it can be fixed. If it can, it is fixed 56 and again checked 52 against the HCs; otherwise an error 60 is generated.

40

| | |
|---|---|
| 41 | The user must start with the "Mouse Training" game if he is a new user |
| 42 | The user must start with games of "Math Readiness". |
| 43 | The user must play "Counting Sticks" before playing "Counting Numbers". |
| 44 | The user must not repeat the same session more than twice. |
| 45 | The user must play games that are appropriate for his age. |
| 46 | The user must play at least one game from each level. |

**Fig. 5.4** A representative table of hard constraints in accordance with a preferred embodiment of the presented method

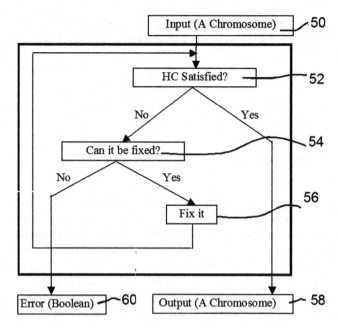

**Fig. 5.5** A scheme for implementation of hard constraints in accordance with a preferred embodiment of the presented method

```
HCFilter (Input : chromosome ch)
begin
    Loop for i from 1 to #HC
        if HC(i) is not satisfied in chromosome ch then
            if HC(i) is fixable then
                fix chromosome ch
                continue loop
            else
                disregard chromosome ch
                return
            end
        end
    end
    accept chromosome ch
    return
end
```

**Fig. 5.6** A pseudo code naïve implementation of a hard constraint filter in accordance with a preferred embodiment of the present method

Figures 5.6 and 5.7 is a pseudo code naïve implementation of an HC filter. The set of HCs are integrated in the system after the best chromosome in the population is chosen. Making an offer for a new session from the chromosome consists of a scoring technique for the set of possible sessions given in a table according to the

- Create the first generation of *I* chromosomes using NDCC.
- Evaluate the fitness of each chromosome using CE with SC penalty.
- While the cost of the best chromosome isn't good enough
  o Remove a portion of the worst chromosomes.
  o Breed new chromosomes with mutations to fill the vacancy.
  o Evaluate the fitness of the new chromosomes using CE with SC penalty.
- Return the best fit chromosome.

**Fig. 5.7** The modified genetic algorithm

values of the chromosome genes. The HCs are checked when matching a session for the chromosome. In case one of the HC is not satisfied, then it is made impossible to choose the corresponding session. For example, if HC 43 is not satisfied, meaning that we find the game "Counting Sticks" appropriate, but the user hasn't played "Counting Numbers" yet, then we mark it impossible to choose "Counting Sticks" by changing its score. We further check from his/her social network for someone who may guide him/her to learn the process for a second round of check. We select from the social network according to the following preference, sisters/brothers, friends, then classmates.

Some HCs can be satisfied before activating the GA, such as HC 41. In this case, we can check if the user is new, skip the GA, and choose the "Mouse Training" game. HC 42 is treated similarly. It should be clear that the description of the embodiments and the attached figures set forth in this specification serves only for a better understanding of the method, without limiting its scope as covered by the Claims.

## 5.3.3  *Computing the Result Evaluation*

The Result Evaluation of a game that was played by a certain user is a weighted sum of his/her performance in various qualities, such as the relative time consumption, the score, the number of "help" button clicks and the usage of the "solve" option, etc. Also including in evaluating the result is the performance of the individuals in the social network of the tested child. This is actually important because a child is not expected to be much better than those in his/her social network, though it is preferred and appreciated to have him/her superior to his/her peers. However, having the peers performing at low quality will reveal the low evaluation result of the tested child. Hence, he/she and his/her peers may need serious and special treatment as a group instead of identifying only the tested child as poor, which is not the right decision. He/She might have passed the evaluation successfully shall he/she be placed in an appropriate environment with qualified peers.

Each quality is measured by a function $Q_i$ for some i. Each quality is multiplied by a weight value w(i), which is preset to be an appropriate importance value of the quality. For instance, the time consumption quality measurement can be formulated as follows (Fig. 5.7):

$$Q_{time} (Result,\ GameInfo) = \frac{Actual\ Playing\ Time - Minimal\ Playing\ Time}{Maximal\ Playing\ Time - Minimal\ Playing\ Time}$$

The quality of "help" button usage can be formulated as follows:

$$Q_{help} (Result,\ GameInfo) = \frac{Actual\ Number\ of\ Requesting\ Help\ Button}{Maximal\ Number\ of\ Helps\ to Request}$$

Some other qualities:

$$Q_{solve} (Result,\ GameInfo) = Click\ on\ Solve\ Button$$
$$Q_{exit} (Result,\ GameInfo) = Click\ on\ Exit\ Button$$
$$Q_{score} (Result,\ GameInfo) = Score$$
$$\vdots$$

**The Fitness Sub-function**

$$Decision\_Value(chromosome) = \left| CE - \left( RE + \delta_{input} \right) \right|$$

$$Final\_Decision = \min_{\forall chromosomes} Decision(chromosome)$$

## 5.4  Assessment Learning

In order to provide further usage of readiness assessment, we discuss plausible learning mechanisms in this section. The motivation of learning aims at information reuse of the assessment results in social aspects; in other words, the assessment of school readiness can be further used if the outcome can be interactively interpreted with a structured network from which a learning protocol focusing on the assessment evolution is incurred. Another reason for this structured learning attributes to the non-deterministic nature of genetic algorithms, that is, we need to gauge the robustness or behavioral patterns of assessment from the software. We expect that the reuse of assessed information by learning aids the optimization processes, leading to enhanced features of the innovative tool presented in the previous section. To enable the aforementioned learning capabilities, we resort to current computational trends in social network [2, 17, 21], which can be categorized to unsupervised learning in consideration of noisy environment in assessment processes. Two possible ways: learning by copying and analyzing group behavior, are described in the sequel.

## 5.4.1  Learning by Copying

In process 12 of Fig. 5.1, the communication between GA (13) and Assessment (14) components can be further analyzed. As presented in the previous section, the evaluation of chromosomes is a sum of measures, and the evolution of chromosomes relates to the computation in choosing the next user session. The concern is that the learning process is somehow mystified by the measure; and this can be alleviated by a copy learning mechanism in social network perspective. The learning process is modeled as a copying process among individuals in serial of sessions, in this case, the genes on chromosomes to be evaluated, in each stage of prompting new sessions after the GA. It is argued that the most effective learning process is to simply copy what is useful and discard what is not (by imposing penalties) among individual components. In regard of assessing user input, it is highly likely that each gene (information) on different chromosomes (combination of choices) evolves differently, and so if genes can be weighted (similar to the notion of "$C(i)$"), the communication between GA and assessment can evolve by copying certain set of genes from previous sessions on pertaining chromosomes. This deviates from the point measure, and actually the learning path can be provided in reference to selection of genes copied and discarded. As simple as it sounds, the argument is influential by automating the evolution of the assessment process. The expected payoff of a known selection can be found by [21] with details of the winning strategies:

$$w_{exp} = w(1 - p_{est})^i + w'(1 - (1 - p_{est})^i)(*)$$

where $w$ is the current weight and acquired $i$ stages ago, $w'$ is the estimated mean weight for all genes, and $p_{est}$ is an estimate of the probability of changing the choice of contents of each chromosome. The copy learning process is to be implemented between GA and assessment module in the core program in Fig. 5.1.

## 5.4.2  Group Behavior

It will be greatly useful if the assessment tool can depict the group behavior of users, as the readiness ability can vary among group of children under different circumstances, leading to the need for administrators to understand the discrep-ancies. The behavior can be structured in network methodology as follows: On individual basis, the genes whose evolutionary patterns are known from previous learning process can be structured as nodes in networks. For example, gene $i$ is linked with gene $j$ if the payoff on certain stage of evaluation is greater than preset threshold, this is analogous to the selection pressure from phenotype to genotype mapping in computational biology. Further, links can be weighted by the payoff in equation (*). The group behavior can be considered as the combination of such networks for all individuals, for example, averaging the weighs of links to produce the group network. As a consequence, the global picture of assessment behaviors can be structured as networks based on assessment stages (process 12) and gene-centered chromosomal level. The visualization of such network for group A can be compared with that for group B, and further explorations can be done on graph similarity search so as to quantify the behaviors and graphical patterns.

Actually, the social network model is used effectively in casting the assessment process into a more realistic and fruitful means leading to robust decisions. In fact, a child is expected to perform at the level of his/her peers. So, to have a successful framework, it is important to identify peers for each child to be tested before the test starts. This will guide the testing by following a success story that relates to one or some of the peers of the child to participate in the test. The social network modeling process starts by identifying the sisters/brothers, friends and classmates for each child to participate in the test. Each one of these three groups is treated separately first starting with sisters/brothers group. The database is searched to identify the history for his/her sisters/brothers; if found, the success and failing cases are identified, if any. Then, the child is testing based on the rich history produced from the already completed testing of his/her sisters/brothers. Further, his/her results are compared against the results achieved by his sisters/brothers to find out if there is improvement in the family because it is anticipated that younger children are mostly affected by the performance of their elder sisters/brothers. Younger children are expected to demonstrate better performance compared to their elder sisters/brothers because they learn from their elders.

The second social network group to be used in driving and focusing the test is the social network of friends (including relatives other than sisters/brothers). Normally, a child is affected by his/her peers with whom he/she spends most of the time playing, especially if he/she is the first child of his/her family. Following a test trend that considers this group of peers may lead to more successful results to the benefit of the child, the family, the school and the community.

The last social network group to consider in the testing is classmates. This is an important group that must be considered in combination with the other group to realize the effect of each group on the tested child. For instance, a child is expected to performance at least as good as his/her sisters/brothers, friends and/or classmates. There are two extreme cases to consider here. First, a child may show degraded performance compared to his/her peers in any of the three groups. Second, a child may demonstrate outstanding performance compared to the other groups. In both cases, we have to isolate the group(s) which contradict the demonstrated performance of the child and run the test using the other group(s) in order to study the effect of the other group(s) on the performance of the child. This will lead to important and vital result whether positive or negative. For the positive performance case, the recommendation may be to move the child to a more elite class when he/she reflects superiority to them; recommend to the parents to watch for his/her friends who may negatively affect the performance at later stages, etc. For the negative performance case, the result may lead to recommend having the child more involved with his/her friends who demonstrated better performance in order to learn from them; his/her sisters/brothers may need to provide further help to improve his/her capabilities; he/she may be moved to a suitable class that fits better the current potential, etc.

## 5.5   The Multi-Sessions Mode

We have also investigated another mode of activation for the software, where we introduce a set of sessions to the user in each round. According to the average result in the set of sessions we create the chromosome and activate the GA to choose a different set of sessions for the user to play. The only addition we introduce to this mode is a pre-determined association of sessions in sets. Each set contains a pre-determined number of sessions that all belong to the same category and are all of the same level.

The added value of this mode of activation is that we avoid singular non-average results that might have occurred. These singular non-average results can be a result of a one-time event that does not indicate anything about the true ability and performance of the user.

Computationally, this mode is better because it requires a fewer number of activations of the GA. The runtime of the GA is one of the heaviest bottlenecks of the software, and reducing the number of times we need to run it is a good optimization step.

## 5.6   Conclusions and Future Enhancements

We have created software that assesses the user's readiness for school in a wide variety of tests. The software is play-based, hence enables the user to use it without having to go through the assessment feeling he/she's being tested. The true novelty of this software is its ability to learn the user behavior and reuse its history during the test course, and thus, providing the appropriate tests and difficulty levels. Further, we incorporated the social network model in the assessment process because we believe that a child is mostly affected by his/her peers whether sisters/brothers, friends or classmates. Expanding the testing and evaluation in this direction has lead to more robust framework because the testing is no more biased; it is comprehensive and considers all aspects related to the tested child. At the end, a child whose performance is low compared to all the three groups of peers raises a red flag for a case that must be carefully studied and should receive special treatment. This way, the early warning may be to the benefit of the child who may improve if degraded performance is noticed at early stage. The use of GA in this software is the perfect natural choice to assess the evolution of the user's understanding and ability to promote his/her performance. Finally, we are currently investigating different approaches to learn the user's performance and improvement, using various learning methods from statistical learning theory. We will also study the influence of the family social network on the assessment of children; the education level of the family may be a good measure to incorporate in the assessment.

# References

1. Berger, L., et al.: First-year maternal employment and child outcomes: Differences across racial and ethnic groups, Child. Youth Serv. Rev. **30**, 365–387 (2008)
2. Centola, D.: The Spread of Behavior in an Online Social Network Experiment. Science, **329**(5996), 1194–1197 (2010)
3. Chambers, L.: (ed.) The Practical Handbook of Genetic Algorithms: Applications, 2nd Edition. CRC Press (2000)
4. Clements, D.H., Nastasi, B.K., Swaminathan, S.: Young children and computers: Crossroad and direction from research, Young Children, **48**, 56–64 (1993)
5. Condron, D.J.: An Early Start: Skill Grouping and Unequal Reading Gains in the Elementary Years, The Socio. Q. **49**, 363–394 (2008)
6. David, H.: Adaptive Learning by Genetic Algorithms: Analytical Results and Applications to Economic Models, Springer (1999)
7. Donald, A.R., Stenner, A.J.: Assessment Issues in the Testing of Children at School Entry, The Future of Child. **15**(1), 15–34 (2005)
8. Downes, T., Reddacliff, C.: Young Children Talking about Computers in their Homes, Australian Computers Education Conference (1996)
9. Duncan, G.J., et al.: School readiness and later achievement, Dev. Psychol. **43**(6), 1428–1446 (2007)
10. Feinstein, L.: Inequality in the Early Cognitive Development of British Children in the 1970 Cohort, Economica. **70**(277), 73–97 (2003)
11. Fogel, D.B.: Evolutionary Computation: Towards a New Philosophy of Machine Intelligence, IEEE Press (2000)
12. Guhn, M., Forer, B.: Translating school readiness assessment into community actions and policy planning–The Early Development Instrument Project, Symposium conducted at the annual meeting of the American Educational Research Association, San Francisco (2006)
13. Hair, E., et al.: Children's school readiness in the ECLS-K: Predictions to academic, health, and social outcomes in first grade, Early Child. Res. Q. **21**, 431–454 (2006)
14. Janus, M., Offord, D.: Development and psychometric properties of the Early Development Instrument (EDI): A measure of children's school readiness, Can. J. Behav. Sci. **39**, 1–22 (2007)
15. Kagan, S.L., Evelyn, M., Sue, B.: Reconsidering children's early development and learning: Toward common views and vocabulary. In Report of the National Education Goals Panel, Goal 1 Technical Planning Group. Washington, DC: Government Printing Office (1995)
16. Klein, S.P., Alony, S.: Immediate and sustained effects of maternal mediating behaviors on young children, J. Early Interv. **17**, 1–17 (1993)
17. Kianmehr, K., Alhajj, R.: Calling Communities Analysis and Identification Using Machine Learning Techniques. Expert Syst. Appl. **36**(3), 6218–6226 (2009)
18. Klein, S.P., Nir-Gal, O.: Humanizing computers for young children: effects of computerized mediation of analogical thinking in kindergartens. J. Comput. Assist. Learn. **8**, 244–254 (1992)
19. Martin, A., Rebecca, M.R., Jeanne, B.-G.: The joint influence of mother and father parenting on child cognitive outcomes at age 5. Early Child. Res. Q. **22**(4), 423–439 (2007)
20. Miller, G.E., Emihovich, C.: The effects of mediated programming instruction on preschool children's self-monitoring. J. Educ. Comput. Res. **2**, 283–299 (1986)
21. Rendell, L., et al.: Why copy others? Insights from the social learning strategies tournament. Science. **328**(5975), 208–213 (2010)
22. Schmitt, L.M.: Theory of Genetic Algorithms, Theor. Comput. Sci. **(259)**, 1–61 (2001)
23. Snow, K.L.: Measuring school readiness: Conceptual and practical considerations, Early Educ. Dev. **17**, 7–41 (2007)
24. Vernon-Feagans, L., Blair, C.: Measurement of school readiness, Early Educ. Dev. **17**, 1–5 (2006)
25. Whitley, D.: A genetic algorithm tutorial. Stat. Comput. **4**, 65–85 (1994)

# Chapter 6
# From Simple Management of Defects to Knowledge Discovery to Optimize Maintenance

**Grégory Claude, Marc Boyer, Gaël Durand, and Florence Sèdes**

## 6.1  Introduction

To ensure the quality of a final product, processing and traceability of defects which occur during its industrial manufacturing process has become an essential activity. Indeed, management of information relative to defects may represent up to 80% of the final product information volume. Therefore, the processing of this mass of information provides a real value-added to, for example, understand scrapping reasons, reduce or even remove this scrapping and anticipate manufacturing issues. A parallel can be drawn with software defects and the numerous follow-up

G. Claude (✉)
Université de Toulouse, Université Paul Sabatier, IRIT UMR 5505
118, route de Narbonne 31062 Toulouse cedex 9, France

Intercim LLC, 1915 Plaza Drive, Eagan, MN 55122, USA

Intercim, 32, rue des Jeûneurs 75002 Paris, France
e-mail: GClaude@intercim.com

F. Sèdes
Université de Toulouse, Université Paul Sabatier, IRIT UMR 5505
118, route de Narbonne 31062 Toulouse cedex 9, France
e-mail: Florence.Sedes@irit.fr

G. Durand
Intercim LLC, 1915 Plaza Drive, Eagan, MN 55122, USA

Intercim, 32, rue des Jeûneurs 75002 Paris, France
e-mail: GDurand@intercim.com

M. Boyer
Université de Toulouse, Université Paul Sabatier, Inserm UMR 825
Place du Docteur Baylac, Pavillon Riser 31059 Toulouse cedex 9, France
e-mail: Marc.Boyer@iut-tlse3.fr

T. Özyer et al. (eds.), *Recent Trends in Information Reuse and Integration*,
DOI 10.1007/978-3-7091-0738-6_6, © Springer-Verlag/Wien 2012

systems for bugs management activities. We can name IBM Rational Clearquest[1] or Atlassian Jira,[2] and other ones, free, such as Mantis Bug Tracker[3] or BugZilla.[4] These software applications offer various features, including bugs creation and tracking in real-time, reporting and environment configuration and customization. Some offer other features such as project planning tools.

However, we can wonder about capitalization of information from all defects processed with such tools, whether software or not. Indeed, these tools are used for defects monitoring, mainly enabling to note them, to follow the process leading to their resolution and to close them. Generated reports are primarily statistics (e.g., the number of created or solved defects during a period, defects in progress sorted by their criticality) and are presented in a graphical form (e.g., histograms, curves, pie charts).

In maintenance activities, we can notice many benefits in taking advantage of the information from defects [16], in particular in aeronautical, aerospace and pharmaceutical industries where notably important workforce causes high rates of defects during the manufacturing process. A first benefit, in the short run and in an online mode, is to make corrective maintenance activities easier by assisting the maintainer in his task of finding a solution to solve a problem. A second benefit, in the long run and in an offline mode, is to prevent the emergence of recurrent defects by highlighting the reasons of their emergence.

A manufacturing cycle of a product is composed with the time spent to foreseen activities and the time spent to unforeseen activities (i.e. the processing of defects). In some industries, aerospace in particular, this second work can represent more than 50% of the time. We think that optimizing maintenance activities is to reduce this time by analyzing and classifying past and solved defects to allow:

- Searching for similar defects to a new created one to know solutions executed in the past to solve the problem.
- Knowing and executing, if possible, the less time-expensive solution to solve a problem.
- Understanding recurrent defects to avoid their emergence.

In this paper, we present an approach which meets our aim of using information from defects to optimize maintenance activities. For this, we propose a modeling of defect on which we rely on. It allows considering both nominal defects (i.e. those that can be foreseen) and non-nominal ones (i.e. those that were not thought, that could not be foreseen). To make it, we particularly focused on defining elements which qualify a defect. Moreover, we can apply it to defects from various maintenances, in fields as different as software or industry. By setting up a framework focused on the defect as a lever of manufacturing process improvement,

---

[1]http://www-01.ibm.com/software/awdtools/clearquest/

[2]http://www.atlassian.com/software/jira/

[3]http://www.mantisbt.org/

[4]http://www.mozilla-europe.org/fr/products/bugzilla/

we want to assist in online correcting, to anticipate it or even to correct the manu-facturing process before it appears. Beyond the simple management of the defect's traceability, i.e. their real-time monitoring which ends with their storage/archiving, we want to take advantage of their information, most often underused information and yet whose capitalization, producing knowledge, can give a real value-added.

The aim of this paper is to propose a framework which takes advantage of the information from defects to reveal the knowledge in order to optimize maintenance activities. First, in Sect. 6.2, we describe the concept of defect, an example of defect resolution and present our context of maintenance. In Sect. 6.3, we define a modeling of defect which is at the base of the approach on defects processing we propose in Sect. 6.4. Section 6.5 presents the elements we use for the implementation of our approach and results. Finally, we conclude and present some perspectives.

## 6.2  Taking Advantage of Information from Defects

In this section, we define the concept of defect by describing our vision of the resolution process. We emphasize the fact that it allows to solve defects but it does not take advantage of the information from them. We give an example of defect to illustrate it. Then, we present our context of maintenance with a state-of-the-art about defect, its representation in the literature and its use in maintenance activities.

### 6.2.1  The Defect Resolution Process

During a manufacturing process, when a variation is noticed between what is defined by this process and what is carried out, a defect is noted in order to materialize this variation and to try to solve it. We can represent a defect resolution as a process during which data about the problem and the solution to be set up is filled in (Fig. 6.1). Once the solution executed or if there was not any solution, the defect is closed and stored. To execute this process, a document, we call it a defect form, is instantiated. Whether paper or electronic, this form serves as a support for a defect. It is composed of fields which store information about the defect, the data filled in by users. Even in a restricted domain, several kinds of defects may emerge so several defect forms templates may exist.

Some details about the objects involved in this activity diagram:

- *Detected defect*: at this stage, user has to choose which defect form template to use to solve the defect. We consider a set of templates because all defects are not solved in the same way according to their type. Several templates are thus defined to allow the user to choose the most appropriate one to solve it.
- *Created defect*: it is an instance of a defect form template in which data must be collected in fields.

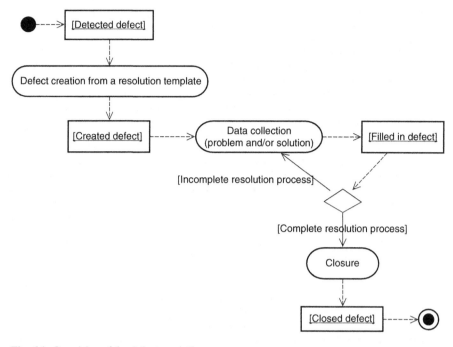

**Fig. 6.1** Our vision of the defect resolution process

- *Filled in defect*: defect description is filled in at several times by several actors (operator(s), manufacturing support, quality assurance, etc) at different times (may take several days to several weeks). However, all actors have an overall vision about the defect. It is not a set of fragmented elements that everyone has to fill in regardless of the other elements but a unique instance completed as the actors go along.
- *Closed defect*: defect resolution process is complete, defect form is closed.

A defect resolution ends with the storage of the defect. We would propose an approach that goes beyond this process, this one being the basis on which we would make an effective defects processing. Without any use of the information contained in the defects database, i.e. without processing extracting knowledge, this resolution process suffers from several issues:

1. *Processing slowness*: fields filling in is an entirely manual activity, no assistance is given to the user to guide him during the defect resolution, whether to fill in fields about the problem or the solution.
2. *Processing cost*: it follows the previous point, the longer the defect resolution takes time, the more expensive the process is, since it blocks the manufacturing process on which the defect is noted, and thus the final product.
3. *Recurrent defects processing*: no assistance to analyze past defects is carried out, no action is taken to avoid recurrent defects emergence.

Fig. 6.2   An example of a defect form

## 6.2.2   An Example of Defect Resolution

Let us consider a manufacturing process during which ten holes are drilled in a metal part constituting an aircraft wing. During the drilling of the 7th, the worker's tool breaks inside, which widens the hole diameter beyond the expected (Fig. 6.2). From this moment, the worker notes a defect in which he informs of the encountered problem. Then, he waits for instructions, from the manufacturing support, on the solution to be implemented to resume this manufacturing process execution. Two hours later, the manufacturing support analyzes this defect. Typically, three solutions are possible.

*First solution: keep the hole as is.* It is a concession that needs to be validated by the design office and will be shared with customer. Because of dozens of defects the design office receive every day, this validation is performed several days after the note of the defect. The manufacturing process is on hold until the defect is closed. Expected delay: 10 days.

*Second solution: repair the hole.* Various standard disposition methods can be used, depending on complementary inspection performed by manufacturing support. This new data is collected. To repair it, it is possible to enlarge the hole to make it cleaner (without residual stress point which could start a tear) and to put a larger rivet. It is also possible to fill the hole to give it a correct diameter. Whatever chosen type of repair, both may require a design change. Obviously, such a repair needs to be detailed in the defect to define precisely the steps to be followed. The manufacturing process is on hold until the defect is closed. Expected delay: 20 days.

*Third solution: scrap the part and build another one.* The simplicity of this approach has to be balanced with the cost of scrap and the cost of delaying all manufacturing processes depending on the availability of this airplane wing part. The decision is usually made in the next three to five days by the factory management.

In any case, the manufacturing process, and thus the current part, is put on hold for at least 10 days, and probably more, delaying the other manufacturing processes that depend on it. If manufacturing cannot wait, and as far as possible, the execution of the manufacturing process and those who depend on it may continue with an alarm on the part mentioning that it must be corrected before the first flight or before customer delivery. Obviously, it also has a high cost: the later the problem is fixed, the higher the cost is.

## *6.2.3 Related Works*

Related works about maintenance activities and defects management can be a quite large domain. In the background section, we present the context of maintenance we consider. Then we take an interest in defects modeling in several works to situate our vision of a defect and we present weaknesses of defects processing systems.

### 6.2.3.1 Background

Defect resolution process is a collaborative work, several users with various profiles and specialties give their experience and knowledge. The document supporting a defect is filled in as the contributions go along. It is a communications medium between the different actors who are in charge of identifying the best solution to be set up to solve it. For this reason, we consider the defect resolution system as a collaborative (or collective) annotation system [7] allowing readers to create annotations in context and so, to share their experience and their knowledge with future readers [23]. Annotations are data given by users. There are two kinds of annotations [24], that we find in a defect resolution system: documents fragments which are information given by only one person such as problem textual description or date, and conversations fragments which mention an exchange between persons on a specific element such as the suggested solution which may be iterative between several users. Unlike traditional annotation systems, in particular web documents, whose annotations are attached to the document because it is available only in a reading mode, defect annotations are directly embedded in the document itself through the fields.

In the literature, maintenance is divided into several categories. In the industrial context, three major categories emerge: the preventive (or proactive) maintenance, the corrective (or reactive) maintenance and the self-maintenance [20]. We can also notice the predictive and holistic (process oriented) ones [4] and the condition-based and intelligent ones [12]. In the software context, Swanson had already talked about the corrective, adaptive and perfective maintenances in the mid 70s [36]. The evolutionary and preventive maintenances have been added to these first categories, still valid nowadays. We do not want to cover all these categories; we focus here on corrective and preventive dimension. The role of corrective maintenance is to solve a new defect's problem so that the element on which it is noted becomes again in conformity with specifications. Preventive maintenance is intended to anticipate and prevent from known defects emergence, so recurrent ones.

### 6.2.3.2   Modeling a Defect

To perform these two maintenances, we need to define a defect, what is the struc-ture of the document which support it, what kind of content can it store. In an organization, all defects are not solved in the same way according to their kind. Several defect forms templates are thus defined to use the most appropriate one. They involve fields which are not only used in one of them but which can be used in every template. Considering nominal defects (i.e. those that can be foreseen), forms involve fields that can be qualified as regular. Indeed, all of them are distinctly named and the data a user filled in is clearly identifiable. Some of these fields define structured data, whose list of values is finite and some others define unstructured data, as free text. Considering non-nominal defects (i.e. those that cannot be foreseen), the form involves non-regular fields. These fields are, for example, comments or description of the first occurrence of a new kind of defect. They define unstructured data. Thus, a defect form is a semi-structured document which involves various fields containing structured and unstructured information. To propose an approach to optimize corrective and preventive maintenance activities, modeling a defect is needed in order to better take advantage of all information about defects and so to reveal knowledge.

In the literature relative to the maintenance, authors are not interested in such a modeling, which would allow taking advantage of information about defects, but rather in the formalization of the maintenance process, which optimizes defect resolution. In industrial maintenance, Waeyenbergh et al. propose a framework, CIBOCOF [37], for the management of maintenance issues. A functional approach is presented by Despujols where various maintenance functions are described: studies, work preparation, scheduling, implementation, supply of raw materials and external services [11]. The TPM reasoning (Total Productive Maintenance) is often cited as a solution [2]. It involves all people at different levels of maintenance with the aim of optimizing the use of each manufacturing tool. In software maintenance, all works converge to a comparable process model [3, 6, 18, 21] whose major steps are (1) software analysis and understanding, (2) choice of a solution or reject of the intervention, (3) solution implementation, (4) intervention closure. Although the formalization of the maintenance process leads to facilitate defect resolution, it does not define how to use information about the defect.

Even if we note a lack of model of defect, we think such a model is essential to use at best information about defects. This model needs to be generic in order to take into account every resolution template and every defect, whether nominal and non-nominal. So, it is necessary to propose a model of descriptor of defect to characterize any defect, to describe the representation of its informational content and its structure. It must consist in representing any defect by a descriptor, i.e. a set of keywords extracted from their data contained in the fields of their form. Working on the descriptors rather than the document itself allows applying techniques which compare two sets of keywords to compute easily the similarity between the two documents. Thus, with a model, it is possible to find documents matching a document query, so to find similar defects to a new one.

### 6.2.3.3 Defects Processing Systems

To perform corrective and preventive maintenance activities, we can note various approaches. Some consider the defects database as a "bag of defects". Others, on the contrary, allow structuring the database, that is to say organize defects in a hierarchical form using different techniques, different algorithms.

As regards preventive maintenance, many approaches and solutions were proposed. We are not going to make a state of the art here; we prefer to refer the reader to the recent comprehensive study [9] which presents a wide range of application fields and numerous defects detection techniques, based on classification principle, nearest neighbors, clustering or even statistic and spectral techniques.

As regards corrective maintenance, we distinguish two approaches. The first one consists in searching relevant defects, because similar to the new defect in some parts, using keywords. For this, the user is prompted to enter one or more keywords. This will start a searching process of these keywords in the solved defects database and thus will select and present similar defects to the user. By using information from one or more of those, and in particular information about solution, the user can propose a solution more easily and more quickly. However, this method is rather addressed to users familiar with the domain and able to provide significant and discriminatory keywords for the search. We can mention another technique similar to the method but really different in its realization. It is based on the knowledge of users dedicated to defects resolution. When they have to solve a new defect, they rely on their own experience, they refer to the way they have already solved this kind of defect. Therefore, a similar defects searching process also exists, but it takes place in the user's mind using his own experience. Obviously, this technique has its limits with the loss, for the company, of these users. The second approach in corrective maintenance consists in organizing documents according to an a priori defined structure to allow the user to browse by selecting different categories. Adapted to the industrial defects management in aerospace, this approach was recently proposed by [16] using a faceted classification [15,28]. It allows finding similar defects, to group them and to set up a feedback between the maintenance department and the design department which defines the templates of industrial manufacturing processes on which defects occur. Nevertheless, a major effort must be made upstream to design a fully usable classification scheme. The use of descriptors, of which we propose a model in this paper, characterizing defects, would make this work easier.

To overcome the limits of these approaches (field (expert) knowledge, upstream considerable work), we want to propose an approach which classifies defects by pointing grouping criteria specified in the model of descriptor of defect we define. So, we can identify various populations, various defects groups that we put in the heart of corrective and preventive maintenance activities. In this way, we use past defects information not only to solve a new one but also to improve manufacturing process quality by feeding back this information from manufacturing teams to design teams, feedback rarely existent within companies [22].

## 6.3   Modeling a Defect

In the related works section, we have noted a lack of model of defect in the literature relative to the maintenance. However, we think such a model is necessary to really take advantage of information about defects. In this section, we propose to address this issue by defining a model as generic as possible in order to represent any defect, nominal or non-nominal, instantiated from any defect form template.

### 6.3.1   Defect Definition

The document which serves as a support for a defect contains data about the characteristics of the detected incident (i.e. the problem) and about the applied protocol to resolve it (i.e. the solution). It involves elements we call "attributes" in which data is contained. We define these attributes by two features, their *structure* and their *descriptive quality* (Table 6.1).

The *structure* feature divides attributes between *constrained* and *loose* attributes. *Constrained* attributes contain accurate, structured data. The set of values these attributes can take is known, beforehand or not. Even if this set is not defined beforehand, the value of such an attribute is not unique among the set of defects and can be found in other ones. *Loose* attributes are unstructured data, among them, we will particularly note the problem description of the defect and its solution description. The value of such an attribute is a free text. Constrained attributes can be compared to closed questions whereas loose attributes may be compared to open ones. We present this first classification in order to emphasize the need to apply

**Table 6.1** Defect attributes features

| Classification of attributes according to their structure | |
| --- | --- |
| Constrained | Structured attribute, with a static or dynamic finite list of values |
| Loose | Unstructured attribute, free text |
| Classification of attributes according to their descriptive quality | |
| Problem descriptive | Attribute concerning problem description |
| Solution descriptive | Attribute concerning solution description |
| Non-descriptive | Attribute not considered as problem nor solution descriptive |

a specific processing to loose attributes to extract from them information as easily handled as from constrained attributes.

Defect forms templates set up many attributes during the resolution process. However, all attributes do not provide the same kind of information. The *descriptive quality* feature divides them into three groups: the *problem descriptive*, the *solution descriptive* and the *non-descriptive* attributes. Thanks to this second categorization, we emphasize attributes which contain relevant data. Therefore, we give the possibility to partition attributes in:

- *Problem descriptive* attributes, which inform about the encountered problem.
- *Solution descriptive* attributes, which explain a way to solve the problem.
- *Non-descriptive* attributes, which do not fit into either of the two previous categories.

An expert must decide the classification of attributes in the proposed partition. Indeed, while the structure feature of an attribute is derived directly from it, its descriptive quality feature is a knowledge that must be given by an expert who well knows the field, the defect form template, because it depends on the context. Of course, the cost of this work increases with the number of attributes.

An attribute can contain data filled in by the user but also data the user did not directly fill in but which arises from the context of the defect emergence, to keep its traceability. This contextual data concerns elements related to the moment when the defect is noted. An example of contextual attribute is the creation date of the defect or the person who noted it. By definition, such an attribute is constrained. Chandola et al. go even beyond our vision since they distinguish contextual defects from other ones [9]. These defects are created in a specific context, have contextual and behavioral attributes and have their own preventive processing. In our case, we do not make this distinction. We do not want to leave out potentially useful information contained in attributes, contextual or not.

## 6.3.2   The Model of Descriptor of Defect

We propose to model a descriptor of defect as a set of attributes, each one having a role in the resolution process and containing data, structured or not (Fig. 6.3). Note

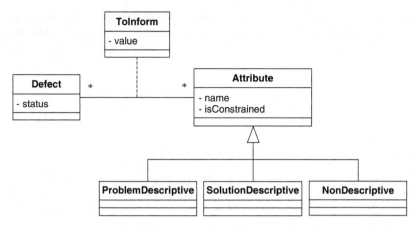

**Fig. 6.3** The proposed model of descriptor of defect

that even if one takes place in a single domain, several forms templates may exist and be used because all kinds of defects are not solved in the same way, using the same steps, the same attributes. However, an attribute may be present in several forms templates. Two defects solved using two different forms templates may therefore have attributes in common.

This model represents our vision of a descriptor of defect as a set of attributes that users fill in. It explicitly states the two features any attribute must have, namely their structure and their descriptive quality, with which we are able to propose an approach to find similar defects to a new one and to make their grouping easier. Whatever the resolution process used, whether nominal or not, a defect is represented by a descriptor instantiated from our model. Indeed, it suffices to identify attributes, these are the elements that contain data about the defect, and to specify their structure and descriptive quality classification. Once this preliminary work is done, the descriptor of defect is automatically instantiated.

## 6.3.3  Model's Evaluation

This model of descriptor of defect was tested on a database containing more than 7,000 real software defects from the bugs' database of the Intercim company, involved in this research project. Intercim offers manufacturing execution and optimization systems for aerospace and pharmaceutical industries. So, defects processing is a problem that arises naturally.

All of the 7,216 defects come from the same form template that involves 8 steps. These steps are not executed sequentially, the succession of steps is defined during the defect resolution, based on the collected data. In this form template, we have identified 91 attributes. Partition according to the structure feature has revealed

20 loose attributes and 71 constrained attributes. One of the experts who defined this process template has indicated the following partition according to the descriptive quality feature: six attributes are problem descriptive among which only one is loose. The major issue in this dataset is that there is not any attribute which describes the solution, no one can be defined as solution descriptive. Nevertheless, it allows testing the model on real data.

Let us take the example given in Sect. 6.2.2 We can note six attributes: *date*, *creator*, *part reference*, *involved part*, *description* and *solution*. Let us identify its descriptor instantiated from the model we propose.

The structure feature distinguishes constrained attributes from loose ones. It defines how the data contained in an attribute must be extracted. In the case of a constrained attribute, a single term is extracted, the data it contains. Thus, a term may be composed of several words. This is interesting for data whose set of words is more meaningful than each separated word (e.g. name and surname or "faceted classification"). In the case of a loose attribute, a specific processing must be applied. Using a set of defects, we can consider that all of them belong to a same domain, which is closed, whose vocabulary is controlled. There is no need to extract semantic information from loose attributes, terms that compose them are sufficient to describe defects.

As presented earlier, the structure feature of an attribute is directly derived from it. It is quite simple to identify the first four attributes as constrained and the last two ones as loose. However, quality descriptive feature can turn out to be more difficult. While it is fairly easy to identify the last attribute as solution descriptive and the previous two ones as problem descriptive, what about the first two attributes, contextual by the way, and the third? For our simplified example, we consider them as non-descriptive. Using a stopwords list to remove the not discriminatory words and the Vector Space Model [34] to weight the extracted terms, we obtain the following descriptor:

- *Problem vector*: {(wing, 0.1), (drill, 0.2), (hole, 0.2), (break, 0.1), (inside, 0.1), (widen, 0.1), (diameter, 0.1), (expected, 0.1)}.
- *Solution vector*: {}.
- *Non-descriptive vector*: {(Christian Durand, 0.33), (28/02/2011, 0.33), (787380, 0.33)}.

This model can also be applied on other defects databases such as Bugzilla. The defect form in which data is collected is composed of 20 fields, including for example:

- *Reporter* (the person who notes the defect).
- *Product and Component* (the kind of defect).
- *Assigned To* (the person responsible for fixing the defect).
- *Summary* (a one-sentence summary of the problem).
- *Keywords* (a list of predefined terms to assign tags to defect).
- *Platform and OS* (the computing environment).
- *Additional Comments* (description about the defect).

All these fields, and also the other ones, can be define as attributes having a structure and a descriptive quality. Except *Summary* and *Additional Comments* which are loose, all attributes are constrained. Moreover, we can imagine that the first one in this list, *Reporter*, is non-descriptive and the following five items are problem descriptive. *Assigned To* could be defined as problem descriptive if we consider the assigned person is an expert in a domain. *Additional Comments* can be problem and solution descriptive because all persons in relation with the defect can add a comment. A comment from the person who notes the defect should be problem descriptive whereas a comment from the person responsible for fixing the defect should be solution descriptive.

The only field which cannot be easily defined as an attribute is *Attachments* (one or more files attached to the defect). Because this is not a field containing a text value, the structure feature is not adapted. However, we can say that it is problem descriptive.

## 6.4   Optimizing Maintenance Using Defect Information

Thanks to this model of descriptor of defect, we can automatically take advantage from information from any defect. Indeed, we are able to propose an approach which reveals knowledge from defect information to make corrective and preventive maintenance activities easier. This approach consists in a sequence of defects grouping based on the attributes' descriptive quality feature. Beforehand, we need to describe how to compare defects by defining the similarity computation between two defects.

### 6.4.1   Similarity Between Defects

We have modeled a descriptor of defect as a set of attributes. Each one's value is a word or a set of words filled in by user (*value* in the class *ToInform*). We propose to summarize defects information as significant words extracted from these attributes. These significant words (*SigWord* in the formulas) are weighted (*w* in the formulas) as in the Vector Space Model [34]. Other models could be used such as the Boolean Model [5] or the Probabilistic Model [31]. In this way, every defect has its own set of significant words. To identify the similarity between two defects, we carry out a similarity computation on their significant words. During the computation, problem and solution descriptive significant words are separated for two reasons. On the one hand, it makes possible to search similarity on one of the two sides and not on the whole defect. On the other hand, each of these two sides uses its own specific vocabulary, even within the same defect. Non-descriptive significant words are not included because we believe that they do not provide additional information.

So, the similarity computation involves a couple of significant words vectors from the descriptor of defect, a problem vector containing significant words about the problem and a solution vector containing significant words about the solution. Thereby, we obtain three similarity computation functions: similarity *sim'* between defects problems (i.e. between significant words in problem vectors), similarity *sim''* between defects solutions (i.e. between significant words in solution vectors) and similarity *sim* between two defects (using *sim'* and *sim''*). Let us:

$V_p^d$ the problem vector $V$ containing the problem descriptive significant words $p$ on the defect $d$,

$$V_p^d = \left\{ (\text{SigWord1}_p^d, w1_p^d), (\text{SigWord2}_p^d, w2_p^d),..., (\text{SigWordK}_p^d, wK_p^d) \right\} \quad (6.1)$$

$V_s^d$ the solution vector $V$ containing the solution descriptive significant words $s$ on the defect $d$,

$$V_s^d = \left\{ (\text{SigWord1}_s^d, w1_s^d), (\text{SigWord2}_s^d, w2_s^d),..., (\text{SigWordL}_s^d, wL_s^d) \right\} \quad (6.2)$$

We define the similarity *sim* between two descriptors of defects as the similarity between their vectors of significant words. Let us take two defects $d1$ and $d2$ which one wants to know the similarity,

$$sim(d1, d2) = sim'(V_p^{d1}, V_p^{d2}) + sim''(V_s^{d1}, V_s^{d2}) \quad (6.3)$$

A similarity computation is generally expressed between 0 (very different defects) and 1 (very similar defects), so we introduce two variables, $\alpha$ and $\beta$, to normalize our computation,

$$sim(d1, d2) = \alpha \left[ sim'(V_p^{d1}, V_p^{d2}) \right] + \beta \left[ sim''(V_s^{d1}, V_s^{d2}) \right], \alpha + \beta = 1 \quad (6.4)$$

Thanks to these two variables, we can search similar defects only on the problem ($\alpha = 1$ and $\beta = 0$), very useful when we want to suggest a solution for a new defect using defects which have a similar problem, or only on the solution ($\alpha = 0$ and $\beta = 1$). We can also search a similarity between defects using all their information (with more importance on the problem: $\alpha > \beta$, or on the solution: $\alpha < \beta$, or the same importance between them: $\alpha = \beta$) but this is not a way which interests us in our approach.

### 6.4.2 Our Approach on Defects Processing to Reveal Knowledge

To reach our aim to assist the user in his solution proposal and to prevent from recurrent defects emergence, we take advantage of information from past, already solved defects. For this, we organize the defects database in a scheme to meet quickly and precisely our searching and similar defects grouping objectives.

At first, as mentioned previously, all defects are not solved in the same way. So, it is meaningless to make comparisons between all defects in a database. What relevant similarity relationship may be established between a documentation defect and a defect which occurs during a drilling? We can imagine that the drilling defect was caused by the documentation defect but it would be difficult to find similarity between defects themselves, between their problem and/or their solution. So, we distinguish numerous defects categories inside a unique defects database. These categories are defined according to various discriminatory elements. The defect form template is probably the merest but other attributes can also have this role.

The interest of creating defects categories is to have a first simple grouping in order to avoid comparing all defects whereas we know it exists distinct populations. The difficulty is to identify discriminatory criteria leading to relevant categories creation. Because of the unstructured information they contain, these criteria cannot be loose attributes, it would require an important work of analysis by an expert or the use of a text processing algorithm that we want to apply only later. So we have to consider only constrained attributes. As we want to facilitate the similar defects searching with a new defect whose solution is obviously unknown, this classification must separate defects only according to their problem. So, we are particularly interested in constrained and problem descriptive attributes. Some of them will define so characteristic values that searching similarity between defects which do not have the same value for one of these attributes does not seem to make sense. In order to not give too much weight to attributes defining categories, their significant words are not involved in the problem vector of the descriptor of the defect.

So, among the set of attributes, only some of them are used to define categories. Thanks to the model of descriptor we propose, we can filter attributes which can serve as defects categories criteria. A first filtering is performed through the structure feature, we only retain constrained attributes. Then, a second filtering is performed through the descriptive quality feature, we only retain problem descriptive attributes. Among the remaining attributes, whose number could be quite large in spite of the filtering step, an expert, who knows the domain well, identifies those that are really discriminatory. This work is necessarily specific to the domain, no attribute can be defined as defects categories criterion by default.

After that, within each category, we identify defects groups in which problems are similar. Comparisons between defects, more precisely between problem vectors of defects, are performed with the similarity formula presented in the previous sub-section. A clustering algorithm uses these comparisons to make problems groups. Then, within each problems group, we identify defects groups in which solutions are similar. In the same manner that problems groups are created, comparisons between solution vectors of defects are performed with the similarity formula and a clustering algorithm uses these comparisons to make solutions groups.

For each problems and solutions group, we define a prototype, i.e. a descriptor of defect automatically constructed which is representative of the group. So, it is made up of a problem vector and a solution one (non-descriptive one is useless since it is not involved in the similarity computation). Each vector is created as follows. Significant words in the same type of vector (problem or solution) of the descriptor of all defects in the group are integrated into the prototype's vector. These significant words are still weighted using the Vector Space Model. Thus, problem vector of a problems group's prototype is used to calculate similarity in comparison with the problem vector of a new defect's descriptor. So, it is not necessary to compare the new defect to all defects that belong to the problems group. With regard to solutions groups, on the one hand, solution vector of the prototype is used to present summarized possible solution. On the other hand, to identify the problems group's solution the most suited to a new problem, problem vector of the prototype and of the descriptor of defect are compared.

### 6.4.3  Approach's Illustration

Let us take again the example given in Sect. 6.2.2. Two criteria defining categories can be identified. We use the defect form template (a constraint and problem descriptive attribute that we do not mentioned before) and the involved part. In this example, its value is "wing". Thanks to these two criteria, we group defects by the resolution template and the involved part. In our example, we consider three resolution templates: *documentation*, *electric* and *mechanic*, and five plane parts: *cockpit*, *fuselage*, *landing gear*, *nose* and *wing*. Defects in the database are thus divided into fifteen categories (Fig. 6.4). So we obtain all defects about a plane wing problem using the same resolution template, the mechanical one (DFT_03). It is the category called *wing mechanic defects*. To search for solutions, we consider the specific category of defects which relate to a problem on the plane wings and which use the same resolution template. In this category, past defects have already been classified in groups (problems groups and in each one, solutions groups) with a clustering algorithm.

To identify similar problems to the current one, the problem vector of the current defect is compared with the problem vector of the prototype of each problems groups. So, we are interesting in groups whose problem is notably described by the significant words *drill* and *hole* and also by *break*, *inside*, *wide*, *diameter* and *expected*. Note that *wing* does not appear since it is involved in the categories definition. We obtain one group with three solutions (Fig. 6.5). Inside this problems group, solutions groups have been created using the significant words *keep*, *repair*, *enlarge*, *fill*, and *scrap*. Their grouping has defined three kinds of solutions to suggest to the user. With this information, the delay of the resolution can be reduced as the search for a solution is instantaneous. The remaining delay would only be the time that the actors take note of the defect and the time of the execution of the solution.

We propose a model of descriptor in an approach which formalizes our vision of a defect and explains the elements involved in the similarity computation, problem

| Defect form template | | |
|---|---|---|
| DFT_01 (documentation) | DFT_02 (electric) | DFT_03 (mechanic) |
| Cockpit documentation defects | Cockpit electric defects | Cockpit mechanic defects |
| Fuselage documentation defects | Fuselage electric defects | Fuselage mechanic defects |
| Landing gear documentation defects | Landing gear electric defects | Landing gear mechanic defects |
| Nose documentation defects | Nose electric defects | Nose mechanic defects |
| Wing documentation defects | Wing electric defects | *Wing mechanic defects* |

Involved part: Cockpit, Fuselage, Landing gear, Nose, Wing

**Fig. 6.4**  An example of defects categories

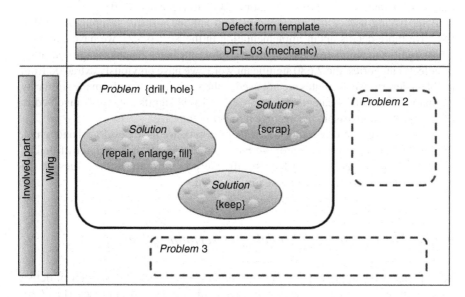

**Fig. 6.5**  An example of defects grouping

and solution descriptive attributes. It identifies the similarity between two defects in order to meet two goals: to assist the user in the defect resolution and to prevent from recurrent defects emergence.

As regards the first objective, we present to the user defects solutions whose problem is similar to the one of the current defect. For this, we do not consider all the past and solved defects but only those belonging to the same category as the current defect. Then, we use problems groups and solutions groups to automatically and quickly find the solutions the most suited to solve the current problem.

As regards the second objective, recurrent defects can be easily deduced using categories and problems groups we defined. Preventing emergence of such defects results from the analysis of large groups by an expert to understand the reasons of their emergence and to set up ways to prevent from reproducing them, by modifying manufacturing process for example.

## 6.5 Experimentation

The model of descriptor of defect and the approach we propose in this paper was tested on the database we have presented in Sect. 6.3.3 containing 7,216 real software defects. All defects concern only one software but various versions of it. They were recorded by persons in charge of quality assurance (not final users) during more than 3 years, from September 2005 to December 2008.

After having used a list of stopwords and applied a suffix stripping algorithm (we give more details about implementation in the Sect. 6.5.2), we have extracted 2,680 significant words from problem descriptive attributes. Thanks to the definition of defects categories, the 2,680 significant words we have previously extracted from problem descriptive attributes are not used at the same time to perform the similarity comparison. In our 59 categories, one reaches 1,098 significant words involved in the similarity comparison, the average number being 213.

In this section, we present the software prototype we have developed and which implements the approach we propose based on the model of descriptor of defect. Main techniques used are detailed. Results are presented and some issues on the approach and on its implementation are discussed.

### 6.5.1 The Operational Model of Descriptor of Defect

To implement our approach, we have instantiated the model of descriptor of defect with additional elements, resulting from our objectives and approach (Fig. 6.6). They are well-known elements in classification like keywords extraction or grouping.

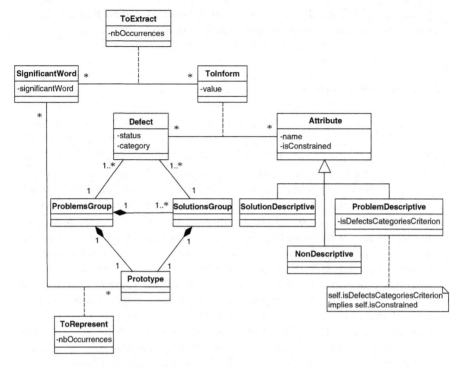

**Fig. 6.6** The operational model of descriptor of defect

We have therefore integrated:

- The fact that an attribute can be used as a defects categories criterion. Only constrained and problem descriptive attributes can have this role. It appears in the class *ProblemDescriptive* with a note to indicate that the boolean property *isDefectsCategoriesCriterion* in this class can be true only if the boolean property *isConstrained* in the class *Attribute* is also true.
- The retrieve of significant words from attributes values. The property *value* in the class *ToInform* is a text containing one significant word if the attribute is constrained and one to several significant words in the case of a loose attribute. One significant word can appear few times in a value, we keep this information with the property *nbOccurrences* in the class *ToExtract*.
- The notion of similarity between descriptors of defect according to the problem or the solution. Similarity computation is performed using the significant words in the descriptors weighted by measures based on the number of occurrences.
- The groups and their prototype. Our approach defines that a defect belongs to only one problems group (class *ProblemsGroup*) and only one solutions group (class *SolutionsGroup*), and a solutions group belongs to a problems group. A solutions group cannot exist without a problems group to be associated. For each group, one prototype (class *Prototype*) is created which is composed

by vectors of weighted significant words as well as a descriptor of defect. A prototype exists only if the group it represents exists.

### 6.5.2 Implementation

Our software prototype, implemented in Java, is divided into several modules to run sequentially. Once the categories are created and assigned to defects, significant words extraction is run. Respecting loose attributes, the text (their *value*) is split into words according to spaces and special characters. We use a list of general English stopwords (containing 671 terms) and a list of field-specific stopwords (which is poorer for the moment) to remove common words. We also apply Porter's suffix stripping algorithm [26, 27],[5] well-suited to Information Retrieval, to produce more relevant groups by considering only root of words. This algorithm applies five successive steps on each word to gradually remove all parts of the word that relate only to its suffix. No linguistic element is used. Preserving only root of words allows processing less significant words during the similarity computation between descriptors and increasing this similarity. In his paper, Porter gives the example of the words "connect", "connected", "connecting", "connection" and "connections" which all have the same meaning. After applying his algorithm, they are all represented by "connect". The words "generalization" and "oscillators" become respectively "gener" and "oscil". To improve the similarity between defects, we could also use an algorithm of dimension reduction such as Singular Value Decomposition (SVD) [35], Principal Components Analysis (PCA) [1], Latent Semantic Analysis (LSA) [10] aka Latent Semantic Indexing (LSI), etc. We are conscious that we leave aside the semantic aspect which could increase the similarity between defects. But this extraction is not the aim of our contribution and we prefer to focus on the development of our approach.

Then, concerning the similarity computation, we chose to use the Vector Space Model with the TF-IDF measure (Term Frequency – Inverse Document Frequency), widespread in Information Retrieval [32, 33] and in Text Mining [13, 14]. This statistical measure evaluates the importance of a word with respect to a document from a corpus of documents and to the corpus itself.

Term Frequency is the number of occurrences of each significant word in one vector of the descriptor. To avoid bias, this number is normalized by dividing it by the number of occurrences of all significant words in the vector.

$$tf_{i,j} = \frac{n_{i,j}}{\sum_k n_{k,j}} \tag{6.5}$$

[5]The founder paper introducing this algorithm was republished in an identical form 26 years later, demonstrating that it is still topical and enhancing its credibility

Where the numerator is the number of occurrences of the significant word $m_i$ in one vector of the descriptor $D_j$ and the denominator is the number of occurrences of all the significant words in this same vector.

Inverse Document Frequency is an indicator of rarity of a significant word in the vector of all descriptors in the corpus. In our approach, a corpus is a category when considering problem vectors and a problems group when considering a solution vectors. The higher the number of vector of descriptors has a significant word is, the smaller the IDF value for this significant word is.

$$idf_i = \log \frac{|D|}{|\{d_j : m_i \in d_j\}|} \tag{6.6}$$

Where the numerator is the number of descriptors in the corpus and the denominator is the number of descriptors having the significant word $m_i$ in the considered vector (problem or solution).

The weight of a significant word is the product of these two measures.

$$w_{ip}^{d_j} = tf_{i,j} \times idf_i \tag{6.7}$$

We use the obtained values to calculate the similarity between descriptors of defects with the cosine measure, according to the problem vector then the solution one.

$$sim(d1, d2) = \sum_{i=1}^{m} (w_{ip}^{d1} \times w_{ip}^{d2}) \Big/ \sqrt{\sum_{i=1}^{m} (w_{ip}^{d1})^2 \times \sum_{i=1}^{m} (w_{ip}^{d2})^2} \tag{6.8}$$

These similarity measures are then used in the hierarchical ascendant classification. This clustering algorithm allows, unlike most of others, not setting a priori the number of groups to obtain but only the similarity threshold. In a category, all defects whose similarity between descriptor's problem vectors is over the threshold constitute a problems group. Solution groups are constituted in the same manner. Finally, a prototype is created for each group.

## 6.5.3   Results

To perform a qualitative evaluation, a study was conducted in a category containing 92 defects. This category possesses the advantages to have enough defects to obtain interesting results by our automatic approach and to not have too many defects to compare these results with those we would like to obtain manually. First, groupings were handmade giving the reference classification. This one reveals that 22 groups have been completed, so the 92 defects were classified into 70 groups.

Then, to identify the threshold value of similarity that produces the most relevant groups, different classifications have been obtained by varying the similarity threshold between 0.9 and 0.3 and compared with the reference classification. To

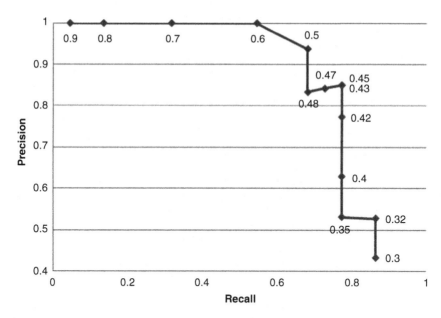

**Fig. 6.7** Precision-recall

assess the quality of classification, we present the precision-recall (Fig. 6.7), and the F-measure (Fig. 6.8) curves. Numbers on the points indicate the threshold value of similarity.

Figure 6.7 shows a curve which presents the proportion of correct groupings made by the algorithm among all the groups the algorithm made (the precision) with respect to the proportion of correct groupings made by the algorithm among all the groups we would like to obtain (the rappel). It shows that the algorithm proposes a best classification for a similarity threshold between 0.43 and 0.5. Outside this range of values, the precision decreases, there is too much noise (i.e. too many incorrect groups), or this is the recall which decreases, there is too much silence (i.e. not enough correct groups).

In the precision-recall curve, the objective is to maximize the precision and the recall. However, it is very difficult to estimate a good similarity threshold since when the precision increases the recall decreases and inversely. The F-measure considers both the precision and the recall. Figure 6.8 shows a curve which represents a weighted average of the precision and the recall. We can easily identify the best similarity threshold between 0.43 and 0.45.

We can specify that the threshold value of similarity which allows obtaining the nearest reference classification is 0.45. Indeed, it's the best value for the precision-recall and it is the point at which the F-measure is the highest. 0.43 could be a value as right as 0.45 but the fall of the F-measure score is really more significant below 0.43 than above 0.45.

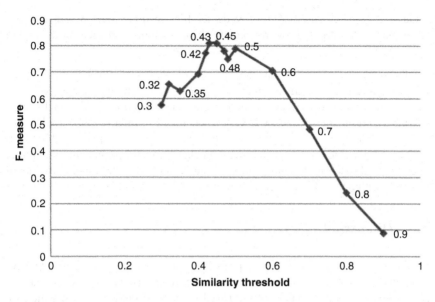

**Fig. 6.8** F-measure

These results point out that our model of descriptor and our defects grouping approach based on this model allows us to propose relevant clusters of defects. In our study, with a similarity threshold of 0.45, we obtain a precision and a recall at about 0.8 and an F-measure upper than 0.8.

Below two examples of groups obtained with a similarity threshold of 0.45. Defects are describing by the problem description, a loose attribute used in defects grouping.

*Group 1*:

- "Layouts-Expressions column do not accept valid expressions in notes panel."
- "Layouts-Expressions column do not accept valid expressions in buyoff panel."
- "Layouts-Expressions do not accept valid expressions in Alert panel."

*Group 2*:

- "Tool & device panel – Edit tools definition layout -> Alias check box does not function as expected."
- "Tool device panel in layouts-> edit tools definition layout -> Value ddl has unnecessary values."
- "Load Order/oper/ step layout -> Load Tool & device panel -> Add row causes unexpected error."
- "Layout -> Tool & Device panel -> Edit tools definition layout does not load the related screen in layout."
- "Tools & Devices – Edit Tools Definition Layout value column should be display as a label and read only."

- "On selecting the Buyoff & tool & device tiles in order, oper & step layouts, it displays a highlighted row for buyoffs & opens tools & devices at the end of page. See Screenshot."

In the reference classification, the last defect does not belong to this second group. This reference classification cannot be achieved with our current algorithm. You can have either a perfect precision (similarity threshold $\geq 0.5$ in Fig. 6.7) but with a significant silence or a very good recall (similarity threshold $\leq 0.3$ in Fig. 6.7) but with significant noise. The use of semantic analysis algorithms based on ontology or automatic classification algorithms based on natural language processing and machine learning, such as neural networks, Bayesian networks or genetic algorithms, could help tend towards this reference classification.

### 6.5.4  Discussion

We have developed a software prototype using a software defects database. We have instantiated our model and implemented our approach. However, our approach is really interesting in the context of industrial manufacturing process, in which we can find many more really similar, recurrent, defects than in software development. Indeed, in the industrial context, a unique manufacturing process of a product is repeated. Tasks are repetitive and so recurrent defects may also be repeated on numerous products. However, in software development, particularly on the database on which we did our tests, it is difficult to find recurrent defects, or at least very similar ones. This is due to the fact that defects concern only one software and are not recorded by final users. In that event, two different users would have recorded the same defect. Nevertheless, a recurrence, really less significant than in the industrial context, can be detected during a version upgrade (non-regression bug) or if a function with a bug is used in several modules. Furthermore, in an industrial manufacturing database, defects will have problem descriptive attributes but also solution descriptive ones. Thus, the validation of our proposition needs to be made on such a database. This new evaluation will reinforce the fact that our model of descriptor of defect is applicable on more than a unique domain.

Moreover, like any clustering algorithm, the hierarchical ascendant classification has its strengths but also its weaknesses [8]. It does not allow overlapping, i.e. a defect is a member of a unique group, in our approach a unique problems group and a unique solutions group. It cannot be a member of several problems or solutions groups with a percentage of belonging for each of them. Nevertheless, this is not a drawback for us. We consider that a defect must belong to only one group. In its single-link version, that we have implemented, this algorithm suffers from the chain effect which can lead to produce heterogeneous groups, with descriptors of defects not all similar by pairs. This drawback can be address implementing the complete-link version. A major weakness of this algorithm for our approach is its quadratic complexity. Even while considering only about 7,000 defects, we can already address performance issues for a category with 1,232 defects (i.e. 758,296 computations of cosine measure on the problem). However, this issue just occurs when we create the classification scheme. Once groups are constituted, we can run the algorithm of similar defects searching to assist in solving a new defect and we can analyze groups without any performance issue. Furthermore, it is possible to optimize this algorithm to solve (or at least heavily reduce) performance issues, using, for example, an incremental version [17, 19, 30].

It could also be interesting to evaluate the quality of the groups we obtain with the clustering algorithm by implementing indicators. Intra-cluster (evaluation in a group) and inter-cluster (evaluation between groups) measures should be involved in these indicators [25, 29].

Finally, as the *Attachments* attribute used in Bugzilla has underlined, the current structure feature reveals a lack. Defining an attribute only as constrained or loose limits the value it contains as a textual form. However, it could be interesting to extend this feature to take into account textual documents and also image, audio and video contents.

## 6.6   Conclusion and Perspectives

We have presented an approach which makes corrective maintenance activities easier by assisting the user in his defect solution proposal thanks to the capitalization of knowledge we revealed from defects. For this, we defined a model of descriptor of defect which separates parts concerning the problem from part concerning the solution. It allows considering defects according to these two sides and thus to improve the relevance of the grouping processing. In addition to the corrective dimension, our approach is suited to the preventive maintenance since an analysis can be made for each defects group in order to identify their emergence causes.

After having worked on the optimization of our algorithm to solve performance issues implementing an incremental version, we plan to do our tests on a manufacturing defects database, our approach being more suited to manufacturing defects than software ones. Moreover, we want to carry out a greater processing of information by taking advantage of the fact that the used vocabulary is controlled, since it is an occupation vocabulary and we foresee to build constrained attributes from looses ones. Eventually, we could take into account the user's profile in the solution ranking. By considering his skills on the one hand, and the way he is used to executing a kind of solution to solve a problem on the other hand, we could re-rank the list of proposed solutions to be adapted to his profile. A solution he is used to failing when he executes it could be depreciated in the list.

## References

1. Abdi, H., Williams, L.J.: Principal components analysis. Wiley Interdiscip. Rev. Comput. Stat. **2**(4), 433–459 (2010)
2. Ahuja, I.P.S., Khamba, J.S.: Total productive maintenance: literature review and directions. Int. J. Qual. Reliab. Manag. **25**(7), 709–756 (2008)

3. Alloui, I.: Conciliating property stability and system evolution through software model analysis. GDR Génie de la Programmation Logicielle 2009, pp. 224–231 (2009)
4. Alsyouf, I.: The role of maintenance in improving companies' productivity and profitability. Int. J. Prod. Econ. **105**(1), 70–78 (2007)
5. Angione, P.V.: On the equivalence of boolean and weighted searching based on the convertibility of query forms. J. Am. Soc. Inform. Sci. **26**(2), 112–124 (1975)
6. Barros, S.: Analyse a priori des conséquences de la modification de systèmes logiciels : de la théorie à la pratique. PhD thesis, Université Paul Sabatier, Toulouse 3 (1997)
7. Cabanac, G., Chevalier, M., Chrisment, C., et al.: Social validation of collective annotations: definition and experiment. J. Am. Soc. Inform. Sci. Tech. **61**(2), 271–287 (2010)
8. Candillier, L.: Contextualisation, visualisation et évaluation en apprentissage non supervisé. PhD thesis, Université Charles de Gaulle, Lille 3 (2006)
9. Chandola, V., Banerjee, A., Kumar, V.: Anomaly detection: a survey. ACM Comput. Surv. **41**(3), 1–58 (2009)
10. Deerwester, S., Dumais, S.T., Furnas, G.W., et al.: Indexing by latent semantic analysis. J. Am. Soc. Inform. Sci. **41**(6), 391–407 (1990)
11. Despujols, A.: Approche fonctionnelle de la maintenance. Techniques de l'Ingénieur, **AG4710**, 1–14 (2004)
12. Dowlatshahi, D.: The role of industrial maintenance in the maquiladora industry: an empirical analysis. Int. J. Prod. Econ. **114**(1), 298–307 (2008)
13. Fatudimu, I.T., Musa, A.G., Ayo, C.K., et al.: Knowledge discovery in online repositories: a text mining approach. Eur. J. Sci. Res. **22**(2), 241–250 (2008)
14. Feldman, R., Fresko, M., Kinar, Y., et al.: Text mining at the term level. In: Zytkow, J.M., Quafafou, M. (eds.) PKDD, Second European Symposium. LNCS, vol. 1510, pp. 65–73. Springer, Heidelberg (1998)
15. Giess, M.D., Wild, P.J., McMahon, C.A.: The generation of faceted classification schemes for use in the organisation of engineering design documents. Int. J. Inform. Manag. **28**(5), 379–390 (2008)
16. Goh, Y.M., Giess, M.D., McMahon, C.A., et al.: From faceted classification to knowledge discovery of semi-structured text records. In: Abraham, A., Hassanien A.-E., Carvalho A.P. de L.F. de, Snasel V. (eds) Foundations of Computational Intelligence Volume 6. Studies in Computational Intelligence, vol. 206, pp. 151–169. Springer, Heidelberg (2009)
17. Gurrutxaga, I., Arbelaitz, O., Martín, J.I., et al.: A stable incremental hierarchical clustering algorithm. In: 11th International Conference on Enterprise Information Systems, pp. 300–304 (2009)
18. Haziza, M., Voidrot, J.F., Minor, E., et al.: Software maintenance: an analysis of industrial needs and constraints. In: Conference on Software Maintenance, pp. 18–26 (1992)
19. He, L.L., Bai, H.T., Sun, J.G., Jin, C.Z.: A general incremental hierarchical clustering method. Comput. Meth. 1303–1307 (2006)
20. Komonen, K.: A cost model of industrial maintenance for profitability analysis and benchmarking. Int. J. Prod. Econ. **79**(1), 15–31 (2002)
21. Lambolez, P.-Y.: Recherche d'informations pour la maintenance logicielle. PhD thesis, Université Paul Sabatier, Toulouse 3 (1994)
22. Levner, E., Zuckerman, D., Meirovich, G.: Total quality management of a production-maintenance system: a network approach. Int. J. Prod. Econ. **56–57**(1), 407–421 (1998)
23. Lortal, G., Lewkowicz, M., Todirascu-Courtier, A.: Annotations: a way to capture experience. In: Gabrys, B., Howlett, R.J., Jain, L.C. (eds.) KES LNCS, vol. 4251, pp. 1131–1138. Springer, Heidelberg (2006)
24. Lortal, G., Lewkowicz, M., Todirascu-Courtier, A.: Des activités d'annotation : de la glose au document. In: Salembier, P., Zacklad, M. (eds.) Annotations dans les Documents pour l'Action, pp. 153–171, Hermes-Lavoisier, Paris (2007)
25. Nguyen, Q.H., Rayward-Smith, V.J.: Internal quality measures for clustering in metric spaces. Int. J. Bus. Intell. Data. Min. **3**(1), 4–29 (2008)
26. Porter, M.F.: An algorithm for suffix stripping. Program: Electronic Library and Information Systems **14**(3), 130–137 (1980)

27. Porter, M.F.: An algorithm for suffix stripping. Program: Electronic Library and Information Systems **40**(3), 211–218 (2006)
28. Prieto-Diaz, R., Freeman, P.: Classifying software for reusability. IEEE Software **4**(1), 6–16 (1987)
29. Raskutti, B., Leckie, C.: An evaluation of criteria for measuring the quality of clusters. In: 16th Joint Conference on Artificial Intelligence, pp. 905–910 (1999)
30. Ribert, A., Ennaji, A., Lecourtier, Y.: An incremental hierarchical clustering. In: Vision Interface Conference, pp. 586–591 (1999)
31. Robertson, S.E., Jones, K.S.: Relevance weighting of search terms. J. Am. Soc. Inform. Sci. **27**(3), 129–146 (1976)
32. Salton, G., Buckley, C.: Term-weighting approaches in automatic text retrieval. Inform. Process. Manag. **24**(5), 513–523 (1988)
33. Salton, G., McGill, M.J.: Introduction du modern information retrieval. McGraw-Hill, New York (1983)
34. Salton, G., Wong, A., Yang, C.S.: A vector space model for automatic indexing. Comm. ACM **18**(11), 613–620 (1975)
35. Stewart, G.W.: On the early history of singular value decomposition. SIAM Rev., Society for Industrial and Applied Mathematics **35**(4), 551–566 (1993)
36. Swanson, B.: The dimensions of maintenance. In: 2nd International Conference on Software Engineering, pp. 492–497. IEEE Computer Society Press (1976)
37. Waeyenbergh, G., Pintelon, L.: Maintenance concept development: a case study. Int. J. Prod. Econ. **89**(3), 395–405 (2004)

# Chapter 7
# Rules for Effective Mapping Between Two Data Environments: Object Database Language and XML

**Tamer N. Jarada, Abdullah M. Elsheikh, Taher Naser, Kelvin Chung, Armen Shimoon, Panagiotis Karampelas, Jon Rokne, Mick Ridley, and Reda Alhajj**

## 7.1 Introduction

The rapid development in automated communication and the diversity of computing platforms necessitated and motivated for the development of platform independent data format that could smoothly provide for portability and extensibility. Intensive research efforts over the past two decades have produced XML as the de facto standard for platform independent sharing of data which is the most valuable commodity for maintaining successful and competitive performance. Data is the most valuable source of knowledge. Once data is acquired, it can be queried to retrieve explicit content and it can be mined to extract and predict implicit content. XML has been embraced as a data model mainly due to its simplicity, readability, and portability, i.e., its ability to be transported over well established protocols, such as HTTP. XML is extremely similar to HTML in structure, making it an ideal data format to be used in conjunction with HTTP. Furthermore, HTML parsers can be easily adapted for dealing with XML data. However, XML serves a purpose different from that of HTML. While the latter is intended for data formatting, the

---

T.N. Jarada (✉) · A.M. Elsheikh · K. Chung · A. Shimoon · J. Rokne
Computer Science Department, University of Calgary, Calgary, AB, Canada

T. Naser · M. Ridley
Department of Computing, School of Computing Informatics and Media, University of Bradford, Bradford, UK

P. Karampelas
Department of Information Technology, Hellenic American University, Manchester, NH, USA

R. Alhajj
Computer Science Department, University of Calgary, Calgary, AB, Canada

Department of Computer Science, Global University, Beirut, Lebanon

Department of Information Technology, Hellenic American University, Manchester, NH, USA
e-mail: alhajj@ucalgary.ca

T. Özyer et al. (eds.), *Recent Trends in Information Reuse and Integration*,
DOI 10.1007/978-3-7091-0738-6_7, © Springer-Verlag/Wien 2012

former specifies and describes structure and context for the data by allowing the user to decide on his/her own tags, structure, nesting, etc. Finally, XML documents can be highly structured, based on an accompanying XML Schema.

Recently, database administrators are facing the challenge of ensuring that their databases can smoothly interface with other heterogeneous systems using XML. However, the most prevalent database system is the relational/object-relational model [19]. This creates some issues, as relational data is inherently different than XML data; there is a semantic gap between the two models; the least to say, the former is flat while the latter provides full flexibility for nested and flat structures. Actually, there exist many legacy databases (such as relational, object-oriented, etc) that are in need of modernization in order to be compatible and competitive in this new era of XML ubiquity. Fortunately, the flexibility of XML has facilitated for an array of options for the storage of XML data, such as relational, object-relational, object-oriented, and native XML databases. Because each option has its own strengths and weaknesses, work must be done to develop standard practices for storing XML data in each of these models. Then, the viability of each alternative can be evaluated and the most appropriate for a given situation would be preferred and adapted.

It has been argued in the literature that the relational (and object-relational) is the most appropriate solution for the storage of XML data, due to its widespread use and significant maturity. However, we argue that the existence of the semantic gap between relational and XML structures forms a major barrier and hence works against the argument. Accordingly, we can further argue that the object-oriented model is more appropriate for hosting XML data, due to its robustness, sufficient level of maturity (ODMG 3.0), and its complementary data structure to XML; in fact, the XML and object-oriented models share more features and hence the semantic gap is very limited if exists. Due to the already mentioned reasons, it is clear that additional work must be done in the field to produce attractive mapping strategies. An investigation of different methodologies is needed in order to evaluate and compare all the options. Inhere we benefit from our rich experience dealing with the three models; we have already investigated and successfully developed different techniques for transforming between XML and each of the relational and the object-oriented paradigms. However, there is a need for a rule driven transformation process that can be augmented and easily adapted to fit as many as possible of the existing object-oriented models though the differences between them are limited and minor but need to be considered carefully in order to produce solutions that are to be accepted by developers and practitioners.

In this paper, we investigate the viability of interfacing object-oriented databases with XML. To do this, we look at an incomplete set of rules analyzed and expanded in our previous work as described in [13] for the conversion of ODL schemas into the widely-accepted XML format, namely XML Schema. Then, we contribute some improvements by making small adjustments that allow for greater flexibility and expansion of the rule set. After completely handled the ODL to XML transformation process, we turn the attention to develop and produce a comprehensive set of rules for mapping XML to ODL. Finally, we propose a set of new rules in order to handle the two way mapping of data.

In order to ensure the effectiveness and correctness of the existing and new rules, we have implemented the rules in a fully working two-way converter between ODL and XML. The implementation is able to handle all the scenarios described in the rules presented in this paper. The implementation helps to show the feasibility of the rule set and the relative ease of the transformation process. We hope that our work can be used as a stepping stone for further investigation into the transformation between ODL and XML Schemas. This study should attract more attention to the area of transformation between ODL and XML.

The rest of this paper is organized as follows. Section 7.2 briefly covers the related work. The set of rules for ODL to XML transformation is described in Sect. 7.3. The XML to ODL conversion rules are presented in Sect. 7.4. The data conversion process is covered in Sect. 7.5. Implementation details are briefly discussed in Sect. 7.6. Section 7.7 is conclusions.

## 7.2  Previous Work

Since XML has become universally adopted as one of the main formats for information exchange and representation on the Internet, the need for data in XML format has dramatically increased [15]. Most of the information however, is stored and maintained by structured databases, thus applications generally convert data into XML for exchange purposes. Benefits for the conversion are: cross platform independence and re-mapping XML data into target applications/databases from anywhere in the world. This situation is routinely found in business situations where most data resides in structured databases and there is a need to transfer such data over the Internet where other departments/clients may access it. Accordingly, the issue of transforming existing databases into XML has been addressed by several research groups, e.g., [7, 8, 12, 14, 16, 17, 21] and [1], ([10] (XML) 1.0 (Fourth Edition). W3C Recommendation 16 August 2006, edited in place 29 Sept 2006, Uppsala Master's Thesis in Computing Science no. 260, 2003). Most of the focus has concentrated on the transformation of relational databases into XML, e.g., [17], [16], [12], ([10] (XML) 1.0 (Fourth Edition). W3C Recommendation 16 August 2006, edited in place 29 Sept 2006, Uppsala Master's Thesis in Computing Science no. 260, 2003). Many tools such as SilkRoute [11] and XPERANTO [5] have been developed to enable the translation of relational data into XML. SilkRoute is a general tool used for viewing and querying relational data in XML. XPERANTO provides a uniform, XML based query interface over an object-relational database that allows users to (re)structure the contents of the database as XML data. Flat Translation (FT), Nesting-based Translation (NeT) and Constraint-based Translation (CoT) are three additional approaches analyzed further next. FT is the most straightforward approach where: (1) tables in a database are mapped to XML elements and; (2) columns in each table are mapped to attributes (in attribute-oriented mode) or elements (in element-oriented mode) in the XML. Attribute-oriented and element-oriented modes are analogous except that the element-oriented mode adds unnecessary ordering semantics to the resulting schema [15]. Since the XML represents the "flat" relational tuples faithfully, this method is called Flat Translational. FT is a simple and effective translation algorithm. FT translates the "flat" relational model to a "flat" XML model in almost a one-to-one manner [15]. Hence, for every tuple in a table, one corresponding element is generated in XML. A shortcoming to this approach is that it does not use a number of basic "non-flat" features (such as "*", "+") provided by XML for data modeling. SQL-X [20] is a graphical tool for generating XML document from relational databases. It provides the user interface for users to edit the query and then the query can be executed later to generate XML documents. Although in their interface, they provide help for users to manage the textual query, the method is not as intuitive as visual query construction. In addition, users need to learn a new query language in order to use Visual SQL-X. Finally, BBQ [18] is a system with strong support for interactive query formulation.

Other approaches exist, such as the work described in [22]. The authors tackle the issue of dealing with legacy relational databases. A reverse engineering approach is adopted to extract a semantically-rich ER model from the given legacy relational database The ER model is then converted into an XML Schema. However, the transformation of object-oriented databases into XML has not received considerable attention, e.g. [8], [14], [21] and [1]. Nevertheless, an approach to conduct the mapping from object-oriented data into XML has been developed in [19]. The authors present a method to transform an existing object-oriented database into XML. Another approach is presented in [13], which offers a set of rules to translate a simple object-oriented schema specified in ODL into a corresponding XML Schema. Finally, it is important to mention that there has been a considerable amount of work done regarding the other direction of the transformation process. A variety of approaches have been proposed for the transformation of XML DTD structures, as well as XML Schemas, into corresponding ODL schemas. Clearly, this is of great benefit for adapting an object-oriented database to store XML data. We argue that it is apparent that a transformation process for extracting the object-oriented data and serving it back as XML data is needed. Therefore, we focus our attention on the object-oriented to XML direction, which serves to complement the XML to object-oriented transformation well.

The work described in [19] establishes the importance of XML as an emerging data format, as well as the fact that most data nowadays resides in relational databases. Therefore, the authors highlighted the need to automate the process of generating XML documents from existing databases. Based on this fact, the paper focuses on the mapping of contents of an existing object-oriented database into XML. In Sect. 7.3, we discuss the work described in [19] in greater detail. Basically, the method maps the contents of object-oriented databases into XML by applying these three key steps:

1. Derive a summary of the object-oriented schema. This is done using an object graph, including nesting and inheritance links found in the object-oriented schema.

2. The object graph is transformed into XML Schema.
3. The object-oriented data is mapped into corresponding XML documents based on the schema produced in step (2).

Although the work described in [19] is only partially related to our area of work, there are some key differences and similarities that should be mentioned. First, this paper takes a *deductive* approach for producing a representation of the object-oriented schema. This is beneficial due to the fact that an explicit object-oriented schema may be difficult, or impossible to acquire. In fact object-oriented databases were widely accepted and pushed hard to replace relational database because they provided smoothly extensibility which was lacking in relational databases until they were expanded into object-relational databases. Further, object-oriented databases could well serve applications where the description and data are incomplete at the initial development stage and become available incrementally at later stages. Eventually, these attractive characteristics are also supported by XML which has found its ways to be accepted as a standard. The key success of XML as compared to object-oriented is portability which cannot be absorbed by the relational or object-relational model. Furthermore, differing object-oriented database models employ various schema formats but they agree on some basics which are part of ODL schema. Hence, the approach taken in this paper should theoretically work with any object-oriented database. Second, in the second step of the transformation process, an intermediate object model is created to represent the schema. We chose to take a similar approach in our implementation. By creating a generic, yet flexible intermediate object representation of the schema, we are able to extend our rule set quite seamlessly, while making changes to the transformation process quite straightforward. Third, the work described in [19] not only deals with schema transformation, but with actual data migration as well. After the corresponding XML Schema is produced, an algorithm is specified for the generation of an XML document based on the generated XML Schema. In our approach described in this paper, we turn the latter process into a set of rules to smoothly address data migration.

We use the work described in [13] as our main area of focus for specifying the rules; eventually, the work described in [13] builds on the initial primitive thus interesting study presented in [9]. The authors address the issue of storage and management of XML data, and then expand a set of existing rules in their effort to define a comprehensive set of rules to help in the transformation from ODL to XML Schema. The authors describe the current situation of XML data in the database community. They mainly recognize the widespread use and status of XML as an upcoming data format; it has been eventually accepted as standard for platform independent data exchange. Furthermore, they make the argument that the ODMG standard [6] has reached a sufficient level of maturity, and therefore, it is a valid and attractive alternative for the storage of XML data. Then, the authors describe the fundamental cornerstone concept: in order to store XML documents in conventional databases (relational, object-relational, and object oriented), a certain amount of translation work in inevitable. Finally, the authors discussed various other options for the storage of XML data, such as native XML and direct file system.

The authors of [13] hypothesize that the global adoption of XML will serve as a catalyst for the acceptance of object-oriented databases as the most appropriate repository for maintaining XML data without sacrificing the semantics of the XML model, which is the case when XML is forced into a flat structure and stored in relational databases. Thus, they believe that more rule based work should be done regarding bi-directional translation between XML documents and object-oriented databases. It is then asserted that ODL is the language of choice for specifying the object-oriented database schema because it is the proposed language standard by ODMG. This will result in high portability between different object-oriented databases. At the end, the authors address the issue that the rules are not complete, and that more work must be done. First, new rules are needed to handle more complex object-oriented modeling constructs. Second, a new set of rules are needed for conversion in the reverse direction (XML Schemas to ODL). Third, some more rules will be needed for data migration. We thoroughly handle all these three aspects in this paper.

## 7.3   Rules for Conversion from ODL Structure to XML Schema

ODL is a well structured language that allows for specifying the definition of an object-oriented schema, including classes, inheritance and attributes. Given an ODL specification, the goal is to produce a corresponding XML schema. The XML schema has only elements and attributes, in addition to allowing for the definition of sequences, keys/keyrefs, cardinality, etc. Inhere, we list the rules that should be applied in order to transform an ODL into XML schema by considering and maintaining as much as possible the characteristics of each model. The process starts from the root, then moves on to the classes and finally considers the inheritance and the attributes. Further, complex attributes are considered in more details by investigating and mapping the details of the underlying domain. Finally, the sample ODL schema shown in Fig. 7.1 will be used as a running example for illustrating the conversion rules. The resulting XML Schema will be shown at the end after all the rules are introduced.

$O2X\_R_1$.   *Base XML document structure:* Since every XML document must have a root element, such an element has to be created with, for example, the database name. One should also associate with the root element an anonymous complex type that includes a special choice element with a maximum occurrence (written as maxOccurs in XML) attribute specified as unbounded.

This rule is quite straightforward. Essentially, XML specifies that all elements within a document must be enclosed within a single root element which is equivalent to the root of the object-oriented hierarchy. Also, a choice element is associated with the root element. This is where the content of the database will be stored. In other words, our transformed XML Schema will have the same overall structure as the original schema. The content model of the root element

```
class author;
class person;

class book (extent books key ISBN)
{
  attribute int ISBN;
  attribute list<int> ratings;
  attribute dictionary<int, int> chapterPageNumbers;
  relationship author writtenby inverse author::write;
};

class author extends person (extent authors key ID)
{
  attribute int ID;
  relationship set<book> write inverse book::writtenby;
  enum sex { MALE, FEMALE };
  attribute sex gender;
};

class person (extent people keys firstName, lastName)
{
  attribute string firstName;
  attribute string lastName;
};
```

**Fig. 7.1** A sample ODL schema

will be declared anonymously, with a choice element of unbounded cardinality. This choice element will be referred to as the root element. The XML schema declaration of the root element (whose name may be arbitrary; we will use the name database here) will be referred as the element declaration, while the schema element will be referred to as the schema root. The additional rules will insert child elements in all of these places. Elements representing classes also have anonymous content models; they are inserted in the root element. The sequence element defining the content model of the class will be referred to as the class root.

$O2X\_R_2$.    *Top-level classes:* There should be in the XML schema one element for each class defined in ODL. Accordingly, each top-level class in the ODL schema is converted into an element with the same name; the latter element is included in the special choice element created by applying rule $O2X\_R_1$. One should also associate with each of the created elements a complex type with a special sequence element. The content of each one of these special sequence elements is specified by invoking the next rules. In summary, this rule regards itself with creating containers for each of the top-level classes. For each class in the ODL schema, a corresponding, like-named container is created, in which its attributes will be stored based on later rules.

In the work described in [9], class types were implemented by declaring anonymous complexTypes containing its attributes and relationships, and then proceeding recursively through them. With the simplifying assumption stated next, we can simplify this to include only the key attributes, as an object in a class is entirely defined by the values of its key attributes. Finally, we have chosen to implement class types in the same manner as relationships (see rules $O2X\_R_5$ and $O2X\_R_6$), also to simplify the implementation, though we lose on the ability to visually differentiate between attributes and relationships.

One additional note to make is how to declare keyrefs for attributes whose type is a list of class type: the name of the keyref would then be given as classType.attribute.item.ref, where classType is the name of the declaring class and attribute is the name of the attribute. The selector XPath is then given the value "./classType/attribute/item". In short, the name of a keyref closely mirrors that of its selector XPath expression.

**Simplifying Assumption:** ODL schemas allow for a key to be any set of properties that is defined for the class. This means that in addition to primitive type attributes, keys may also be complex types, classes, structures, unions, or even relationships. However, the implementation of any type beyond the primitive ones increases the complexity of the transformation process substantially. In this version of our work, we have chosen to adopt a simplifying Assumption. This assumption allows us to make progress in our implementation, while avoiding some tricky technical details beyond the scope of our research. The simplifying assumption can be stated as follows: All classes must have at least one key attribute, and every key must be of primitive or enumerated type.

$O2X\_R_3$.    *Attributes with primitive domains:* This rule states that each basic data type attribute is mapped into a like-named element within the container described in rule $R_2$. In other words, the attributes specified in ODL with basic data types (string, short, date, float, etc.) are translated into atomic elements with the same name. These elements are included in the corresponding special sequence element that has already been produced by rule $O2X\_R_2$, and their data types should, if possible, be the same, although users are allowed to indicate another compatible data type.

The handling of attributes of primitive type is accomplished as follows: an element is created with the same name as the corresponding attribute name, the value of the type attribute for the new element is specified as the XSD analogue of its ODL type. Refer to Table 7.1 for a listing of type mappings ([24]: Data Types Second Edition. http://www.w3.org/TR/xmlschema-2/).

Note that not all ODL primitive types have appropriate XSD analogues: though all numeric types are easily mapped to their XSD counterparts, the ODL char and octet types do not have XSD counterparts. In other words, we have noted that there exist primitive types in ODL for which there is no easy XSD analogue. For char, one workaround is to create a new simpleType called char.type, restricting xsd:string by adding in a length element with the value attribute set to 1 and the fixed attribute set to true (to disallow the empty string). A workaround for octet is to either restrict xsd:hexBinary as above, or use xsd:byte or xsd:unsignedByte. Care must be taken to ensure that any new simpleTypes added for primitive types must only be added once per transformation. In Fig. 7.1, the attributes "ISBN" of class "book" and "ID" of class "author" are both of primitive (integer) type.

Consideration should also be taken as to what the intention of the value may be. For example, a string attribute may be representing a URI, for which a conversion

**Table 7.1**  Primitive type mappings

| ODL | XSD |
| --- | --- |
| Boolean | xsd:boolean |
| Double | xsd:double |
| Float | xsd:float |
| Int | xsd:int |
| Long | xsd:long |
| Short | xsd:short |
| String | xsd:string |
| Unsigned long | xsd:unsignedLong |
| Unsigned short | xsd:unsignedShort |

to xsd:anyURI may be more apt than xsd:string. However, this may be beyond the scope of what an automated tool can do, i.e., user involvement is required to guarantee that the intended domains are specified appropriately.

$O2X\_R_4$.   *Key attributes:* It is possible for a class to have explicitly specified key attributes for which a special key element has to be declared as a sub-element of the element that has resulted from rule $O2X\_R_1$. This special element should also have a name attribute whose value should be the same as the name of the class that the key attribute belongs to, with the letter $K$ (abbreviating the word Key) appended to it. The xpath attribute of the selector (sub)element belonging to the special key element should contain a Xpath expression like: `.//class_name_which_the_key_attribute_belongs_to`. Nevertheless, users should have the option to define the value of the Xpath expression. The xpath attribute of the (sub)element field of the special key element should contain a Xpath expression like: keyattribute_name.

In this rule, key attributes are declared as key elements within the root XML document structure (element from rule $O2X\_R_1$). It then defines a convention for naming the key, and creating the Xpath expressions for the selector and field elements within the key.

We also consider compound keys in order to have this rule more general and comprehensive. In other words, as ODL allows for compound keys, we extend this to compound keys by adding additional fields. The complete rule is thus as follows:

For each class, a key element is added to the element declaration. The name of the key, if classType is the name of the class, is given as classType.key. The value of the xpath attribute of the selector child element is given as "./classType".

For each key attribute in the class, a field element is added as a child to the key element, with its xpath attribute given the name of the key attribute as its value. In Fig. 7.1, we can see two single-part keys, "ISBN", and "ID"; the class person makes use of a compound key, using two strings, firstName and lastName.

$O2X\_R_5$.   *Multiple values specified by collection type (one-to-many relationship):* Each relationship that contains any of the three keywords set, list or bag, is converted into an element with the same name. This element is included in the special sequence element already produced by rule $O2X\_R_2$ and, since the cardinality of the relationship is greater than one, it is necessary to explicitly associate with this attribute the attribute maxOccurs with value unbounded. Further, this element should include an anonymous complex data type with a special element complexContent. Moreover, this special element should include a special restricted element that contains an attribute with the name of the key attribute of the referred class. Additionally, a special keyref element has also to be declared as a subelement of the element that has been produced by rule $O2X\_R_1$. This special element should also have a name attribute with its value specified as: `class_name_which_the_relationship_belongs_to_Ref_class _name_referred_to`. This special element should also include a refer attribute with the value: `key_element_name_of_referred_class`. The xpath attribute of the selector (sub)element belonging to this special element should contain a Xpath expression like: `.//class_name_which_the_ relationship_belongs_to/relationnship_name`. Nevertheless, users should have the possibility to define the value of the Xpath expression. The xpath attribute of the (sub)element field of this special element should contain a Xpath expression like: `@key_attribute_name_of_the_referred_ class`.

This rule concerns itself with handling attributes which have multiple values specified in a collection. First, a like-named element is created within the container, which has an unbounded number of occurrences. Within this element, the corresponding related element is declared. Then, a corresponding keyref element is declared within the root document element from rule $O2X\_R_1$. This element follows a naming convention, and specifies the appropriate selector and field Xpath statements.

$O2X\_R_6$.    *One-to-One relationships:* Each relationship that does not involve any of the three keywords set, list or bag, is converted into an element with the same name. The translation is similar to that specified in rule $O2X\_R_5$, with three exceptions: (1) the attribute maxOccurs (with value unbounded) should not be associated with the created element; (2) the user should define its data type, although a default data type may be used for this matter; (3) in the declaration of the special keyref element, the xpath attributes of the selector and field (sub) elements will have other values; so, the xpath attribute of the selector element should contain a Xpath expression like: `class_name_which_the_ relationship_belongs_to`. Nevertheless, users should have the ability to define the value of the latter Xpath expression. The xpath attribute of the field element should contain a Xpath expression like: `//relationship_name`.

Similar to rule $O2X\_R_5$, a like-named element is created within the class container. This like-named element simply takes on the type of the referred attribute. Finally, it is worth mentioning that in the work described in [9], one-to-one and one-to-many relationships are handled differently; one-to-one relationships were treated as attributes, and were unable to handle compound keys. To visually separate relationships from attributes, we have decided to adopt for one-to-one relationships the same method used in one-to-many relationships. The complete rule is articulated as follows:

If a relationship is declared, a new element with the name of its relationship is declared, which is to be added to the class root. If this is a one-to-many relationship (i.e., the relationship is of a list type), the maxOccurs attribute of this element is set to unbounded.

The relationship element is of an anonymous complexType. For each key attribute in the related class, an attribute with the name of the related key attribute is added to the complexType, with its type matching that of the type of the related key attribute.

Finally, we add a keyref element to the element declaration. If the name of the relationship is "relationship", the name of the declaring class is classType, and the name of the related class is relatedType, then the name of the keyref is classType.relationship.ref and the value of its refer attribute is relatedType.key. The xpath attribute of the selector element child is given as "./classType". For each key attribute in the class, a field element is added as a child to the key element. The xpath attribute therein is given the value "@attribute", where attribute is the name of the key attribute.

Care must be taken to ensure that the field children of the keyref are declared in the same order as that of the related key. In Fig. 7.1, we have an example of both one-to-one and one-to-many relationships, namely "writtenby" of class "book", and "write" of class "author", respectively.

$O2X\_R_7$.   *Lists:* A simple approach is taken for lists. Essentially, a like-named element is placed in the class container for any list attribute in the ODL schema. Then, a simple type with an embedded list element is declared, based on the data type of the ODL list. In other words, each attribute of a class that contains the keyword "list" in its definition is converted into an element with the same name; the new element is included in the special sequence element that has resulted from rule $O2X\_R_2$. This element should include an anonymous simple data type. Moreover, the simple data type should include a special list element with an itemType attribute whose value could be indicated by the user, or defined by default (e.g., string).

The work described in [9] implements list types by creating an (anonymous) XSD simpleType, and using XML Schema's own list datatype as its analogue. However, this method is flawed; firstly, it cannot deal with lists of any type other than primitive or enumerated type. Furthermore, it does not handle primitive types appropriately: according to the XML Schema specification ([23]: Structures Second Edition. http://www.w3.org/TR/xmlschema-1/; [24]: Data Types Second Edition. http://www.w3.org/TR/xmlschema-2/), the list type is represented using a space-separated sequence of atomic datatypes (which correspond to ODL primitive types). This means that lists of the ODL string type may be misrepresented if these strings contain spaces. To avoid the use of special encoding techniques for differentiating between spaces within the strings themselves and delimitations between list items, we use the following approach.

If an attribute is of a list type, which, for our purpose, could be specified as list, set, bag, sequence, or array, a new element is created with an anonymous complexType as its content model. This complexType will have an unbounded number of child elements named item. This element type is then inserted into the class root.

The content model of the item element will depend on the type of the corresponding list element. For instance, if the attribute is a list of lists, the item element will again contain an unbounded item child. In Fig. 7.1, the attribute "ratings" of the class "book" is of list type.

$O2X\_R_8$. *Complex data type attributes:* This rule takes advantage of the hierarchical nature of XML and describes an elegant way of dealing with complex data types specified in ODL. Each complex attribute is created within its corresponding container created in rule $O2X\_R_2$. Then, the appropriate other rules can be applied to this complex data structure in order to populate the newly created element. In other words, each attribute with a complex data type is converted into an element with the same name. This element should be included in the special sequence element that has resulted from rule $O2X\_R_2$. However, this special element should include an anonymous complex type with a special sequence element in which the elements corresponding to the conversion of the attributes integrated in the class (that defines the complex type of the attribute being converted) are declared. Of course, the rule to be applied is specified according to the data type of each specific attribute. For instance, if the data type is simple then we should apply rule $O2X\_R_3$ again; if it is complex, we should apply rule $O2X\_R_8$ once more, and so on.

$O2X\_R_9$. *Enumerations:* This rule describes how to handle anonymous enumerations. A like-named element is placed in the class container produced by rule $O2X\_R_2$. Within this element, enumeration elements are specified and should conform to the type of the enumeration in the ODL schema. In other words, each attribute (of a class in ODL) that contains the keyword enum in its definition is converted into an element with the same name; the element should be included in the special sequence element that has resulted from rule $O2X\_R_2$. This element should include an anonymous simple data type. Moreover, the simple data type should include a special restriction element with a base attribute, whose value could be specified by the user, or defined by default (e.g., string). The restriction element should also include, for each one of the values of the enum data type, a special enumeration element with the same value in its value attribute.

Enumerated types may be declared in any of three scopes: globally, on the same level as classes, locally within the scope of a class, or anonymously declared within an attribute. We declare a new XSD simpleType, restricting xsd:string. However, instead of inserting it into the attribute declaration, we insert the simpleType into the schema root. As is necessary under XML Schema, each enumerated type is given a name. Given the enumerated type enumType:

1. If enumType is globally declared, the name enumType.enum is given.
2. If enumType is locally declared in class classType, the name classType.enumType.enum is given.
3. If enumType is declared anonymously in attribute of class classType, the name classType.attribute.enum is given.

Note that there is no ambiguity if there exists a class with an attribute and enumerated type declared with the same name, even if the attribute is not of the enumerated type. Furthermore, we note that a locally declared or anonymous enumerated type may not share the same name as a globally declared one in ODL.

If an attribute is of enumerated type, the transformation will create an element with the attribute name as its name and our assigned type name as the value of its type attribute, as done with attributes of primitive type. This element will be inserted into the class root, as with all the other properties.

In Fig. 7.1, we see an example of enumerated types in the class author. It has a locally declared enum called "sex", which is the type of the attribute "gender".

$O2X\_R_{10}$.   *Dictionary Types:*  A commonly used construct in the object-oriented paradigm is dictionaries. A dictionary allows a user to query a list of associations, searching for an entry that matches the search criteria. The structure of XML lends itself well to handle dictionaries. ODL defines a dictionary as a simple list of a structured type called Association [6]. Each Association, in turn, defines a key and a value. As such, if an attribute is of dictionary type, the transformation will create an element with an attribute as its name and an anonymous complexType as its content model. This anonymous complexType will have an unbounded number of children of elements named item. The content model of the item element is also an anonymous complexType, which has two child elements (either as an all or as a sequence; we have chosen sequence for stylistic reasons), named key and value.

The content models of the key and value elements depend on the key and value types of the dictionary; this is handled in a manner similar to that of lists. If the key and value types involve classes, keyrefs will also be added accordingly.

In Fig. 7.1, we have included an example of a dictionary type. In our sample, we use it to associate chapters with page numbers, in the same way as they are associated in a table of contents. In class book, we define a dictionary attribute named chapterPageNumbers.

$O2X\_R_{11}$.   *Inheritance:*   No object-oriented schema transformation process would be complete without support for inheritance. One of the main attractions of the object-oriented model is the ability to abstract common class characteristics into superclasses and allowing slight variation in subclasses. Because the content models of classes are anonymously declared, we cannot use the XML Schema analogue of subclassing (deriving new complexTypes from existing complex-Types by extension) here. Instead, we choose to do the following:

If a class extends another class, then we must deal with all inherited and new properties as with any other class. As XML does not support explicit inheritance, it is interpreted as nesting. We handle this in one of two ways based on the user preference. First, we may produce a nested XML structure where the element that corresponds to a subclass is extended to include all the definition of the superclass. Second, we may produce a flat XML structure using key/keyref as follows. The selector element in the superclass key element has its xpath attribute appended with "|./subclass", where subclass is the name of the subclass. Similarly, the selector element in any keyref element referencing the superclass key must have its xpath attribute appended in the same way. This process is repeated for every superclass in the class hierarchy.

Note that by the simplifying assumption stated above under rule $O2X\_R_2$, the subclass must declare its own key, which may be any combination of new or inherited attributes, even the same key attributes as that of its superclass.

To illustrate the functionality of inheritance, we have inherited the "person" class in the "author" class in Fig. 7.1.

$O2X\_R_{12}$. *Structures and Unions:*   The implementations of structured and union types are relatively straightforward. However, there is one major obstacle preventing the full implementation of these types, along with classes that do not have an extent, have an extent but no key, or otherwise violate the simplifying assumption stated above. These cases are to be considered in a future extension of our work.

Structure is implemented as complexTypes to be added to the schema root, with a naming scheme similar to that of enumerated types, but with the ".struct" extension. These complexTypes have as children a sequence (or all) element containing its members, whose types are then recursively resolved. With this, attributes of structured type may be resolved as with primitive types.

Union is also implemented as complexTypes. In ODL, union contains a discriminator member sharing the name of the union type. This discriminator must be of primitive or enumerated type, which will be the first child in our sequence of children. The second child is a choice element containing each member of the union. Again, union type elements are added to the schema root with a naming scheme similar to that of enumerated types, but with the ".union" extension. Similar to structures and enumerated types, attributes of union type can then be coded in the same way as primitive types. However, the problem with a full implementation is the cyclic dependency problem: a structure may declare a member of a structure type, which may declare a member of the original structure type. With it, we cannot guarantee that our transformation will terminate. In fact, the simplifying assumption was introduced as a response to tackle the cyclic dependency.

$O2X\_R_{13}$. *Dates, Times, Intervals, and Time-stamps:* ODL has a way of codifying dates, times, intervals, and timestamps, using the structured types date, time, interval, and timestamp, as well as their corresponding class type counterparts Date, Time, Interval, and Timestamp [6]. These types can be treated specially by converting them into the XSD types xsd:date, xsd:time, xsd:duration, and xsd:datetime, respectively. However, a proper translation between these two types must be defined, owing to the possible fine-print differences between them.

$O2X\_R_{14}$. *Null Objects:* We have not explored whether the value of a property may be null, as ODL does not explicitly define whether this is possible. However, in the event that it is, we have an implementation for it.

Elements declaring classes must set the nillable attribute to true. Elements representing attributes of list and dictionary type attributes must also set the nillable attribute of their item, key, or value children, if those types are not of primitive or enumerated type – it is believed that attributes of these types will never be null. In addition, elements representing relationships and attributes of class type must set the use attribute of each attribute child to optional.

Note that, unfortunately, there is no way in XML Schema to enforce more granular conditions in which attributes are required, optional, or prohibited: it is the intention that, in the last case, all attributes are prohibited if we have an xsi:nil attribute with value true (representing a null value), and required if this attribute is absent. (Other XML Schema languages, such as RELAX NG, can enforce this requirement).

$O2X\_R_{15}$.   *Interfaces:*   Though interfaces are normally ignored because they are defining only the behavior of an object (which would be lost in the XML transformation), interfaces may also declare enumerated types, structures, and unions, which can then be used by any class implementing any of the mentioned types. Further exploration of this is needed.

**The Extended Simplifying Assumption:**   We will focus on relaxing the Simplifying Assumption to include more types. We have proposed the Extended Simplifying Assumption, which is defined as follows. A type in ODL is said to satisfy the Extended Simplifying Assumption if it is any of the following: (1) primitive or enumerated type (including string); (2) fixed<N, M>, string<N>, date, time, interval, or timestamp; (3) a fixed-length array of a type conforming to the Extended Simplifying Assumption; (4) a structured type for which every member conforms to the Extended Simplifying Assumption; (5) a class for which all key attributes conform to the Extended Simplifying Assumption; (6) a to-one relationship to a class where all key attributes of the related class conform to the Extended Simplifying Assumption.

Note that the first condition, along with a non-recursive form of the fifth condition, makes up the original Simplifying Assumption.

Note that the fixed-length array depends on an alternate implementation, using $N$ distinctly named child elements instead of one child (item) repeated $N$ times. If we choose to add the last two conditions, we must also come up with a better name-mangling scheme, in the event of two key portions being classes or relationships whose keys have a common name.

The reason that the Simplifying Assumption may be extended as such is due to the fact that these types may be easily keyed (directly for the first two conditions, and via fields that can be more than one level deep in the remaining conditions), and thus easily keyrefed. Similarly, we have excluded other types because of limitations regarding what is allowed in a selector XPath (this is the reason why the alternate fixed-length array implementation is needed), or the nature of the type in question (for example, lists are excluded because its length may vary).

It is our belief that the Simplifying Assumption may never be entirely removed; we have not dealt with the remaining types of keys that may be permitted in ODL.

As a result of applying the rules described above to the sample ODL schema in Fig. 7.1, the following XML Schema is produced:

```
<?xml version="1.0" encoding="UTF-8" standalone="no"?>
<xsd:schema xmlns:xsd="http://www.w3.org/2001/XMLSchema">
 <xsd:element name="database">
  <xsd:complexType>
   <xsd:choice maxOccurs="unbounded">
    <xsd:element name="person">
     <xsd:complexType>
      <xsd:sequence>
       <xsd:element name="firstName" type="xsd:string"/>
       <xsd:element name="lastName" type="xsd:string"/>
      </xsd:sequence>
     </xsd:complexType>
    </xsd:element>
    <xsd:element name="book">
     <xsd:complexType>
      <xsd:sequence>
       <xsd:element name="ratings">
        <xsd:complexType>
         <xsd:sequence maxOccurs="unbounded">
          <xsd:element name="item" type="xsd:int"/>
         </xsd:sequence>
        </xsd:complexType>
       </xsd:element>
       <xsd:element name="ISBN" type="xsd:int"/>
       <xsd:element name="chapterPageNumbers">
        <xsd:complexType>
         <xsd:sequence maxOccurs="unbounded">
          <xsd:element name="item">
           <xsd:complexType>
            <xsd:sequence>
             <xsd:element name="key" type="xsd:int"/>
             <xsd:element name="value" type="xsd:int"/>
            </xsd:sequence>
           </xsd:complexType>
          </xsd:element>
         </xsd:sequence>
        </xsd:complexType>
       </xsd:element>
       <xsd:element name="writtenby">
        <xsd:complexType>
         <xsd:attribute name="ID" type="xsd:int"/>
        </xsd:complexType>
       </xsd:element>
      </xsd:sequence>
     </xsd:complexType>
    </xsd:element>
    <xsd:element name="author">
```

```
  <xsd:complexType>
   <xsd:sequence>
    <xsd:element name="firstName" type="xsd:string"/>
    <xsd:element name="ID" type="xsd:int"/>
    <xsd:element name="lastName" type="xsd:string"/>
    <xsd:element name="gender" type="author.sex.enum"/>
    <xsd:element maxOccurs="unbounded" name="write">
     <xsd:complexType>
      <xsd:attribute name="ISBN" type="xsd:int"/>
     </xsd:complexType>
    </xsd:element>
   </xsd:sequence>
  </xsd:complexType>
 </xsd:element>
 </xsd:choice>
</xsd:complexType>
<xsd:key name="person.key">
<xsd:selector xpath="./person|./author"/>
 <xsd:field xpath="firstName"/>
 <xsd:field xpath="lastName"/>
</xsd:key>
<xsd:key name="book.key">
 <xsd:selector xpath="./book"/>
 <xsd:field xpath="ISBN"/>
</xsd:key>
<xsd:keyref name="book.writtenby.ref" refer="author.key">
 <xsd:selector xpath="./book/writtenby"/>
 <xsd:field xpath="@ID"/>
</xsd:keyref>
<xsd:key name="author.key">
 <xsd:selector xpath="./author"/>
 <xsd:field xpath="ID"/>
</xsd:key>
<xsd:keyref name="author.write.ref" refer="book.key">
 <xsd:selector xpath="./author/write"/>
 <xsd:field xpath="@ISBN"/>
</xsd:keyref>
</xsd:element>
<xsd:simpleType base="xsd:string" name="author.sex.enum">
 <xsd:restriction>
  <xsd:enumeration value="FEMALE"/>
  <xsd:enumeration value="MALE"/>
 </xsd:restriction>
</xsd:simpleType>
</xsd:schema>
```

Finally, our transformer is tested against a suite of 25 test inputs, all of which have been verified to create XML Schemas that pass the XML Schema validator ([25] (REC (20010502) version, as amended) Checking Service. http://www.w3.org/ 2001/03/webdata/xsv).

## 7.4   Rules for Converting XML Schema into ODL

An XML schema is characterized by having elements and attributes; elements may
be nested or may reflect a flat structure that simulates the nesting using key/keyref
referencing pair. The task accomplished by the rules presented in this section may
be articulated as follows: for a given XML schema (whether flat or nested), produce
a corresponding ODL structure. As ODL allows for inheritance which is not part
of XML by definition, we have two choices; either we produce a one level class
hierarchy where all classes are under the root or we produce a multilevel class
hierarchy by applying an optimization rule on the one level hierarchy to minimize
overlap and redundancy between the classes. Conversion of flat XML involves one
more step to produce a class hierarchy that incorporates nesting in a more natural
way. In other words, the key/keyref pair lead to extending one class to include all
the definition of the referenced class.

$X2O\_R_1$   *Create the main classes:* Every object-oriented schema must have
classes to hold definition for the structure of the objects to be populated in the
database. To create the main classes, it is important to start by locating in the
XML schema all first level elements; these are elements not nested within other
elements. For each first level element create a corresponding main class with the
same name as the element.

$X2O\_R_2$   *Create the nested classes:* Two alternatives are possible because XML
allows for both flat and nesting; even the two choices may coexist in the same
XML schema. Therefore, it is important to trace the nesting of elements or the
references specified using key/keyref in order to find the appropriate additional
classes to add to the class hierarchy. We need to recursively trace each element $E$
for which a class $c$ has been created and apply the following two rules.

$R_{X2O\_2.1}$   *Nested classes from nested XML schema:* If element $E$ has a subelement $E_s$ such that $E_s$ has complex structure (includes some subelements) then
create a new class $c_s$ corresponding to $E_s$.

$X2O\_R_{2.2}$   *Nested classes from nested XML schema:* If element $E$ has keyref
specification which references another element $E_k$ then add to the definition
of class $c$ a new non-primitive attribute with the name $A_{Ek}$ and specify the
domain of $A_{Ek}$ as the class that corresponds to element $E_k$; create the latter
class if it does not exist.

$X2O\_R_3$   *Define primitive attributes for simple elements:* For each simple element found in any of the already processed elements $E$, create in the corresponding class $c$ a primitive attribute with the same name and domain as the
simple element.

$X2O\_R_4$   *Define primitive attributes for attributes in elements:* For each attribute
found in any of the already processed elements $E$, create in the corresponding
class $c$ a primitive attribute with the same name and domain as the attribute in $E$.

*X2O_R₅*    *Define the keys in classes:* If element $E$ has a key which may be one or more attributes/elements then create a key for the corresponding class $c$; the key for class $c$ must include all the attributes that correspond to the components of the key of $E$.

*X2O_R₆*    *Define non-primitive attributes in classes:* Both XML and ODL allow for complex domains to be specified. Hence the mapping into non-primitive attributes is almost straight forward. It would have been more challenging and would require more effort if the destination of the mapping was the relational model which does not allow attributes with complex types.

> *X2O_R₆.₁*    *Relationships:* For every element reflecting a relationship type inside element $E$, create a corresponding attribute in class $c$ that corresponds to $E$. The destination for the relationship should be set equivalent to the definition of the corresponding element in $E$ by specifying the type of the relationship attribute to be an existing class and the linking attribute in the latter class should be determined.

> *X2O_R₆.₂*    *Collections:* Both simple and complex domains may be allowed to have collections (set, list, bag or array) as the value; this may be found explicitly specified in XML using "maxOccurs". Also relationships may be specified to connect to a collection of values. For all these cases, determine the collection type from the definition of the corresponding element and reflect that into the definition of the attribute or relationship in the given class.

> *X2O_R₆.₃*    *Enumeration:* Locate all specifications of enumerated type in the XML schema. Each of these enumerated types is defined with an element $E$; accordingly, locate the correspond class $c$ and define in class $c$ the same enumerated type found in $E$.

> *X2O_R₆.₄*    *Dictionary:* For elements specified to have the dictionary type in the XML schema, find the components of the dictionary type and define it accordingly for the corresponding attributes in the ODL schema.

*X2O_R₇*    *Decide on attribute that may have null value:* By tracing the XML elements, it is possible to find out the "minOccurs" clause specified for some elements. For every attribute $A$ in the ODL schema, if the element that corresponds to $A$ has its "minOccurs" specified as "0" then attribute $A$ is allowed to have the value "null" and this should be reflected into the definition of attribute $A$ in ODL.

*X2O_R₈*    *Define interfaces in classes:* One of the key features of object-oriented paradigm is encapsulation and message passing. This is facilitated by providing at minimum two methods per attribute to allow for setting/modifying and getting/returning its value, respectively. Accordingly, for each attribute $A$ defined in ODL, generate two methods *get()* and *set(X)*; these methods will have very simple code; the former will display the value of attribute $A$ in the receiving object and the latter will replace in the receiving object the existing value of $A$ with $X$.

### 7.4.1   Enforcing Inheritance into the Hierarchy

The outcome from the XML to ODL conversion would not immediately produce any inheritance information; it is a one level hierarchy with all classes listed as direct children of the root. However, the inheritance information is implicitly present in the hierarchy and simulated by duplication of information in various classes. Hence, our target is to enforce reusability instead of duplication; this is possible by deriving a list of superclasses for each given class $c$ to increase inheritance instead of duplication. We benefit from our experience in optimizing the class hierarchy for a more comprehensive set of cases as described in our work described in [4] which work fine by properly placing in the class hierarchy all virtual classes produced by our object query language defined in [2] and [3].

Non-inherited (locally defined) characteristics of class $c$ include both locally defined attributes and locally defined behavior, denoted $L_{attributes}(c)$ and $L_{behavior}(c)$, respectively. The inherited characteristics of class $c$ include both all attributes and behavior in the superclasses of class $C$, denoted $(W_{attributes}(c) - L_{attributes}(c))$ and $(W_{behavior}(c) - L_{behavior}(c))$, respectively. The sizes of inherited and locally defined characteristics are inversely proportional, i.e., as one increases the other decreases and vice versa. Our aim is to maximize inherited and hence minimize locally defined characteristics. To achieve this, we adjust the list of superclasses of class $c$ to include classes that maximize inherited and minimize locally defined characteristics. Further, we argue that enforcing multiple inheritance in a data model helps in increasing reusability by providing the flexibility of adjusting superclasses when possible in a way to increase the characteristics that class $c$ inherits by allowing more than one class to appear in the inheritance list of $c$, and hence decrease locally defined characteristics of class $c$.

The XML to ODL conversion process produces classes for which superclasses are not specified. Hence superclasses must be found for the produced classes. For this purpose, we have to define some rules capable of handling the process by investigating and adjusting the list of superclasses for a given class. We define rules that investigates the possibility of maximizing reusability by adding some existing classes to the list of superclasses of class $c$, denoted $C_p(c)$, or by pushing class $c$ into the list of superclasses of some existing classes. Here, we have to consider and investigate all classes in the hierarchy because every class $c$ produced by the conversion has no specified subclasses, denoted $C_b(c)$, i.e., $C_b(c)$ is empty. The process is not very challenging because all the classes we have are coming from the same XML schema and hence there is no attribute naming problem as attributes correspond to elements; in case of duplication, duplicated attributes will have same name and domain in all classes in which they appear. Therefore, it will be straightforward to find the overlap between classes.

$X2O\_R_9$   *Find potential superclasses:* For each pair of classes $c_1$ and $c_2$ produced as the result of applying rules $X2O\_R_1$ to $X2O\_R_8$ in Sect. 7.4 above, if $L_{attributes}(c_1) \bigcap L_{attributes}(c_2)$ is not empty then created a new class $c_{1\_2}$ to include all attributes in $L_{attributes}(c_1) \bigcap L_{attributes}(c_2)$. Further, $L_{behavior}(c_{1\_2})$

is set to include $L_{behavior}(c_1) \bigcap L_{behavior}(c_2)$ as every class has in its local behavior only pairs of methods, one pair per attribute to set/modify and get/return the value of the attribute.

## 7.5  Rules for Data Conversion

Mapping the schema will not be complete, useful and acceptable without mapping the data. The data mapping process involves two main rules. First, one object (root element) is generated for each root element (object) and attributes (elements) with primitive domains are mapped directly. Second, all complex domains are mapped after their containers are ready to hold them; this second process is handled separately because in object-oriented databases we cannot reference a non-existing object; so we have to make the object skeleton ready first and then reference it in the second phase. The other way around is very similar, but in one pass to produce instantiations for elements with values for their primitive and non-primitive components specified. In the remainder of this section, we first present the rules for producing objects and then we present the rule that maps objects into instantiation of elements.

$X2O\_D\_R_1$    *Creating objects from documents:* Recall that we have created one class $c$ for every element $E$ of complex type. For each instantiation $I_E$ of element $E$, create a corresponding object $O_c$ in the local instances of class $c$, denoted $L_{instances}(c)$; object $o_c$ will have a new unique identifier and the values of its primitive attributes will be taken from the corresponding instantiation $I_E$.

$X2O\_D\_R_2$    *Deciding on values of non-primitive attributes for objects:* This rules has immediate benefit from the outcome of rule $X2OD\_R_1$; this rule uses the object identifiers created by rule $X2OD\_R_1$ to specify the values of non-primitive attributes. For every object $o_c$ which has been created by rule $X2OD\_R_1$, if class $c$ of object $o_c$ has non-primitive attributes then for each non-primitive attribute $A$ in class $c$, locate object(s) $o_I$ that corresponds to the instantiation of the element that corresponds to attribute $A$ and set the value of attribute $A$ in object $o_c$ to object(s) $o_I$. The latter value may be single or collection of values; for the collection of values case, the values $o_I$ are arranged to fit the specification of the collection type whether set, list, bag or array.

$O2X\_D\_R_1$    *Creating documents from objects:* For each element $E$, locates its corresponding class $c$. Some subelements from $E$ may have their definition exists in superclass(es) of $c$ because of the inheritance characteristic specified for class $c$ but missing in XML. If this is the case, then start with elements $E$ that correspond to classes $c$ with no subclasses; produce one instantiation of $E$ for each object $o_c$ in $c$; each instantiation will utilize from $o_c$ in $c$ the values that are part of $o_c$ due to the inheritance relationship between class $c$ and its superclasses. For each non-leaf class $c_L$ which has a corresponding element $E_L$, if $c_L$ has some objects (in $L_{instances}(c_L)$) then for each object in $L_{instances}(c_L)$ create a corresponding instantiation for element $E_L$.

## 7.6    Implementation Details

In the process of devising the rules enumerated above for both sides of the two-way mapping, including the schema and the data, we have also prepared a reference implementation written in Java. This reference implementation contains a lexer and parser to accept input, as well as a transformer to do the actual XML and ODL transformations. The parser creates a set of objects representing ODL classes, attributes, relationships, and enumerated types; it is also capable of producing XML schemas and XML documents; it is set to error if the simplifying assumption is violated. Our reference parser all parts of ODL that comply with the corresponding rules specified in Sect. 7.3. It is also capable of parsing all parts of XML schema that satisfy the requirements for the conversion rules presented in Sect. 7.4.

The core of the transformer are the ODLTransformer and XMLTransformer classes. As the process incorporated in each of these two classes is the reverse of the process in the other class, we will only describe the ODL to XML transformation process as specified in the ODLTransformer class.

After lexing and parsing the input ODL file, the ODL object representing the ODL schema, an object of type ODLSchema, is passed into the ODLTransformer, which proceeds to build a DOM tree. This DOM tree can then be further manipulated using XSL Transformations. In our case, we simply choose to output the DOM tree as-is, which then is our output XML Schema file.

ODLTransformer depends on a set of ItemBuilders to create the content models of each class type. There are separate ItemBuilders for primitive types, enumerated types, class types (which are also used in relationships), list types, and dictionary types. These ItemBuilders will also create the necessary keyrefs, though ODL-Transformer itself holds the responsibility of creating keys as well as simpleTypes representing enumerated types.

## 7.7    Conclusions

The need to convert between the object-oriented database model and the XML database model has been covered in this paper. It is eventually more attractive than the XML to relational conversion because the two models covered in this paper have more commonalities and hence the semantic gap almost disappears to the benefit of the transformation process both way. The mapping from ODL to XML involves the conversion between an object schema, in the form of ODL, to an XML schema, using XML Schema, and then the conversion of the data itself following the schemas that have been created. The conversion of an object-oriented schema to an XML schema is a nontrivial process, requiring both human intervention and automated tools. Though we may make rules representing general guidelines on how to covert between ODL and XSD, they can only cover a limited subset of the ODL grammar. However, these rules cover a sufficiently large portion of ODL so that human intervention is kept to a minimum. Finally, the converted XML schema may also impose conditions on how the data is to be transformed as well, as with the case with date and time datatypes. The conversion from XML to ODL is more challenging as we need to decide on the inheritance for ODL; we could successfully handle this vital property and requirement by invoking an optimization process that minimizes the overlap between classes, redundancy in other words. We argue that the rules that we have defined should cover the vast majority of the needs of most object-oriented database users and administrators.

# References

1. Ahmad, U., et al.: An integrated approach for extraction of objects from XML and transformation to heterogeneous object oriented databases. In: Proc. of ICEIS, pp. 445–449 (2003)
2. Alhajj, R., Arkun, M.E.: A query model for object-oriented database systems. In: Proc. of the IEEE International Conference on Data Engineering, Vienna (1993)
3. Alhajj, R., Polat, F.: Closure maintenance in an object-oriented query model. In: Proc. of the ACM International Conference on Information and Knowledge Management, Maryland (1994)
4. Alhajj, R., Polat, F., Yilmaz, C.: Views as first-class citizens in object-oriented databases. VLDB Journal 14(2), 155–169 (2005)
5. Carey, M., et al.: XPERANTO: Publishing object-relational data as XML. In: Proc. of WebDB (2000)
6. Cattell, R.G.G., Barry, D., et al.: The Object Data Standard: ODMG 3.0. Academic Press, London (2000)
7. Cheng, J., Xu, J.: XML and DB2. In: Proc. of IEEE ICDE, San Diego, pp. 569–573 (2000)
8. Chung, T.-S., Park, S., Han, S.-Y., Kim, H.-J.: Extracting object-oriented database schemas from XML DTDs using inheritance. In: Proc. of the International Conference on Electronic Commerce and Web Technologies, pp. 49–59 (2001)
9. De Sousa, A., Pereiera, J., Carvalho, J.: Mapping rules to convert from ODL to XML-SCHEMA. In: Proc. of the International Conference of the Chilean Computer Science Society, IEEE Computer Society (2002)
10. Extensible Markup Language (XML) 1.0 (Fourth Edition). W3C Recommendation 16 August 2006, edited in place 29 Sept 2006, Uppsala Master's Thesis in Computing Science no. 260, 2003.
11. Fernandez, M.F., Tan, W.C., Suciu, D.: SilkRoute: Trading between relational and XML. In: Proc. of WWW (2000)
12. Fong, J., Pang, F., Bloor, C.: Converting relational database into XML document. In: Proc. of the International Workshop on Electronic Business Hubs, pp. 61–65 (2001)
13. Jarada, T.N., Chung, K., Shimoon, A., Karampelas, P., Alhajj, R., Rokne, J.: Mapping rules for converting from ODL to XML schemas. In: Proc. of International Conference on Information Integration and Web-based Applications & Services, Paris (2010)
14. Johansson, T., Heggbrenna, R.: Importing XML Schema into an Object-Oriented Database Mediator System, Uppsala Master's Theses in Computing Science no. 260 (2004)
15. Lee, D., Mani, M., Chiu, F., Chu, W.W.: NeT & CoT: Translating relational schemas to XML schemas using semantic constraints. In: Proc. of ACM International Conference on Information and Knowledge Management (2002)
16. Liu, C., Vincent, M.W., Liu, J., Guo, M.: A Virtual XML Database Engine for Relational Databases. In: Xsym, Vol. 2824, pp. 37–51 (2003)
17. Lo, A., Alhajj, R., Barker, K.: VIREX: visual relational to XML conversion tool. J. Vis. Lang. Comput. 17(1), 25–45 (2006)

18. Munroe, K.D., Papakonstantinou, Y.: BBQ: A visual interface for browsing and querying of XML. In: Proc. of IFIP Working Conf. on Visual Database Systems, pp. 277–296 (2000)
19. Naser, T., Kianmehr, K., Alhajj, R., Ridley, M.J.: Transforming Object-Oriented Database into XML. In: Proc. of IEEE IRI, pp. 600–605 (2007)
20. Orsini, R., Pagotto, M.: Visual SQL-X: A Graphical Tool for Producing XML Documents from Relational Databases. In: Proc. of the International Conference on World Wide Web, Hong Kong (2001)
21. Toth, D., Valenta, M.: Using Object And Object-Oriented Technologies for XML-native Database Systems. In: Proc. of the Dateso Annual International Workshop on Databases, TExts, Specifications and Objects (2006)
22. Wang, C., Lo, A., Alhajj, R., Barker, K.: Converting Legacy Relational Database into XML Database through Reverse Engineering. In: Proc. of ICEIS (2004)
23. XML Schema Part 1: Structures Second Edition. http://www.w3.org/TR/xmlschema-1/
24. XML Schema Part 2: Data Types Second Edition. http://www.w3.org/TR/xmlschema-2/
25. XML Schema (REC (20010502) version, as amended) Checking Service. http://www.w3.org/2001/03/webdata/xsv

# Chapter 8
# Feature Selection for Highly Imbalanced Software Measurement Data

**Taghi M. Khoshgoftaar, Kehan Gao, and Jason Van Hulse**

## 8.1 Introduction

Many data sets include an overabundance of features (i.e., attributes). However, not all features make the same contributions to the class. Selecting a subset of features that are most relevant to the class attribute is a necessary step to create a more meaningful and usable model [11]. In addition, for a two-group classification problem, class imbalance is frequently encountered [7, 34]. Class imbalance refers to the cases where the examples of one class are significantly outnumbered by the examples of the other class in a data set. For instance, in software quality classification, fault-prone (*fp*) modules typically are much less common than not-fault-prone (*nfp*) modules. Traditional classification algorithms attempt to improve classification accuracy without considering the relative significance of the different classes, resulting in a large number of misclassifications from minority class (e.g., *fp*) to majority class (e.g., *nfp*). This type of misclassification is extremely severe in some domains such as software quality assurance, implying a lost opportunity to correct a faulty module prior to deployment and operation. A variety of techniques have been proposed to counter the problems associated with class imbalance [30].

In this study, we investigate six filter-based feature ranking techniques to select a subset of attributes and one data sampling technique to deal with the class imbalance problem. The six filter-based feature ranking techniques are chi-square (CS), information gain (IG), gain ratio (GR), two types of ReliefF (RF and RFW), and symmetrical uncertainty (SU). The data sampling technique used is random undersampling (RUS). The main contribution of this study is that we propose a

T.M. Khoshgoftaar (✉) · J.V. Hulse
Florida Atlantic University, Boca Raton, FL 33431, USA
e-mail: taghi@cse.fau.edu; jvanhulse@gmail.com

K. Gao
Eastern Connecticut State University, Willimantic, CT 06226, USA
e-mail: gaok@easternct.edu

T. Özyer et al. (eds.), *Recent Trends in Information Reuse and Integration*,
DOI 10.1007/978-3-7091-0738-6_8, © Springer-Verlag/Wien 2012

novel strategy to combine a feature ranking technique and data sampling. We first use RUS to balance the two classes of a data set. Then, we apply one of the feature ranking techniques to the sampled data and rank the features according to their predictive powers. In order to eliminate the biased outcome due to using RUS once, we repeat this process $k$ times and aggregate the $k$ different rankings using the mean (average). Finally, the best set of features is selected from either the original data or the sampled data to form the training data set.

The experiments of this study were carried out in the context of software quality modeling. Software quality and reliability can be improved many ways during the software development process. One effective method is to utilize software metrics and defect data collected during the software development life cycle and build defect predictors using data mining algorithms to estimate the quality of target program modules [15, 22]. Typically, a software quality estimation model is trained using software metrics and defect data collected from prior development experiences of the organization, and then the trained model is applied to the project under development. Such a strategy allows practitioners to intelligently allocate project resources and focus more on the potentially problematic modules. A common but unfortunate practice among researchers and practitioners is to simply train a defect prediction model using a commonly used learner and with the available set of software metrics without regard to the quality of the underlying software measurement data [15, 22, 25]. In contrast, we advocate a careful evaluation of the quality of data prior to defect prediction modeling.

To validate the effectiveness of our proposed method, we applied this repetitive feature selection technique to three groups of software data sets, each containing three separate releases. The class distributions of the three groups of data are different, with 92–94%, 85–88%, and 71–76% not-fault-prone ($nfp$) modules, respectively. We also compared our proposed method to two other approaches – (1) using a feature ranking technique alone and (2) using the data sampling and feature ranking techniques together once. In addition, two scenarios were considered for each of the approaches. One scenario refers to the training data being formed by the original data with the selected attributes, while the other scenario refers to the training data being formed by the sampled data with the selected attributes. Therefore, for a given feature ranking technique and a given data sampling method, we had a total of six different strategies to produce training data. After feature selection, five different classifiers were applied to the training data sets with the attributes selected by the six strategies. The experimental results demonstrate that our proposed repetitive feature selection method performs, on average, significantly better than the other two approaches. Moreover, the comparison results show that the relative strength of the repetitive feature selection over other methods becomes evident when a training data set is highly imbalanced (i.e., the percentage of $nfp$ examples is very high). For the two scenarios associated with the original and sampled data, the results show that the training data extracted from the original data performed better than the one formed from the sampled data. To our knowledge, no similar studies have ever been done in related literature.

The remainder of the paper is organized as follows. Section 8.2 presents related work. The six filter-based feature ranking techniques and our newly proposed repetitive feature selection method, as well as the five classifiers and the associated classification performance metric used in the study are described in Sect. 8.3. A case study is provided in Sect. 8.4. Finally, conclusions and future work are summarized in Sect. 8.5.

## 8.2   Related Work

Generally, feature selection is divided into two groups, *wrapper*-based and *filter*-based feature selections. For a wrapper-based technique, a learner is involved in the feature selection process. The potential problems of a wrapper-based technique lie in its high computational cost and a risk of overfitting to the data. On the other hand, for a filter-based technique, instead of using a learner, the intrinsic characteristics of the data are used to assess feature(s). Therefore, the filter method is computationally faster compared to the wrapper technique, and the selected features are determined by the data and its characteristics rather than an external learner.

Many extensive studies on feature selection have been done in data mining and machine learning during recent years. [23] did a comprehensive survey of feature selection algorithms and presented an integrated approach to intelligent feature selection. Another good review on various aspects of the attribute selection problem was done by [11]. They outlined the main techniques and approaches used in attribute selection, including feature construction, feature ranking, multivariate feature selection, efficient search methods, and feature validity assessment methods. [12] investigated six attribute selection techniques that produce ranked lists of attributes and applied them to 15 standard machine learning data sets from the UCI collection. The experimental results showed no single best approach for all situations. However, a wrapper-based approach was the best overall attribute selection schema in terms of accuracy if speed of execution was not a factor. Otherwise, CFS (correlation-based feature selection), CNS (consistency-based subset evaluation), and RLF (Relief) had the best overall performance.

Numerous variations of feature selection have been developed and employed in a range of fields [8, 14, 16, 27]. [16] introduced methods for feature selection based on support vector machines (SVM). [14] investigated the importance of attribute selection in judging the qualification of patients for cardiac pacemaker implantation. [27] applied feature subset selection with three filter-based models and two wrapper-based models to five software engineering data sets. In the context of text mining, [8] investigated multiple filter-based feature ranking techniques.

Our research group recently studied various feature selection techniques including filter-based and wrapper-based methods [9, 10, 18, 19, 31] and applied them to a variety of software data sets. The results demonstrate that the performances of the classification models were maintained or even improved when over 85 percent of the features were eliminated from the original data sets.

The class imbalance problem is observed in various domains [7, 17, 34]. To overcome the difficulties associated with learning from imbalanced data, various techniques have been developed. Data sampling is the primary approach for handling class imbalance, and it involves balancing the relative class distributions of the given data set. For a two-group problem such as predicting program modules as either *fp* or *nfp*, there are two types of data sampling approaches: majority undersampling and minority oversampling [3, 5, 28]. In this study we consider an effective data sampling approach, random undersampling, for addressing the class imbalance problem in defect prediction modeling. Due to limited space, we could not consider other data sampling techniques [3, 5, 28] in this study.

Imbalanced data can also be dealt with directly when creating the classifier. This is easier with some classifiers than with others. For example, Naïve Bayes produces a probability that an instance belongs to a given class, and one can then adjust the threshold at which an instance is put into a certain class and solve the problem of attempting to classify everything as being in the majority class to improve accuracy. A related approach is cost-sensitive classification [6]. However, the costs of different types of misclassification errors are not easy to obtain or estimate during the modeling process. Future work will consider these approaches in the context of this study.

While considerable work has been done for feature selection and data sampling separately, limited research can be found on investigating them both together, particularly in the software engineering field. [4] studied data row pruning (data sampling) and data column pruning (feature selection) in the context of software cost/effort estimation. However, the data sampling in their study does not focus on the class imbalance problem, and unlike this study, the classification models are meant for non-binary problems. Moreover, they focused on the case in which data sampling is used prior to feature selection. [24] introduced the concept of active feature selection, and investigated a selective sampling approach to active feature selection in a filter model setting. However, the purpose of data sampling in their work was for reducing the data set size instead of addressing class imbalance. In this paper, we use both feature selection and data sampling together as a data preprocessing step to form the training data.

## 8.3 Methodology

### *8.3.1 Filter-Based Feature Ranking Techniques*

The process of feature ranking involves scoring each feature according to a particular method, allowing the selection of the best set of features. The six standard filter-based feature ranking techniques used in this work include: chi-square (CS), information gain (IG), gain ratio (GR), two types of ReliefF (RF and RFW), and symmetrical uncertainty (SU). The chi-square – $\chi^2$ (CS) test [26] is used to examine the distribution of the class as it relates to the values of the feature in question. The null hypothesis is that there is no correlation; each value is as likely to have instances in any one class as any other class. Given the null hypothesis, the $\chi^2$ statistic measures how far away the actual value is from the expected value:

$$\chi^2 = \sum_{i=1}^{r} \sum_{j=1}^{n_c} \frac{(O_{i,j} - E_{i,j})^2}{E_{i,j}}$$

where $r$ is the number of different values of the feature, $n_c$ is the number of classes (in this work, $n_c = 2$), $O_{i,j}$ is the observed number of instances with value $i$ which are in class $j$, and $E_{i,j}$ is the expected number of instances with value $i$ and class $j$. The larger this $\chi^2$ statistic, the more likely it is that the distribution of values and classes are dependent; that is, the feature is relevant to the class.

Information gain, gain ratio, and symmetrical uncertainty are measures based on the concept of entropy from information theory [32]. Information gain (IG) is the information provided about the target class attribute Y, given the value of another attribute X. IG measures the decrease of the weighted average impurity of the partitions compared to the impurity of the complete set of data. A drawback of IG is that it tends to prefer attributes with a larger number of possible values, i.e., if one attribute has a larger number of values, it will appear to gain more information than those with fewer values, even if it is actually no more informative. One strategy to solve this problem is to use the gain ratio (GR), which penalizes multiple-valued attributes. Symmetrical uncertainty (SU) is another way to overcome the problem of IG's bias toward attributes with more values, doing so by dividing by the sum of the entropies of X and Y.

Relief is an instance-based feature ranking technique introduced by [21]. ReliefF is an extension of the Relief algorithm that can handle noise and multiclass data sets, and is implemented in the WEKA[1] tool [32]. When the `WeightByDistance` (weight nearest neighbors by their distance) parameter is set as default (false), the algorithm is referred to as RF; when the parameter is set to "true", the algorithm is referred to as RFW.

## 8.3.2  The Repetitive Feature Selection Approach

The proposed method is designed to deal with feature selection for imbalanced data. Algorithm 1 presents the procedure of this new approach. It consists of two basic steps:

---

[1]WEKA (Waikato Environment for Knowledge Analysis) is a popular suite of machine learning software written in Java, developed at the University of Waikato. WEKA is free software available under the GNU General Public License. In this study, all experiments and algorithms were implemented in the WEKA tool.

**input** :
1. Data set $D$ with features $F^j$, $j = 1, \ldots, m$;
2. Each instance $\mathbf{x} \in D$ is assigned to one of two classes $c(\mathbf{x}) \in \{fp, nfp\}$;
3. Filter-based feature ranking technique $\omega \in \{CS, GR, IG, RF, RFW, SU\}$ ;
4. Data sampling technique: RUS (random undersampling) ;
5. A predefined threshold: number (percentage) of the features to be selected.
**output**:
Ranking $\mathbb{R} = \{r^1, \ldots, r^m\}$ where $r^j$ represents the rank for attribute $F^j$.

**for** $i = 1$ *to* $k$ **do**
$\quad$| $\quad$Use RUS to balance $D$ and get the balanced data $D_i$ ;
$\quad$| $\quad$Employ $\omega$ to rank features $F^j$ on $D_i$, and get new rankings $\omega_i(F^j)$, $j = 1, \ldots, m$ ;
**end**
Create feature ranking $\mathbb{R}$ by combining the $k$ different rankings $\{\omega_i(F^j)|i = 1, \ldots, k\}$ $\forall j$
with mean (average).
Select features according to feature ranking $\mathbb{R}$ and a predefined threshold.

**Algorithm 8.1:** Repetitive Feature Selection Algorithm

1. Using the random undersampling (RUS) technique to balance data. RUS creates the balanced data by randomly removing examples from the majority class. After sampling, the ratio between the majority (*nfp*) examples and minority (*fp*) examples is 50–50 in this study.
2. Applying a filter-based feature ranking technique to the sampled (balanced) data and ranking all the features according to their predictive powers (scores).

In order to alleviate the biased results generated due to the sampling process, we repeat the two steps $k$ times ($k = 10$ in this study) and aggregate $k$ rankings using the mean (average). Finally, the best set of attributes is selected.

### 8.3.3 Classification Algorithms

The software defect prediction models are built using five different classification algorithms, including Naïve Bayes (NB) [32], Multilayer Perceptron (MLP) [13], Support Vector Machine (SVM) [29], Logistic Regression (LR) [32], and K-nearest-neighbors (KNN) [32]. These learners were selected for two key reasons: (1) they do not have a built-in feature selection capability, and (2) they are commonly used in both software engineering and data mining domains [15, 22, 25].

The WEKA data mining tool [32] is used to instantiate the different classifiers. Generally the default parameter settings for the different learners are used (for NB and LR), except for the below-mentioned changes. A preliminary investigation in the context of this study indicated that the modified parameter settings are appropriate.

• In the case of MLP, the `hiddenLayers` parameter was changed to "three" to define a network with one hidden layer containing three nodes, and the `validationSetSize` parameter was changed to "ten" to cause the classifier to leave 10% of the training data aside for use as a validation set to determine when to stop the iterative training process.

- For the KNN learner, the `distanceWeighting` parameter was set to "Weight by 1/distance", and the kNN parameter was set to "5".
- In the case of SVM, two changes were made: the `complexity constant c` was set to "5.0", and `build Logistic Models` was set to "true". A linear kernel was used by default.

### 8.3.4   Classification Performance Metric

In this study, we use the Area Under the ROC (receiver operating characteristic) curve (i.e., AUC) to evaluate classification models. The ROC curve graphs true positive rates versus the false positive rates (the positive class is synonymous with the minority class). Traditional performance metrics for classifier evaluation consider only the default decision threshold of 0.5. ROC curves illustrate the performance across all decision thresholds. A classifier that provides a large area under the curve is preferable over a classifier with a smaller area under the curve. A perfect classifier provides an AUC that equals 1. AUC is one of the most widely used single numeric measures that provides a general idea of the predictive potential of the classifier. It has also been shown that AUC is of lower variance and is more reliable than other performance metrics such as precision, recall, and F-measure [15].

## 8.4   A Case Study

### 8.4.1   Data Sets

The case study data is obtained from the publicly available PROMISE software project data repository [2]. We study the software measurement data sets for the Java-based Eclipse project[2] [35]. The software metrics and defect data are aggregated at the software packages level; hence, a program module is a Java package in Eclipse. We consider three releases of the Eclipse system, where the releases are denoted as 2.0, 2.1, and 3.0 [35].

Each system release contains the following information [35]: name of the package for which the metrics are collected (*name*); number of defects reported six months prior to release (*pre-release defects*); number of defects reported six

---

[2]Full data sets can be found at http://www.st.cs.uni-saarland.de/softevo/bug-data/eclipse/

months after release (*post-release defects*); a set of complexity metrics computed for classes or methods and aggregated by using average, maximum, and total at the package level (*complexity metrics*); and structure of abstract syntax tree(s) of the package consisting of the node size, type, and frequency (*structure of abstract syntax tree(s)*).

In our case study, we modify the original data by: (1) removing all non-numeric attributes, including the package names, and (2) converting the post-release defects attribute to a binary class attribute, where fault-prone (*fp*) is the minority class and not-fault-prone (*nfp*) is the majority class. A program module's membership in a given class is determined by a post-release defects threshold, *thd*. A program module (package) with *thd* or more post-release defects is labeled as *fp*, while those with fewer than *thd* defects are labeled as *nfp*.

For a given system release we consider three values for *thd*: for releases 2.0 and 3.0, *thd* = {10, 5, 3}, and for release 2.1, *thd* = {5, 4, 2}. This results in three groups of data sets, as shown in Table 8.1. A different set of thresholds is chosen for release 2.1 because we wanted to maintain relatively similar class distributions for the three data sets in a given group. Table 8.1 presents key details about the different data sets used in our study. All data sets contain 209 software attributes, which include 208 independent (predictor) attributes and one dependent attribute (*fp* or *nfp* label). The three groups of data sets exhibit different class distributions with respect to the *fp* and *nfp* modules, where Eclipse-I is relatively the most imbalanced and Eclipse-III is the least imbalanced.

## 8.4.2 Experiment Design

The main purpose of the experiment is to compare the performance of the classification models when using different feature selection strategies. As mentioned previously, we have three different feature selection approaches: (1) feature ranking technique used alone, (2) data sampling and feature ranking techniques used

**Table 8.1** Data characteristics

| Data set | Rel. | thd | #Attr. | #Inst. | #fp | %fp | #nfp | %nfp |
|---|---|---|---|---|---|---|---|---|
|  | 2.0 | 10 | 209 | 377 | 23 | 6.1 | 354 | 93.9 |
| Eclipse-I | 2.1 | 5 | 209 | 434 | 34 | 7.8 | 400 | 92.2 |
|  | 3.0 | 10 | 209 | 661 | 41 | 6.2 | 620 | 93.8 |
|  | 2.0 | 5 | 209 | 377 | 52 | 13.8 | 325 | 86.2 |
| Eclipse-II | 2.1 | 4 | 209 | 434 | 50 | 11.5 | 384 | 88.5 |
|  | 3.0 | 5 | 209 | 661 | 98 | 14.8 | 563 | 85.2 |
|  | 2.0 | 3 | 209 | 377 | 101 | 26.8 | 276 | 73.2 |
| Eclipse-III | 2.1 | 2 | 209 | 434 | 125 | 28.8 | 309 | 71.2 |
|  | 3.0 | 3 | 209 | 661 | 157 | 23.8 | 504 | 76.2 |

together only once, and (3) the repetitive process of data sampling followed by feature ranking. Each approach contains two scenarios: (1) the training data set extracted from the original data, and (2) the training data set extracted from sampled data. Therefore, there are six strategies considered in the experiment.

- **Strategy 1**: Feature ranking technique used alone and the training data sets are formed from the original data with the selected attributes (denoted **NS-o**).
- **Strategy 2**: Data sampling followed by feature ranking and the training data sets are formed from the original data with the selected attributes (denoted **nonRep-o**).
- **Strategy 3**: A repetitive process of data sampling followed by feature ranking and the training data sets are formed from the original data with the selected attributes (denoted **Rep-o**).
- **Strategy 4**: Feature ranking technique used alone and the training data sets are formed from a sampled data with the selected attributes (denoted **NS-s**).
- **Strategy 5**: Data sampling followed by feature ranking and the training data sets are formed from a sampled data with the selected attributes (denoted **nonRep-s**).
- **Strategy 6**: A repetitive process of data sampling followed by feature ranking and the training data sets are formed from a sampled data with the selected attributes (denoted **Rep-s**).

The six strategies are illustrated in Fig. 8.1, where S1 through S6 represents the six strategies described above respectively.

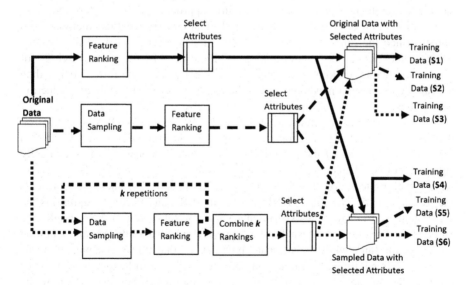

**Fig. 8.1** Six feature selection strategies

### 8.4.3  Results and Analysis

During the experiment, one parameter needs to be determined in advance, that is, how many features will be selected for modeling? In this paper, we choose $\lceil \log_2 n \rceil$ attributes that have the highest scores, where $n$ is the number of the independent attributes in the original data set. For the three groups of Eclipse data $n = 208$, so $\lceil \log_2 n \rceil = 8$. We select $\lceil \log_2 n \rceil$ attributes, because (1) related literature does not provide guidance on the appropriate number of features to select; and (2) one of our recent empirical studies [20] recommended using $\lceil \log_2 n \rceil$ features when employing WEKA to build random forests learners for binary classification in general and imbalanced data sets in particular. Although we use different learners here, a preliminary study showed that $\lceil \log_2 n \rceil$ is still appropriate for various learners.

After feature selection, we applied five different classifiers to the training data sets with the selected attributes, and we used AUC to evaluate the performance of the classifications. All the results are reported in Table 8.2, which contains three subtables, each representing the results for each group of the data sets. In the experiments, ten runs of five fold cross-validation were performed. The values presented in the tables represent the average AUC for every classification model constructed over the ten runs of five fold cross-validation. We also present (1) the average performance (last column of the tables) of each feature selection approach (NS, nonRep, and Rep) for each learner over the six feature ranking techniques and across two scenarios (original and sampled), and (2) the average performance (last row of the tables) of each feature ranking technique for each scenario (original and sampled) over the five learners and across three approaches (NS, nonRep, and Rep). The results demonstrate that (1) for each of the learners, the repetitive method (Rep) outperformed the other two approaches (NS and nonRep) based on the average performances across six feature ranking techniques and two scenarios; and (2) for each of the feature ranking techniques, the original scenario performed better than the sampled scenario on average across the five learners and three feature selection approaches.

We also performed a three-way ANalysis Of VAriance (ANOVA) F test [1] on the classification performance for each group of the data sets (Eclipse-I, -II, and -III) separately to examine if the performance difference (better/worse) is statistically significant or not. Three factors are designed as follows: Factor A represents the six feature ranking techniques (CS, IG, GR, RF, RFW, and SU), Factor B represents the five classifiers (NB, MLP, KNN, SVM, and LR), and Factor C represents the six strategies designed in the experiment (NS-o, nonRep-o, Rep-o, NS-s, nonRep-s, and Rep-s). The null hypothesis for the ANOVA test is that all the group population means are the same while the alternate hypothesis is that at least one pair of means is different. In addition, the interactions between each pair of factors, as well as for all three factors, are also examined in the test. Tables 8.3, 8.4, and 8.5 show the ANOVA results for Eclipse-I, -II, and -III, respectively. The $p$-value is less than the typical cutoff, 0.05, for each main factor and most interactions of each group of data sets except for interaction A×B×C on Eclipse-II and Eclipse-III, where $p$-value is

**Table 8.2** Classification performance in terms of AUC

| Learner | Feature Selection | Ranking technique | | | | | | | | | | | | Avg. |
|---|---|---|---|---|---|---|---|---|---|---|---|---|---|---|
| | | CS | | GR | | IG | | RF | | RFW | | SU | | |
| | | Original | Sampled | Original | Sampled | Original | Sampled | Original | Sampled | Original | Sampled | Original | Sampled | |
| (a) Eclipse-I | NS | 0.8355 | 0.8426 | 0.8152 | 0.8211 | 0.8527 | 0.8607 | 0.8174 | 0.8205 | 0.8132 | 0.8104 | 0.8323 | 0.8339 | 0.8296 |
| NB | nonRep | 0.8471 | 0.8529 | 0.8460 | 0.8520 | 0.8476 | 0.8512 | 0.8488 | 0.8409 | 0.8126 | 0.8137 | 0.8458 | 0.8508 | 0.8425 |
| | Rep | 0.8542 | 0.8548 | 0.8553 | 0.8572 | 0.8589 | 0.8554 | 0.8858 | 0.8791 | 0.8603 | 0.8561 | 0.8610 | 0.8592 | 0.8614 |
| | NS | 0.8539 | 0.8486 | 0.8094 | 0.8118 | 0.8750 | 0.8624 | 0.8228 | 0.8127 | 0.8356 | 0.7993 | 0.8307 | 0.8365 | 0.8332 |
| | nonRep | 0.8609 | 0.8559 | 0.8551 | 0.8570 | 0.8594 | 0.8491 | 0.8352 | 0.8254 | 0.8330 | 0.8013 | 0.8562 | 0.8516 | 0.8450 |
| MLP | Rep | 0.8700 | 0.8637 | 0.8782 | 0.8701 | 0.8717 | 0.8675 | 0.8432 | 0.8532 | 0.8604 | 0.8420 | 0.8799 | 0.8697 | 0.8641 |
| | NS | 0.8829 | 0.8703 | 0.8337 | 0.8200 | 0.9003 | 0.8893 | 0.8275 | 0.7807 | 0.7752 | 0.7630 | 0.8689 | 0.8554 | 0.8389 |
| | nonRep | 0.8841 | 0.8649 | 0.8751 | 0.8618 | 0.8799 | 0.8632 | 0.8771 | 0.8411 | 0.8199 | 0.7987 | 0.8806 | 0.8624 | 0.8591 |
| KNN | Rep | 0.8919 | 0.8773 | 0.8940 | 0.8822 | 0.8966 | 0.8811 | 0.8931 | 0.8736 | 0.8797 | 0.8559 | 0.8982 | 0.8867 | 0.8842 |
| | NS | 0.8911 | 0.8880 | 0.8604 | 0.8601 | 0.9029 | 0.9014 | 0.8519 | 0.8467 | 0.8523 | 0.8240 | 0.8820 | 0.8830 | 0.8703 |
| | nonRep | 0.8958 | 0.8860 | 0.8946 | 0.8830 | 0.8965 | 0.8854 | 0.8768 | 0.8527 | 0.8569 | 0.8238 | 0.8943 | 0.8873 | 0.8778 |
| SVM | Rep | 0.9038 | 0.8963 | 0.9077 | 0.9002 | 0.9076 | 0.8997 | 0.8957 | 0.8805 | 0.8923 | 0.8719 | 0.9098 | 0.9031 | 0.8974 |
| | NS | 0.8628 | 0.8307 | 0.8458 | 0.7834 | 0.8716 | 0.8381 | 0.8631 | 0.8379 | 0.8538 | 0.8242 | 0.8573 | 0.8026 | 0.8393 |
| | nonRep | 0.8629 | 0.8032 | 0.8652 | 0.8165 | 0.8589 | 0.8034 | 0.8861 | 0.8350 | 0.8582 | 0.8083 | 0.8700 | 0.8138 | 0.8401 |
| LR | Rep | 0.8630 | 0.8247 | 0.8714 | 0.8183 | 0.8643 | 0.8268 | 0.8938 | 0.8386 | 0.8904 | 0.8388 | 0.8717 | 0.8227 | 0.8520 |
| | Avg. | 0.8707 | 0.8573 | 0.8605 | 0.8463 | 0.8762 | 0.8623 | 0.8612 | 0.8412 | 0.8463 | 0.8221 | 0.8692 | 0.8546 | 0.8557 |

(continued)

**Table 8.2** (Continued)

| Learner | Feature Selection | Ranking technique | | | | | | | | | | | | Avg. |
|---|---|---|---|---|---|---|---|---|---|---|---|---|---|---|
| | | CS | | GR | | IG | | RF | | RFW | | SU | | |
| | | Original | Sampled | Original | Sampled | Original | Sampled | Original | Sampled | Original | Sampled | Original | Sampled | |
| Eclipse II | | | | | | | | | | | | | | |
| NB | NS | 0.8504 | 0.8511 | 0.8260 | 0.8246 | 0.8565 | 0.8548 | 0.8262 | 0.8269 | 0.8214 | 0.8142 | 0.8484 | 0.8479 | 0.8374 |
| | nonRep | 0.8486 | 0.8534 | 0.8367 | 0.8432 | 0.8499 | 0.8527 | 0.8518 | 0.8483 | 0.8351 | 0.8341 | 0.8484 | 0.8512 | 0.8461 |
| | Rep | 0.8533 | 0.8524 | 0.8478 | 0.8490 | 0.8526 | 0.8517 | 0.8737 | 0.8736 | 0.8699 | 0.8663 | 0.8534 | 0.8522 | 0.8580 |
| MLP | NS | 0.8823 | 0.8893 | 0.8576 | 0.8538 | 0.8881 | 0.8926 | 0.8456 | 0.8344 | 0.8276 | 0.8219 | 0.8793 | 0.8837 | 0.8630 |
| | nonRep | 0.8865 | 0.8872 | 0.8760 | 0.8731 | 0.8857 | 0.8892 | 0.8706 | 0.8556 | 0.8632 | 0.8406 | 0.8889 | 0.8858 | 0.8752 |
| | Rep | 0.8931 | 0.8961 | 0.8873 | 0.8909 | 0.8907 | 0.8939 | 0.8741 | 0.8774 | 0.8764 | 0.8772 | 0.8923 | 0.8958 | 0.8871 |
| KNN | NS | 0.8912 | 0.8867 | 0.8536 | 0.8527 | 0.8927 | 0.8906 | 0.7854 | 0.7868 | 0.7684 | 0.7601 | 0.8867 | 0.8833 | 0.8448 |
| | nonRep | 0.8861 | 0.8864 | 0.8747 | 0.8719 | 0.8852 | 0.8876 | 0.8569 | 0.8412 | 0.8393 | 0.8235 | 0.8848 | 0.8865 | 0.8687 |
| | Rep | 0.8906 | 0.8908 | 0.8826 | 0.8847 | 0.8890 | 0.8907 | 0.8709 | 0.8684 | 0.8734 | 0.8652 | 0.8886 | 0.8903 | 0.8821 |
| SVM | NS | 0.9179 | 0.9121 | 0.8913 | 0.8829 | 0.9197 | 0.9143 | 0.8730 | 0.8582 | 0.8364 | 0.8369 | 0.9145 | 0.9061 | 0.8886 |
| | nonRep | 0.9185 | 0.9104 | 0.9077 | 0.8966 | 0.9185 | 0.9104 | 0.9024 | 0.8724 | 0.8818 | 0.8566 | 0.9197 | 0.9092 | 0.9003 |
| | Rep | 0.9206 | 0.9163 | 0.9172 | 0.9118 | 0.9210 | 0.9167 | 0.9073 | 0.8962 | 0.9043 | 0.8910 | 0.9201 | 0.9153 | 0.9115 |
| LR | NS | 0.8999 | 0.8836 | 0.8872 | 0.8710 | 0.9024 | 0.8882 | 0.8711 | 0.8678 | 0.8529 | 0.8486 | 0.8987 | 0.8859 | 0.8798 |
| | nonRep | 0.9067 | 0.8897 | 0.9023 | 0.8841 | 0.9104 | 0.8891 | 0.9012 | 0.8798 | 0.8860 | 0.8681 | 0.9097 | 0.8864 | 0.8928 |
| | Rep | 0.9119 | 0.8890 | 0.9083 | 0.8878 | 0.9084 | 0.8890 | 0.9054 | 0.8984 | 0.9023 | 0.8959 | 0.9101 | 0.8936 | 0.9000 |
| | Avg | 0.8905 | 0.8863 | 0.8771 | 0.8719 | 0.8914 | 0.8874 | 0.8677 | 0.8590 | 0.8559 | 0.8467 | 0.8896 | 0.8849 | 0.8757 |

Eclipse III

| | | | | | | | | | | | | | | |
|---|---|---|---|---|---|---|---|---|---|---|---|---|---|---|
| NB | NS | 0.7877 | 0.7891 | 0.7682 | 0.7688 | 0.7889 | 0.7908 | 0.7879 | 0.7855 | 0.8032 | 0.8005 | 0.7854 | 0.7863 | 0.7869 |
| | nonRep | 0.7830 | 0.7848 | 0.7668 | 0.7679 | 0.7842 | 0.7860 | 0.8032 | 0.8020 | 0.8044 | 0.8061 | 0.7801 | 0.7825 | 0.7876 |
| | Rep | 0.7857 | 0.7867 | 0.7753 | 0.7795 | 0.7872 | 0.7882 | 0.8079 | 0.8102 | 0.8065 | 0.8080 | 0.7844 | 0.7843 | 0.7920 |
| MLP | NS | 0.8561 | 0.8530 | 0.8257 | 0.8244 | 0.8568 | 0.8538 | 0.8141 | 0.8036 | 0.8301 | 0.8224 | 0.8528 | 0.8513 | 0.8370 |
| | nonRep | 0.8568 | 0.8520 | 0.8326 | 0.8255 | 0.8584 | 0.8524 | 0.8467 | 0.8307 | 0.8449 | 0.8330 | 0.8537 | 0.8496 | 0.8447 |
| | Rep | 0.8539 | 0.8534 | 0.8362 | 0.8228 | 0.8568 | 0.8545 | 0.8488 | 0.8418 | 0.8511 | 0.8398 | 0.8523 | 0.8514 | 0.8469 |
| KNN | NS | 0.8618 | 0.8560 | 0.8419 | 0.8398 | 0.8600 | 0.8554 | 0.7530 | 0.7451 | 0.7605 | 0.7506 | 0.8562 | 0.8545 | 0.8196 |
| | nonRep | 0.8531 | 0.8513 | 0.8122 | 0.7980 | 0.8547 | 0.8510 | 0.7896 | 0.7742 | 0.7882 | 0.7756 | 0.8517 | 0.8478 | 0.8206 |
| | Rep | 0.8562 | 0.8535 | 0.8126 | 0.8009 | 0.8558 | 0.8536 | 0.8058 | 0.7987 | 0.8070 | 0.8022 | 0.8579 | 0.8527 | 0.8297 |
| SVM | NS | 0.8852 | 0.8771 | 0.8669 | 0.8620 | 0.8861 | 0.8782 | 0.8424 | 0.8239 | 0.8559 | 0.8368 | 0.8851 | 0.8776 | 0.8648 |
| | nonRep | 0.8847 | 0.8741 | 0.8650 | 0.8561 | 0.8849 | 0.8742 | 0.8706 | 0.8509 | 0.8728 | 0.8546 | 0.8821 | 0.8745 | 0.8704 |
| | Rep | 0.8852 | 0.8777 | 0.8714 | 0.8657 | 0.8862 | 0.8790 | 0.8735 | 0.8612 | 0.8751 | 0.8626 | 0.8858 | 0.8784 | 0.8752 |
| LR | NS | 0.8871 | 0.8822 | 0.8645 | 0.8614 | 0.8865 | 0.8823 | 0.8406 | 0.8387 | 0.8536 | 0.8524 | 0.8831 | 0.8784 | 0.8676 |
| | nonRep | 0.8830 | 0.8779 | 0.8632 | 0.8575 | 0.8826 | 0.8774 | 0.8713 | 0.8687 | 0.8737 | 0.8732 | 0.8825 | 0.8770 | 0.8740 |
| | Rep | 0.8859 | 0.8807 | 0.8703 | 0.8689 | 0.8866 | 0.8817 | 0.8762 | 0.8772 | 0.8758 | 0.8776 | 0.8849 | 0.8807 | 0.8789 |
| | Avg | 0.8537 | 0.8500 | 0.8315 | 0.8266 | 0.8544 | 0.8506 | 0.8288 | 0.8208 | 0.8335 | 0.8264 | 0.8519 | 0.8485 | 0.8397 |

**Table 8.3** Three-way ANOVA for Eclipse-I

| Source | Sum Sq. | d.f. | Mean Sq. | F | $p$-value |
|---|---|---|---|---|---|
| A | 0.7026 | 5 | 0.1405 | 80.54 | 0 |
| B | 1.1252 | 4 | 0.2813 | 161.22 | 0 |
| C | 1.1915 | 5 | 0.2383 | 136.57 | 0 |
| A×B | 0.5096 | 20 | 0.0255 | 14.60 | 0 |
| A×C | 0.4970 | 25 | 0.0199 | 11.39 | 0 |
| B×C | 0.4695 | 20 | 0.0235 | 13.45 | 0 |
| A×B×C | 0.2261 | 100 | 0.0023 | 1.30 | 0.026 |
| Error | 9.1082 | 5220 | 0.0017 | | |
| Total | 13.8297 | 5399 | | | |

**Table 8.4** Three-way ANOVA for Eclipse-II

| Source | Sum sq. | d.f. | Mean sq. | F | $p$-value |
|---|---|---|---|---|---|
| A | 1.1086 | 5 | 0.2217 | 179.24 | 0 |
| B | 1.8920 | 4 | 0.4730 | 382.37 | 0 |
| C | 0.6201 | 5 | 0.1240 | 100.25 | 0 |
| A×B | 0.4035 | 20 | 0.0202 | 16.31 | 0 |
| A×C | 0.5166 | 25 | 0.0207 | 16.71 | 0 |
| B×C | 0.0954 | 20 | 0.0048 | 3.85 | 0 |
| A×B×C | 0.1341 | 100 | 0.0013 | 1.08 | 0.268 |
| Error | 6.4572 | 5220 | 0.0012 | | |
| Total | 11.2274 | 5399 | | | |

**Table 8.5** Three-way ANOVA for Eclipse-III

| Source | Sum sq. | d.f. | Mean sq. | F | $p$-value |
|---|---|---|---|---|---|
| A | 0.7653 | 5 | 0.1531 | 187.80 | 0 |
| B | 5.3294 | 4 | 1.3324 | 1634.83 | 0 |
| C | 0.1166 | 5 | 0.0233 | 28.61 | 0 |
| A×B | 0.9619 | 20 | 0.0481 | 59.01 | 0 |
| A×C | 0.2493 | 25 | 0.0100 | 12.24 | 0 |
| B×C | 0.0323 | 20 | 0.0016 | 1.98 | 0.006 |
| A×B×C | 0.0980 | 100 | 0.0010 | 1.20 | 0.085 |
| Error | 4.2542 | 5220 | 0.0008 | | |
| Total | 11.8069 | 5399 | | | |

greater than 0.05 for both groups of data. This means that for each main factor, the alternate hypothesis is accepted, namely, at least two group means are significantly different from each other. In addition, the pairwise interactions significantly affect classification performance in each group of data sets. In other words, changing value of one factor will significantly influence the value of the other factor, and vice versa.

We further carried out a multiple comparison test [1] on each main factor and the interactions (A×C and B×C) with Tukey's honestly significant difference criterion. Since this study is more interested in comparing the performances of the six different feature selection strategies, we only present the pairwise interactions that involve the six strategies (Factor C). Note that for all the ANOVA and multiple comparison tests, the significance level $\alpha$ was set to 0.05.

Figures 8.2–8.4 show the multiple comparisons on the three groups of data sets, each with five subfigures representing Factor A, B, and C, and interaction A×C and B×C, respectively. The figures display graphs with each group mean represented by a symbol (◦) and 95% confidence interval as a line around the symbol. Two means are significantly different if their intervals are disjoint, and are not significantly different if their intervals overlap. Matlab was used to construct the ANOVA models and perform the multiple comparisons presented in this work, and the assumptions for constructing ANOVA models were validated. From these figures we can see the following points:

- Among the six standard filter-based feature selection methods, IG performed best, followed by CS and SU; the other three techniques, GR, RF, and RFW, performed significantly worse than the first three. This is true for all three groups of data sets. However, the significance levels are a bit different with respect to the different class distribution of each group of data sets. For example, the distinction between the pair of IG and CS can be clearly seen in Fig. 8.2a but not in Fig. 8.3a and Fig. 8.4a.
- For the five classifiers, the performance of LR was quite different from set to set. For example, LR performed worst on Eclipse-I (see Fig. 8.2b) but performed best on Eclipse-III (see Fig. 8.4b). The other four classifiers showed relatively stable performance across different data sets; SVM always performed very well, followed by MLP and KNN; NB performed worst.
- For the six feature selection strategies, which are made up of the three feature selection approaches (NS, nonRep, and Rep) each being combined with two scenarios (original and sampled), we found that when the scenario is fixed, the three approaches ranked in terms of their performance from best to worst are Rep, nonRep and NS; when the feature selection approach is fixed, the original scenario performed better than the sampled case. These patterns are reflected by all three groups of data sets. However, a slight difference exists between groups of data. For instance, nonRep-s significantly outperformed NS-s for Eclipse-I (see Fig. 8.2c) but insignificantly outperformed NS-s for Eclipse-III see Fig. 8.4c.
- There are 36 groups for interaction A×C; these are formed by each of six feature ranking techniques being combined with six feature selection strategies. The group means demonstrate that in addition to the two main factors, the classification performance is greatly influenced by their interactions. For example, for Eclipse-1, IG significantly outperformed GR for NS-o, but this pattern was not observed when nonRep-o and Rep-o strategies were adopted (see Fig. 8.2d). Also, the figures display different performance distributions of the six filter-based rankers with respect to the different feature selection strategies. When data sampling was not involved in feature selection (NS-o/NS-s), the six rankers exhibit quite different performances. On the other hand, the six rankers demonstrate relatively consistent performances when data sampling

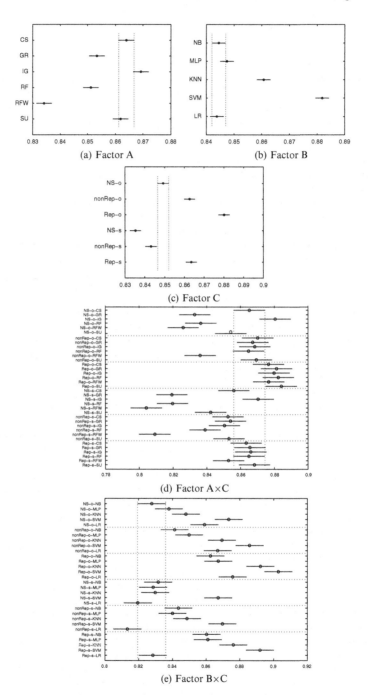

**Fig. 8.2** Eclipse-I: Multiple comparison. (**a**) Factor A (**b**) Factor B (**c**) Factor A (**d**) Factor A×C (**e**) Factor B×C

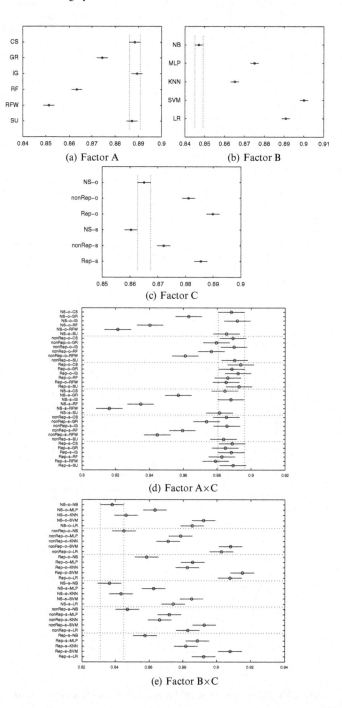

**Fig. 8.3** Eclipse-II: Multiple comparison. (**a**) Factor A (**b**) Factor B (**c**) Factor C (**d**) Factor A×C (**e**) Factor B×C

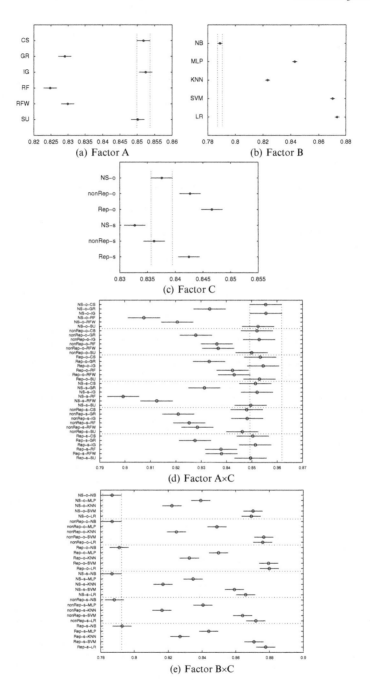

**Fig. 8.4** Eclipse-III: Multiple comparison. (**a**) Factor A (**b**) Factor B (**c**) Factor C (**d**) Factor A×C (**e**) Factor B×C

was performed with feature selection (nonRep-o/nonRep-s and Rep-o/Rep-s), especially when the repetitive method (Rep-o/Rep-s) was adopted.
- For interaction B×C, there are 30 groups that are formed by each of five learners being combined with six feature selection strategies. The group means show that the two main factors played an important role in the classification performance and their interactions did too. For instance, SVM performed slightly better than LR when the NS-o strategy was employed but worse than LR when the NS-s strategy was used (see Fig. 8.4e).
- One point that is clearly observed in the multiple comparisons is that the performance advantage of the repetitive feature selection approach over other methods becomes increasingly dramatic as the class imbalance becomes more serious (i.e., the percentage of *fp* examples becomes increasing low). This may imply that our proposed method is especially appropriate when classification occurs for highly imbalanced data (e.g., the percentage of *fp* examples is less than 10%).

### 8.4.4   Defect Prediction with All Metrics

In addition to the primary experiment discussed above, we also compared defect prediction models built on smaller subsets of attributes (using feature selection) to those built with a complete set of attributes (no feature selection). Table 8.6 shows the classification performances of the five learners using the original data with 208 software metrics. Each value presented in the table is the average AUC value over the ten runs of five fold cross-validation outcomes and across the three data sets of a given group. The results demonstrate that for the MLP, KNN, SVM and LR learners, all feature selection strategies performed on average significantly better than the case where no feature selection was used. For the NB learner, the Rep strategy outperformed the case where all attributes were used, while the nonRep and NS strategies resulted in similar or slightly worse performance.

The above result is of particular importance to a software quality analyst, since a useful model can be built using only a select few software metrics. This provides a less-cumbersome and clearer insight into the software quality trends of the target project, as compared to analyzing the model with respect to a large set of metrics. Based on empirical evidence, we recommend against the common practice of using the complete available set of software metrics for defect prediction modeling.

**Table 8.6**  Classification performance on full data sets

|             | NB     | MLP    | KNN    | SVM    | LR     |
|-------------|--------|--------|--------|--------|--------|
| Eclipse-I   | 0.8495 | 0.7829 | 0.7235 | 0.8028 | 0.6750 |
| Eclipse-II  | 0.8293 | 0.8082 | 0.7572 | 0.8614 | 0.7210 |
| Eclipse-III | 0.7871 | 0.7852 | 0.7565 | 0.8354 | 0.6897 |

### 8.4.5   Threats to Validity

A typical software development project is very human intensive, which can affect many aspects of the development process, including software quality and defect occurrence. Consequently, software engineering research that utilizes controlled experiments for evaluating the usefulness of empirical models is not practical. The case study presented in this paper is an empirical software engineering effort, for which the software engineering community demands that its subject have the following characteristics [33]: (1) developed by a group, and not by an individual; (2) be a large, industry-sized project, and not a toy problem; (3) developed by professionals, and not by students; and (4) developed in an industry/government organization setting, and not in a laboratory.

The software systems that are used in our case study were developed by professionals in large software development organizations using established software development processes and management practices. The software was developed to address real-world problems. We note that our case studies fulfill all of the above criteria specified by the software engineering community.

Threats to external validity are conditions that limit generalization of case study results. The analysis and conclusion presented in this article are based upon the metrics and defect data obtained from nine software data sets. The same analysis for another software system or learner may provide different results which is a potential threat in all empirical software engineering research. However, we place our emphasis on the process of comparing the different feature selection techniques considered in this study. Our comparative analysis can easily be applied to another software system. Moreover, since all our final conclusions are based on ten runs of five fold cross-validation and statistical tests for significance, our findings are grounded in using sound methods.

Threats to internal validity are unaccounted for influences on the experiments that may affect case study results. Poor fault proneness estimates can be caused by a wide variety of factors, including measurement errors while collecting and recording software metrics, modeling errors due to the unskilled use of software applications, errors in model-selection during the modeling process, and the presence of outliers and noise in the training data set. Measurement errors are inherent to the data collection effort. In our comparative study, a common model-building and model-evaluation approach is used for all feature selection techniques and classifiers considered. Moreover, the experiments and statistical analysis was performed by only one skilled person in order to keep modeling errors to a minimum.

## 8.5   Conclusion

Feature selection for highly imbalanced data is a major challenge facing data mining. In this paper, we proposed a repetitive feature selection method to deal with the problem. The outline of this novel approach is that we sample the data $k$ times and each time we apply a given feature selection technique to the sampled data and find a ranking of the features; we combine or aggregate the $k$ different rankings by using mean (average). In order to evaluate the effectiveness of the proposed method, we applied this feature selection technique to three groups of software data sets, each group having three separate releases. We used five different classifiers to build a number of classification models with the selected attributes. We also compared the new method to two other approaches – (1) feature ranking used alone and (2) data sampling and feature ranking used together but only once. Moreover, two scenarios were examined for each of the three approaches. One scenario is the training data set consisting of original data with the selected attributes and the other scenario is the training data set consisting of sampled data with the selected attributes. The experimental results demonstrate that (1) data sampling improves feature selection when the data sets have unequal example size of the two classes; (2) our proposed repetitive method performed significantly better than the other two approaches on average; (3) the performance strength of our proposed method over other methods becomes increasingly evident as the class imbalance becomes more serious; and (4) for a given feature selection approach, the original scenario usually resulted in better performance than the sampled scenario; The conclusions obtained in this paper regarding the effectiveness of the proposed feature selection approach are based on the experiments conducted on the data sets from this specific software system. Future work will involve more experiments on the data sets from different domains. Moreover, different data sampling techniques and feature ranking methods can also be considered in the repetitive feature selection process. Also, although in this paper the $k$ parameter used in the proposed method is fixed at 10, future work could consider a policy which would not use a single value for the $k$ parameter but rather would stop the algorithm when some stability in the results is reached.

# References

1. Berenson, M.L., Goldstein, M., Levine, D.: Intermediate Statistical Methods and Applications: A Computer Package Approach Prentice-Hall, Englewood Cliffs, NJ, 2 edition (1983)
2. Boetticher, G., Menzies, T., Ostrand, T.: Promise repository of empirical software engineering data (2007)
3. Chawla, N.V., Bowyer, K.W., Hall, L.O., Kegelmeyer, P.W.: SMOTE: Synthetic minority over-sampling technique. J. Artif. Intell. Res. **16**, 321–357 (2002)
4. Chen, Z., Menzies, T., Port, D., Boehm, B.: Finding the right data for software cost modeling. IEEE Software. **22**(6), 38–46 (2005)
5. Cieslak, D.A., Chawla, N.V., Striegel, A.: Combating imbalance in network intrusion datasets. In: Proceedings of 2006 IEEE International Conference on Granular Computing, pp. 732–737, Athens, Georgia (2006)
6. Elkan, C.: The foundations of cost-sensitive learning. In: Proceedings of the Seventeenth International Conference on Machine Learning, pp. 239–246 (2001)
7. Engen, V., Vincent, J., Phalp, K.: Enhancing network based intrusion detection for imbalanced data. Int. J. Knowl. Base. Intell. Eng. Syst. **12**(5-6), 357–367 (2008)

8. Forman, G.: An extensive empirical study of feature selection metrics for text classification. J. Mach. Learn. Res. **3**, 1289–1305 (2003)

9. Gao, K., Khoshgoftaar, T.M., Van Hulse, J.: An evaluation of sampling on filter-based feature selection methods. In: Proceedings of the 23rd International Florida Artificial Intelligence Research Society Conference, p. 416–421, Daytona Beach, FL, USA (2010)

10. Gao, K., Koshogoftaar, T.M., Napolitano, A.: Exploring software quality classification with a wrapper-based feature ranking technique. In: Proceedings of 21st IEEE International Conference on Tools with Artificial Intelligence, pp. 67–74, Newark, NJ (2009)

11. Guyon, I., Elisseeff, A.: An introduction to variable and feature selection. J. Mach. Learn. Res. **3**, 1157–1182 (2003)

12. Hall, M.A., Holmes, G.: Benchmarking attribute selection techniques for discrete class data mining. IEEE Trans. Knowl. Data Eng. **15**(6), 1437–1447 (2003)

13. Haykin, S.: Neural Networks: A Comprehensive Foundation Prentice-Hall, (2 edn.) NJ, USA (1998)

14. Ilczuk, G., Mlynarski, R., Kargul, W., Wakulicz-Deja, A.: New feature selection methods for qualification of the patients for cardiac pacemaker implantation. Comput. Cardiol. **34**(2-3), 423–426 (2007)

15. Jiang, Y., Lin, J., Cukic, B., Menzies, T.: Variance analysis in software fault prediction models. In: Proceedings of the 20th IEEE International Symposium on Software Reliability Engineering, pp. 99–108, Bangalore-Mysore, India (2009)

16. Jong, K., Marchiori, E., Sebag, M., van der Vaart, A.: Feature selection in proteomic pattern data with support vector machines. In: Proceedings of the 2004 IEEE Symposium on Computational Intelligence in Bioinformatics and Computational Biology (2004)

17. Kamal, A.H., Zhu, X., Pandya, A.S., Hsu, S., Shoaib, M.: The impact of gene selection on imbalanced microarray expression data. In: Proceedings of the 1st International Conference on Bioinformatics and Computational Biology; Lecture Notes in Bioinformatics; Vol. 5462, pp. 259–269, New Orleans, LA (2009)

18. Khoshgoftaar, T.M., Bullard, L.A., Geo, K.: Attribute selection using rough sets in software quality classification. Int. J. Reliab. Qual. Saf. Eng. **16**(1), 73–89 (2009)

19. Khoshgoftaar, T.M., Gao, K.: A novel software metric selection technique using the area under roc curves. In: Proceedings of the 22nd International Conference on Software Engineering and Knowledge Engineering, pp. 203–208, San Francisco, CA (2010)

20. Khoshgoftaar, T.M., Golawala, M., Van Hulse, J.: An empirical study of learning from imbalanced data using random forest. In: Proceedings of the 19th IEEE International Conference on Tools with Artificial Intelligence, Vol. 2, pp. 310–317, Washington, DC, USA (2007)

21. Kira, K., Rendell, L.A.: A practical approach to feature selection. In: Proceedings of 9th International Workshop on Machine Learning, pp. 249–256 (1992)

22. Lessmann, S., Baesens, B., Mues, C., Pietsch, S.: Benchmarking classification models for software defect prediction: A proposed framework and novel findings. IEEE Trans. Software Eng. **34**(4), 485–496 (2008)

23. Liu, H., Yu, L.: Toward integrating feature selection algorithms for classification and clustering. IEEE Trans. Knowl. Data Eng. **17**(4), 491–502 (2005)

24. Liu, H., Motoda, H., Yu, L.: A selective sampling approach to active feature selection. Artif. Intell. **159**(1-2), 49–74 (2004)

25. Menzies, T., Greenwald, J., Frank, A.: Data mining static code attributes to learn defect predictors. IEEE Trans. Software Eng. **33**(1), 2–13 (2007)

26. Plackett, R.L.: Karl pearson and the chi-squared test. Int. Stat. Rev. **51**(1), 59–72 (1983)

27. Rodriguez, D., Ruiz, R., Cuadrado-Gallego, J., Aguilar-Ruiz, J.: Detecting fault modules applying feature selection to classifiers. In: Proceedings of 8th IEEE International Conference on Information Reuse and Integration, pp. 667–672, Las Vegas, Nevada (2007)

28. Seiffert, C., Khoshgoftaar, T.M., Van Hulse, J.: Improving software-quality predictions with data sampling and boosting. Part A: Systems and Humans, IEEE Trans. Syst. Man Cybern. **39**(6), 1283–1294 (2009)

29. Shawe-Taylor, J., Cristianini, N.: (2 edn.) Support Vector Machines, Cambridge University Press, (2000)
30. Van Hulse, J., Khoshgoftaar, T.M., Napolitano, A.: Experimental perspectives on learning from imbalanced data. In: Proceedings of the 24th International Conference on Machine Learning, pp. 935–942, Corvallis, OR, USA (2007)
31. Wang, H., Khoshgoftaar, T.M., Gao, K., Seliya, N.: Mining data from multiple software development projects. In: Proceedings of the 3rd IEEE International Workshop Mining Multiple Information Sources, pp. 551–557, Miami, FL (2009)
32. Witten, I.H., Frank, E.: Data Mining: Practical Machine Learning Tools and Techniques Morgan Kaufmann, (2 edn.) (2005)
33. Wohlin, C., Runeson, P., Host, M., Ohlsson, M.C., Regnell, B., Wesslen, A.: Experimentation in Software Engineering: An Introduction Kluwer International Series in Software Engineering. Kluwer Academic Publishers, Boston, MA (2000)
34. Zhao, Z.M., Li, X., Chen, L., Aihara, K.: Protein classification with imbalanced data. Proteins: Structure, Function, and Bioinformatics, **70**(4), 1125–1132 (2007)
35. Zimmermann, T., Premraj, R., Zeller, A.: Predicting defects for eclipse. In: Proceedings of the 29th International Conference on Software Engineering Workshops, pp. 76, Washington, DC, USA, IEEE Computer Society (2007)

# Chapter 9
# Long-Term Time Series Prediction Using $k$-NN Based LS-SVM Framework with Multi-Value Integration

**Zifang Huang and Mei-Ling Shyu**

## 9.1 Introduction

Time series modeling and prediction are very attractive topics, which play an important role in many fields such as transportation prediction [4], power prediction [13, 18], and health care study [7]. The purpose of time series prediction is to forecast the values of data points ahead of time, where long-term time series prediction is to make the predictions multi-step ahead. The prediction process is commonly performed by observing and modeling the past values, and assuming that the future values will follow the same trend. When the prediction horizon increases, the uncertainty of the future trend also increases, rendering a more challenging prediction problem. Researchers have dedicated their effort to study how to extract as much knowledge as possible from the past values, and how to better utilize such knowledge for long-term time series prediction. There has been previous research work in order to tackle this challenge based on some classical time series prediction approaches, such as exponential smoothing [12], linear regression [14], autoregressive integrated moving average (ARIMA) [33], support vector machines (SVM) [25], artificial neural networks (ANN) [10, 33], and fuzzy logic [10].

In general, there are two ways to utilize a single value prediction model to predict multi-step ahead. The first way is to train one prediction model by optimizing the performance at the next time step, and iterate the same model using the previously predicted values as a part of the input to generate prediction for a higher horizon.

Z. Huang (✉)
Department of Electrical and Computer Engineering, University of Miami, Coral Gables, FL 33124, USA
e-mail: z.huang3@umiami.edu

M.-L. Shyu
Department of Electrical and Computer Engineering, University of Miami, Coral Gables, FL 33124, USA
e-mail: shyu@miami.edu

T. Özyer et al. (eds.), *Recent Trends in Information Reuse and Integration*, DOI 10.1007/978-3-7091-0738-6_9, © Springer-Verlag/Wien 2012

The approaches in this category are called recursive approaches which the prediction over a high horizon is obtained by recursively using earlier predicted values as the input. However, these approaches suffer from the error propagation problem. The other way is to train one prediction model for each prediction horizon by optimizing the performance at the each time step. The approaches in this category are direct approaches [6, 21]. Direct approaches need to train multiple models, so it takes a longer time in the training step; while it does not suffer from the error accumulation problem. Direct approaches usually outperform the recursive approaches on the prediction accuracy aspect. A multi-input multi-output local learning (LL-MIMO) approach [1,30], which is used as a comparison approach in the experiment, predicts the future values as a whole simultaneously. However, it could still be decomposed into multiple models which belong to the direct approach.

For long-term time series prediction [6, 15, 18, 20, 26], much effort has been put on deriving variations of least squares support vector machine (LS-SVM) [8,29] and ANN [32] approaches, which are fundamental approaches for nonlinear classification and function estimation and are successfully applied to time series prediction. The most frequently used kernels in LS-SVM are linear kernel, polynomial kernel, Radial Basis Function (RBF) kernel, and Multilayer Perceptron (MLP) kernel. The kernel trick extends the LS-SVM theory to a nonlinear technique without an explicit construction of the nonlinear mapping function. Composite kernels [11] have been studied, which combine both global kernel (e.g., polynomial kernel) and local kernel (e.g., RBF kernel) to balance the characteristics of fitting and generalization of these two kernels. Meanwhile, a prior knowledge based Green's kernel [5] was used in chaotic time series prediction by using the concept of matched filters. However, there was only one round of experiment on one dataset in the paper, which showed that Green's kernel outperforms the rest of the kernels. The generality of its application is not proved. For further improving long-term prediction performance, an input feature selection strategy was also combined with LS-SVM [28] and ANN [23]. By using the features selectively, it better utilizes the information from the past, and reduces the computational complexity of the predictor. However, there is not much research effort that focuses on choosing the instances adaptively from the training dataset to reduce the input dataset for training a prediction model.

Wavelet methods are widely used in noise removal in both one-dimensional signals and image data. Also, there has been increasing interest in wavelet transform for time series prediction. Many approaches have been proposed for time-series filtering and prediction by combining wavelet transform with prediction models, such as neural networks [16, 19], Kalman filtering [3], and autoregressive models [27]. In [24], it demonstrates how multiresolution prediction can capture short-range and long-term dependencies by combining a wavelet denoising technique and a wavelet predictive method. [31] proposed a direct modeling approach for long-term non-linear time series predictions by introducing multiresolution wavelet-based non-linear auto-regressive moving average (NARMA) models.

In this chapter, the direct approach is adopted in our proposed $k$-NN based LS-SVM framework with multi-value integration of the long-term time series prediction. A $k$-NN based instance selection strategy is utilized to obtain a smaller training dataset to train an LS-SVM regressor for each given testing instance. A new distance function, which integrates the Euclidean distance and the dissimilarity of the trend of a time series, is used in the $k$-NN approach. This idea was firstly introduced in our previous work [8]. The $k$-NN method here is used to reduce the size of the training dataset; while it was considered for the input feature selection in [28].

The rest of the chapter is organized as follows. In Sect. 9.2, we briefly introduce the LS-SVM method. Section 9.3 presents the proposed $k$-NN based LS-SVM framework with multi-value integration. The experiments and discussions of the experimental results are presented in Sect. 9.4. The chapter is concluded in Sect. 9.5.

## 9.2 An Overview of the LS-SVM Method

The LS-SVM method is commonly used as a nonlinear regressor [29]. In the primal weight space, let $x \in \mathbb{R}^p$, $y \in \mathbb{R}$, and $\varphi(x) : \mathbb{R}^p \rightarrow \mathbb{R}^{p_h}$ be a mapping function which maps the input vector $x$ at dimension $p$ to a high dimensional feature space at dimension $p_h$. The LS-SVM model is described as:

$$y(x) = w^T \varphi(x) + b, \tag{9.1}$$

where $w$ is a weight vector and $b$ is a bias. Given a training dataset $\{x_j, y_j\}_{j=1}^N$ where $N$ is the number of the training instances, let $e_j$ be an error variable, $e =< e_1; \ldots; e_N >$, and $\gamma$ be a regularization parameter, the optimization problem can be formulated in the primal weight space as follows.

$$\min_{w,b,e} \mathscr{J}_P(w, e) = \frac{1}{2} w^T w + \gamma \frac{1}{2} \sum_{j=1}^N e_j^2,$$

$$\text{subject to } y_j = w^T \varphi(x_j) + b + e_j, j = 1, \ldots, N. \tag{9.2}$$

However, when $w$ becomes infinite dimensional, it is unsolvable in the primal weight space. To address this issue, the solution is to construct the Lagrangian as in (9.3) and solve it in the dual space.

$$\mathscr{L}(w, b, e; \alpha) = \mathscr{J}_P(w, e) - \sum_{j=1}^N \alpha_j \{w^T \varphi(x_j) + b + e_j - y_j\}, \tag{9.3}$$

where $\alpha_j$ are Lagrange multipliers. The dual problem is derived as in (9.4) by eliminating the variables $w$ and $e$ under the conditions for optimality.

$$\begin{bmatrix} 0 & 1_v^T \\ 1_v & \Omega + I/\gamma \end{bmatrix} \begin{bmatrix} b \\ \alpha \end{bmatrix} = \begin{bmatrix} 0 \\ y \end{bmatrix}, \tag{9.4}$$

where $y =< y_1; \dots; y_N >$, $1_v =< 1; \dots; 1 >$, $\alpha =< \alpha_1; \dots; \alpha_N >$, and the kernel trick is

$$\Omega_{ij} = \varphi(x_i)^T \varphi(x_j) = K(x_i, x_j), \qquad i, j = 1, \dots, N. \qquad (9.5)$$

The function estimation result is given in (9.6).

$$y(x) = \sum_{j=1}^{N} \alpha_j K(x, x_j) + b. \qquad (9.6)$$

In this study, the RBF kernel, $K(x, x_j) = exp(-\|x - x_j\|_2^2/\sigma^2)$, is used. There are two tuning parameters, namely $\gamma$ and $\sigma$. $\gamma$ is the regularization parameter, determining the trade-off between the training error minimization and smoothness. $\sigma$ is the bandwidth. The values of these two parameters are selected empirically in the experiments. More details about the LS-SVM method could be found in [29].

## 9.3 The Proposed $k$-NN Based LS-SVM Framework

The time series studied in this chapter is a one-dimensional numerical sequence. Long-term time series prediction is to predict the future $n$ values based on the past $p$ observations, where $n > 1$ and $p \geq 1$. As shown in (9.7), we adopt the direct prediction strategy in the proposed framework, and it trains one model $f_i$ for each prediction horizon $i$.

$$\overline{x}_{t+i} = f_i(x_{t-p+1}, \dots, x_{t-1}, x_t), \qquad 1 \leq i \leq n. \qquad (9.7)$$

Given a time series $d$ for training with *Tlength* data points. The sliding window approach is used to re-align the one-dimensional time series into a matrix. The matrix, denoted as $D$, is the training dataset, where each row in the matrix is one training instance. For the prediction problem described above, the size of the sliding window is set to $p + n$. The size of $D$ is $(Tlength - p - n + 1) \times (p + n)$, where the first $p$ columns are the input for training the models, and the $(p + i)th$ column is the output for the model $f_i$. For example, if we have a training time series $d : x_1, x_2, x_3, x_4, x_5, x_6, x_7, x_8$, where $Tlength = 8$. Set $p = 2$ and $n = 3$. The corresponding matrix $D$ is shown in Table 9.1. The size of $D$ is $(Tlength - p - n + 1) \times (p + n) = 4 \times 5$, where the first 2 columns are the input for training the models, and the $(2 + i)th$ column is the output for the model $f_i$.

The most common way of training a prediction model is to use the whole training dataset as the input and treat every instance in the dataset equally. It is observed that training from the instances, which have similar inputs, could render a better model, and it better captures the correlation between the inputs and the corresponding outputs. Therefore, instead of using the whole available training

**Table 9.1** An example for constructing training dataset

| $x_1$ | $x_2$ | $x_3$ | $x_4$ | $x_5$ |
|---|---|---|---|---|
| $x_2$ | $x_3$ | $x_4$ | $x_5$ | $x_6$ |
| $x_3$ | $x_4$ | $x_5$ | $x_6$ | $x_7$ |
| $x_4$ | $x_5$ | $x_6$ | $x_7$ | $x_8$ |

**Fig. 9.1** The proposed $k$-NN based LS-SVM framework

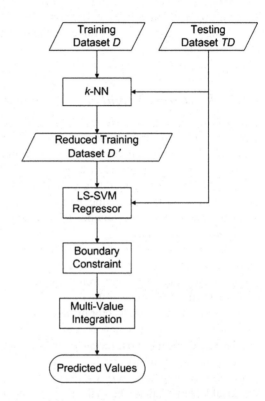

dataset, we dynamically select training instances which are closer to the testing instance to train an LS-SVM model for each testing instance. So the proposed approach is a lazy algorithm. It waits until a testing instance comes to train a prediction model and to do the prediction. The seen part of the testing instance is only the input of the testing instance, and the output is what we predict which is unseen. The training dataset is obtained from the most recent *Tlength* data points, which facilitates the prediction model to be adapted to the latest pattern. Even though the instance selection process will take some time, it reduces the size of the input training data for LS-SVM, which significantly decreases the complexity of training the LS-SVM regressor.

The proposed $k$-NN based LS-SVM framework is shown in Fig. 9.1, which contains five steps. First, $k$-nearest neighbors are selected from the training dataset $D$ for each instance in the testing dataset $TD$, using the distance measure in (9.11) defined in Sect. 9.3.1. The selected $k$ instances form the training dataset $D'$ for LS-SVM method. Second, an LS-SVM regressor is trained for each prediction horizon from $D'$. Third, the testing instance is used as the input to the obtained LS-SVM regressors, and the regressors will return a series of predicted values. Fourth, bound the predicted values within a dynamically calculated range. Finally, the last step is to integrate multiple predicted values as the $n$-step ahead prediction value for that testing instance. This step requires the LS-SVM regressors to generate a prediction over a horizon $n'$ which is larger than the required prediction horizon $n$, i.e., $n' > n$. Accordingly, $n'$ regressors are trained at the second step, and $n'$ predicted values are obtained at the third step.

### 9.3.1  k-Nearest Neighbors Component

$k$-NN is a classification method in pattern recognition. Recently, some studies applied $k$-NN approach in regression by returning the average value of the $k$-nearest neighbors. In our framework, we use $k$-NN method to select $k$ instances from the training dataset to generate a smaller dataset for training LS-SVM regressor. The closest $k$ instances to the input testing instance are selected. Usually, $k$ is much smaller than the number of instances in the training dataset, which is ($Tlength - p - n + 1$).

A new distance metric is defined for evaluating the distance of time series segment, which incorporates both the Euclidean distance and the dissimilarity of the trend of a time series. The trend of a time series is described by a vector which is the first order difference of the time series.

The length of the input vector in each instance is $p$, while we use the most recent $p_1$ values as the input to this $k$-Nearest Neighbors component, where $p_1 \leq p$. Similarly, we use the most recent $p_2$ values as the input to the following LS-SVM Regressor component, where $p_2 \leq p$ and $p = MAX(p_1, p_2)$.

Let the training time series $d$ be $< x_{t+1}, x_{t+2}, \ldots, x_{t+Tlength-1}, x_{t+Tlength} >$. We could derive that the $j$th row in $D$ is $<x_{t+j}, x_{t+j+1}, \ldots, x_{t+j+p-1}, x_{t+j+p}, \ldots, x_{t+j+p+n-1}>$, where $1 \leq j \leq (Tlength - p - n + 1)$. The first $p$ values in the row is the corresponding input vector: $<x_{t+j}, x_{t+j+1}, \ldots, x_{t+j+p-1}>$. The input vector to $k$-Nearest Neighbors component is $<x_{t+j+p-p_1}, x_{t+j+p-p_1+1}, \ldots, x_{t+j+p-1}>$, which is the last $p_1$ values of the input vector to the system. Its first order difference is

$$< z_{t+j+p-p_1}, \ldots, z_{t+j+p-2} > = < x_{t+j+p-p_1+1} - x_{t+j+p-p_1}, \ldots,$$
$$x_{t+j+p-1} - x_{t+j+p-2} > \tag{9.8}$$

The size of $<z_{t+j+p-p_1}, \ldots, z_{t+j+p-2}>$ is $1 \times (p_1 - 1)$.

Given a testing input vector $<x_{T+1}, \ldots, x_{T+p-1}, x_{T+p}>$, which starts at time point $T + 1$. The last $p_1$ values will be used here, i.e., $<x_{T+p-p_1+1}, \ldots, x_{T+p-1}, x_{T+p}>$. We first calculate its Euclidean distance with each training instance, denoted as $E(j)$ according to (9.9).

$$E(j) = \sqrt{(x_{T+p-p_1+1} - x_{t+j+p-p_1})^2 + \ldots + (x_{T+p} - x_{t+j+p-1})^2}. \quad (9.9)$$

The first order difference of the testing input vector is $<z_{T+p-p_1+1}, \ldots, z_{T+p-1}>$ $= <x_{T+p-p_1+2} - x_{T+p-p_1+1}, \ldots, x_{T+p} - x_{T+p-1}>$, whose size is $1 \times (p_1 - 1)$ as well. Calculate the Euclidean distance between the differential testing input vector and each differential training input vector, denoted as $F(j)$, according to (9.10).

$$F(j) = \sqrt{(z_{T+p-p_1+1} - z_{t+j+p-p_1})^2 + \ldots + (z_{T+p-1} - z_{t+j+p-2})^2}. \quad (9.10)$$

The length of vector $E$ and vector $F$ is $(Tlength - p - n + 1)$. We use a combination of the normalized $E$ and $F$ as the distance metric for the $k$-NN method. The distance $Dis$ is defined by (9.11).

$$Dis(j) = \frac{E(j) - MIN(E)}{MAX(E) - MIN(E)} + \frac{F(j) - MIN(F)}{MAX(F)\text{-}MIN(F)}, \quad (9.11)$$

where $MAX(E)$, $MIN(E)$, $MAX(F)$, and $MIN(F)$ are the maximum and minimum values for $E$ and $F$, respectively. $k$ instances corresponding to the smallest distance measures are selected as a reduced training dataset for LS-SVM.

Here, we give a simple example to demonstrate how to select the instances. Let the training time series $d$ be $< 1, 2, 4, 6, 5, 3, 1, 1 >$, where $Tlength = 8$. Set $p = 2$ and $n = 3$. The corresponding matrix $D$ is shown in Table 9.2.

Set $p_1 = 2, k = 2$. Given a testing input vector: $<2, 3>$, $E(j)$ is calculated, $i.e.,$, the Euclidean distance between the testing input vector and each training instance.

$$E(1) = \sqrt{(2-1)^2 + (3-2)^2} = \sqrt{2},$$

$$E(2) = \sqrt{(2-2)^2 + (3-4)^2} = 1,$$

$$E(3) = \sqrt{(2-4)^2 + (3-6)^2} = \sqrt{13},$$

$$E(4) = \sqrt{(2-6)^2 + (3-5)^2} = \sqrt{20}.$$

The first order difference of the testing input vector is $<3-2> = <1>$. Calculate $F(j)$, $i.e.,$, the Euclidean distance between the differential testing input vector and each differential training input vector.

**Table 9.2** An example for instance selection

| 1 | 2 | 4 | 6 | 5 |
|---|---|---|---|---|
| 2 | 4 | 6 | 5 | 3 |
| 4 | 6 | 5 | 3 | 1 |
| 6 | 5 | 3 | 1 | 1 |

$$F(1) = \sqrt{(1 - (2 - 1))^2} = 0,$$

$$F(2) = \sqrt{(1 - (4 - 2))^2} = 1,$$

$$F(3) = \sqrt{(1 - (6 - 4))^2} = 1,$$

$$F(4) = \sqrt{(1 - (5 - 6))^2} = 2.$$

Hence, we have $MAX(E) = \sqrt{20}$, $MIN(E) = 1$, $MAX(F) = 2$, and $MIN(F) = 0$. Calculate $Dis(j)$.

$$Dis(1) = \frac{\sqrt{2} - 1}{\sqrt{20} - 1} + \frac{0 - 0}{2 - 0} = 0.12,$$

$$Dis(2) = \frac{1 - 1}{\sqrt{20} - 1} + \frac{1 - 0}{2 - 0} = 0.5,$$

$$Dis(3) = \frac{\sqrt{13} - 1}{\sqrt{20} - 1} + \frac{1 - 0}{2 - 0} = 1.25,$$

$$Dis(4) = \frac{\sqrt{20} - 1}{\sqrt{20} - 1} + \frac{2 - 0}{2 - 0} = 2.$$

$k = 2$ instances corresponding to the smallest distance are selected as the reduced training dataset. In this example, the first two instances, with distance measure at 0.12 and 0.5 respectively, are selected as the training dataset to train LS-SVM.

### 9.3.2 Boundary Constraint Component

When the prediction horizon is very large, the prediction model might not be able to capture the changing patterns of the time series very well. The prediction model is fitted to the training data, and by carrying on the old patterns it might generate a value which is too small or too large. For some real world time series data, their values should fall within a certain reasonable range. Also, we do not expect sharp pulses in the prediction result. Therefore, a *Boundary Constraint* component is used to bound the predicted values rendered by LS-SVM regressors. We set the upper and lower bounds, denoted as $UpB$ and $LowB$ respectively, based on the values in the training time series data $d$.

$$UpB = MAX(d) + 0.02 \times STD(d);$$

$$LowB = MIN(d) - 0.02 \times STD(d).$$

Here, *MAX(d)* and *MIN(d)* are the maximum and minimum values of the training time series data $d$, and *STD(d)* is the standard deviation of $d$. Thus the boundaries are set to the maximum and minimum values of the training time series data with a moderate extension, where 0.02 is selected empirically. If a predicted value is larger than the upper bound or smaller than the lower bound, we reset the value to the upper bound or lower bound value, respectively. Otherwise, the predicted value is rendered as the final output. Therefore, the final output of the system is limited to the range of $[MIN(d) - 0.02 \times STD(d), MAX(d) + 0.02 \times STD(d)]$. The boundary is set dynamically according the updating training time series data, and it is domain knowledge free.

### 9.3.3 Multi-Value Integration Component

Various methods have been proposed on combining prediction results from individual models. Contrary opinions arose on which types of the models should be combined, different methods with distinct nature or very similar models. Some studies [2] believe that it is meaningless to combine the models which are not significantly different, because the models access to the same information set and capture similar patterns. While some other researchers [34] claim that it is also important to combine forecasts from very similar models. We agree when multiple models are utilized for estimation and prediction, the model uncertainty could generally be reduced.

In the proposed framework, we first generate a prediction over a horizon $n'$. Each predicted value is corresponding to a unique LS-SVM regressor. Given a testing input vector starting at time point $T + 1$ with length $p$, $<x_{T+1}, \ldots, x_{T+p-1}, x_{T+p}>$, the predicted values could be denoted as $<\overline{x}_{T+p+1}, \ldots, \overline{x}_{T+p+n}, \ldots, \overline{x}_{T+p+n'}>$, where the prediction value at time $T + p + n$ is the expected output. Notice that the prediction of the time series value at time $T + p + n$ is also done in the previous $n' - n$ testing instances. Instead of simply taking $\overline{x}_{T+p+n}$ as the output for prediction at horizon $n$, we integrate the previous $h$ predictions of the time series value at time $T + p + n$, where $0 \leq h \leq n' - n$, tog ether with the current prediction $\overline{x}_{T+p+n}$ by calculating the mean value of these $h + 1$ predictions. In order to better illustrate how the approach works, an example is given in Fig. 9.2.

| 1 | $x_1$ | $x_2$ | $\overline{x}_3$ | $\overline{x}_4$ | $\overline{x}_5$ |
| 2 | $x_2$ | $x_3$ | $\overline{x}'_4$ | $\overline{x}'_5$ | $\overline{x}'_6$ |
| 3 | $x_3$ | $x_4$ | $\overline{x}''_5$ | $\overline{x}''_6$ | $\overline{x}''_7$ |
| 4 | $x_4$ | $x_5$ | $\overline{x}'''_6$ | $\overline{x}'''_7$ | $\overline{x}'''_8$ |

**Fig. 9.2** An example for *multi-value integration* component

In this example, let $p = 2$, $n' = 3$, $n = 1$, and $T = 3$, which means the prediction takes the past 2 values ($p = 2$) as the input vector, and predicts the next 3 values ($n' = 3$), while the goal is to predict the next value ($n = 1$). The current testing instance is instance 4 which starts with $x_{T+1} = x_4$. $h$ can be an integer within the range of $[0, 2]$.

- If $h = 0$, it renders the prediction value at time $T + p + n = 6$, i.e., $\overline{x}_6'''$ as the final prediction value at horizon $n = 1$.
- If $h = 1$, it calculates the mean of the previous one prediction of the time series value at time $T + p + n = 6$ and the current prediction $\overline{x}_6'''$ as the final output, which is $(\overline{x}_6'' + \overline{x}_6''')/2$.
- If $h = 2$, it calculates the mean of the previous two predictions of the time series value at time $T + p + n = 6$ and the current prediction $\overline{x}_6'''$ as the final output, which is $(\overline{x}_6' + \overline{x}_6'' + \overline{x}_6''')/3$.

Because the values used for integration distribute diagonally in the matrix as shown in Fig. 9.2, we also name this method as diagonal multi-value integration.

**Parameters**

In our proposed framework, a set of parameters is selected empirically, including the length of the training time series *Tlength*, prediction horizon $n$ and $n'$, the length of the input vector for *k-Nearest Neighbors* component $p_1$, the length of the input vector for LS-SVM model $p_2$, parameter $k$ in the $k$-NN approach, two parameters used in LS-SVM with the RBF kernel (namely, $\gamma$ and $\sigma$), and the parameter $h$ in *Multi-Value Integration* component. $p$ can be specified once both $p_1$ and $p_2$ are given, $p = MAX(p_1, p_2)$.

## 9.4 Experiments and Results

In this section, we present our experimental results on three types of datasets. The datasets are briefly described in Sect. 9.4.1. Section 9.4.2 introduces the two measurements used to evaluate the performance. Comparison experimental results with the traditional LS-SVM approach [22,29] and the LL-MIMO algorithm [1] are shown in Sect. 9.4.3. The experiments were conducted on an Intel Core 2 machine with two 2.66 GHz CPUs and 3.25 GB of RAM.

### 9.4.1 Datasets

Three types of datasets are used in the comparative experiments, including the Mackey-Glass time series benchmark, four time series provided by NNGC1 competition, and a chaotic laser time series. The datasets exhibit diverse patterns.

The Mackey-Glass time series [17] is generated by the following delayed differential equation:

$$\frac{dx(t)}{dt} = \frac{ax(t-\tau)}{1 + x(t-\tau)^{10}} - bx(t).$$ (9.12)

This time series is commonly used to evaluate and compare the performance of time series prediction approaches [6, 15, 20]. In our experiment, 2,201 data points were generated with an initial value $x(0) = 1.2$, where $a = 0.2$, $b = 0.1$, and $\tau = 17$ by using the 4th order Runge-Kutta method. We used the last 2,000 data points of the time series in the experiments.

NNGC1 competition [4] provided diverse non-stationary, heteroscedastic transportation time series data with different structures and frequencies. The datasets are frequently used in the publications as well. The four time series used in our experiments are the four longest series collected hourly. The length of each time series is 1,742.

The chaotic laser time series is a univariate time record of a single observed quantity, measured in a physics laboratory experiment. It comprises measurements of the intensity pulsations of a single-mode Far-Infrared-Laser $NH_3$ in a chaotic state [9]. It has been used in benchmarking studies after the time series prediction competition organized by the Santa Fe Institute. The length of the time series is 1,000.

### 9.4.2 Error Measurements

We employed two error measurements, root mean squared error (*RMSE*) and symmetric mean absolute percentage error (*SMAPE*), to evaluate the performance of the prediction models.

Let $X$ be the real time series, and $\overline{X}$ be the predicted time series obtained at prediction horizon $n$. The length of both $X$ and $\overline{X}$ is $m$. *RMSE* is the square root of the variance, which is defined in (9.13). *RMSE* is closely related to the value range of the time series data.

$$RMSE = \sqrt{\frac{\sum_{t=1}^{m}(x_t - \overline{x}_t)^2}{m}}.$$ (9.13)

In order to compare the performance over different datasets, we calculate *SMAPE*, which is based on relative errors as defined in (9.14). *SMAPE* is the mean value of the difference between $x_t$ and $\overline{x}_t$ divided by their average. It has a lower bound and an upper bound at 0 and 2, respectively.

$$SMAPE = \frac{1}{m}\sum_{t=1}^{m}\frac{|x_t - \overline{x}_t|}{(x_t + \overline{x}_t)/2}.$$ (9.14)

### 9.4.3   Experimental Results

Two existing methods are used in the performance comparison, traditional LS-SVM [22, 29] and LL-MIMO approach [1]. Due to the high complexity of the traditional LS-SVM approach, we set the prediction horizon $n$ to 20 for the comparative experiments. For the proposed framework, $n'$ is set to 30 for all the datasets. Grid searching was done to tune the rest of the parameters one by one within a preset value range, including the length of the training time series $Tlength$, the length of the input vector for $k$-*Nearest Neighbors* component $p_1$, the length of the input vector for LS-SVM model $p_2$, parameter $k$ in the $k$-NN approach, $\gamma$ and $\sigma$ used in LS-SVM with the RBF kernel, and the parameter $h$ in *Multi-Value Integration* component. For the comparison purpose, the parameters required in LS-SVM and LL-MIMO are set to be the same values as the ones used in the proposed framework for a specific time series. For the traditional LS-SVM approach, it has parameters $Tlength$, $p$, $\gamma$, and $\sigma$. In the case of the LL-MIMO approach, $Tlength$, $p$, and $k$ need to be set.

Theoretically, $Tlength$ can be any positive integer as long as it is smaller than the length of the given time series, $p_1$ and $p_2$ can be within the range of $[1, Tlength - n']$, $k$ can be within the range of $[1, Tlength - p - n + 1]$, $\gamma$ and $\sigma$ can be any positive numbers, and $h$ can be within the range of $[0, n' - n]$. However, it is not practical to consider every possible combination of the parameters to compare the performance of the approaches. Table 9.3 lists the value ranges of the parameters. It takes time cost into consideration for setting the ranges.

The search intervals of $Tlength$, $k$ and $h$ are set to 100, 10, and 1, respectively. $p_1$ and $p_2$ are set to the values among $(3, 6, 10, 15, 20, 25, 30, 35, 40, 45, 50)$. The searching grid for $\gamma$ is $(0.02, 1, 5, 10, 17, 20, 30, 50, 100)$, and for $\sigma$ is $(0.02, 1, 5, 10, 17, 20, 30, 50)$. Since there is no guideline on how to set the best parameters for different datasets, we tried to find the (near)-optimal set of parameters using an iterative way. That is, the parameters are set empirically. The ranges can be larger or smaller than the given ones in Table 9.3. If the ranges are larger, we might be able to find a better (near)-optimal set of parameters, but it will be more time consuming. On the other hand, if the ranges are smaller, the searching process will be faster, but it limits the chance of finding the optimal. We select the values of the parameters corresponding to the lowest *RMSE*. Table 9.4 shows the selected parameter values for each time series dataset. The parameter values for each time series dataset varies due to the distinct characteristics of each dataset as expected. From Table 9.4 we can observe that the length of the training time series is not the longer the better. An explanation to that is the pattern of a time series is changing, and a longer training time series does not necessarily represent the latest pattern. The optimal $k$ values

**Table 9.3** Value ranges for parameter tuning

| $Tlength$ | $p_1$ | $p_2$ | $k$ | $\gamma$ | $\sigma$ | $h$ |
|---|---|---|---|---|---|---|
| [500,1000] | [3,50] | [3,50] | [50,150] | [0.02,100] | [0.02,50] | [0,10] |

**Table 9.4** Preset parameters values

| Dataset | Tlength | $p_1$ | $p_2$ | $k$ | $\gamma$ | $\sigma$ | $h$ |
|---|---|---|---|---|---|---|---|
| Mackey-Glass | 700 | 15 | 30 | 80 | 30 | 50 | 1 |
| NNGC1-1 | 600 | 20 | 20 | 70 | 10 | 10 | 2 |
| NNGC1-2 | 600 | 20 | 20 | 60 | 5 | 10 | 1 |
| NNGC1-3 | 600 | 35 | 25 | 110 | 10 | 10 | 2 |
| NNGC1-4 | 600 | 30 | 20 | 70 | 5 | 10 | 4 |
| Chaotic laser | 700 | 6 | 25 | 70 | 20 | 17 | 1 |

**Table 9.5** Performance in terms of *RMSE*

| Dataset | LL-MIMO | LS-SVM | $k$-NN based LS-SVM |
|---|---|---|---|
| Mackey-Glass | 0.0735 | 0.0080 | 0.0015 |
| NNGC1-1 | 6594.2 | 4039.9 | 3590.6 |
| NNGC1-2 | 155.84 | 113.1 | 100.27 |
| NNGC1-3 | 7462.1 | 4771.5 | 4399.6 |
| NNGC1-4 | 2394.7 | 1,683 | 1450.8 |
| Chaotic laser | 20.528 | 5.4761 | 2.1454 |

**Table 9.6** Performance in terms of *SMAPE*

| Dataset | LL-MIMO | LS-SVM | $k$-NN based LS-SVM |
|---|---|---|---|
| Mackey-Glass | 0.0709 | 0.0067 | 0.0013 |
| NNGC1-1 | 0.4323 | 0.1978 | 0.1748 |
| NNGC1-2 | 0.3913 | 0.2764 | 0.2188 |
| NNGC1-3 | 0.3706 | 0.1677 | 0.1510 |
| NNGC1-4 | 0.2882 | 0.1604 | 0.1339 |
| Chaotic laser | 0.2731 | 0.0968 | 0.0339 |

are relevantly small, and it proves that $k$-nearest neighbors component improves the performance of the prediction model in terms of *RMSE*, by creating a smaller training dataset. Meanwhile, $h$ is always greater than 0, so it is helpful to utilized some additional predicted values to generate the final prediction result.

The comparison results on six time series at prediction horizon $n = 20$ are reported in Table 9.5 and Table 9.6, in terms of *RMSE* and *SMAPE*, respectively. As we could see from the results, the proposed approach can always achieve the lowest prediction errors in terms of both *RMSE* and *SMAPE*. All models perform well in predicting the Mackey-Glass time series dataset. It is a synthetic data series without any noise, and has an almost periodic pattern, thus the prediction errors are quite small. Datasets provided by NNGC1 are collected from daily traffic, it contains noise, and the data itself represents a much more complicated model, which render more challenge for prediction, while the proposed approach generates promising prediction results on this real world data. The prediction performance on the chaotic laser time series, which was obtained in a physics laboratory experiment, is pretty good as well. A column plot of Table 9.6 is presented in Fig. 9.3 to show the comparison results in terms of *SMAPE* more clearly.

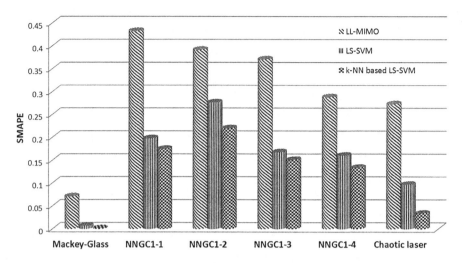

**Fig. 9.3** Performance in terms of *SMAPE*

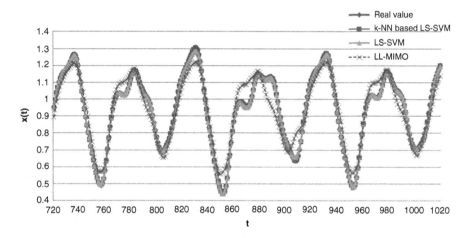

**Fig. 9.4** Prediction results for Machey–Glass

We present some of the 20 steps ahead prediction results by all three approaches for each of the three types of data, including results for Mackey-Glass dataset, NNGC1-4 dataset (representing NNGC1 data), and Chaotic laser dataset. Figure 9.4 shows some of the 20 steps ahead prediction results for time series Mackey-Glass. Due to the limited space, only the first 300 predicted data points out of the 1,281 testing data are shown in the plot together with the corresponding real values. The predicted time series by LS-SVM and our framework are very close to the real time series, while our framework achieved lower error measures.

Figure 9.5 shows some of prediction results for time series NNGC1-4. The first 150 predicted data points out of the 1,123 testing data points are shown in the plot.

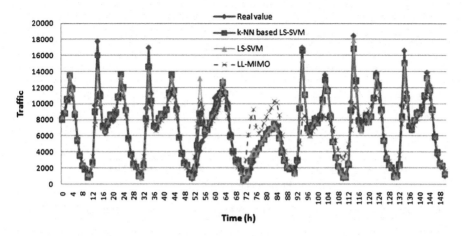

**Fig. 9.5**  Prediction results for NNGC1-4

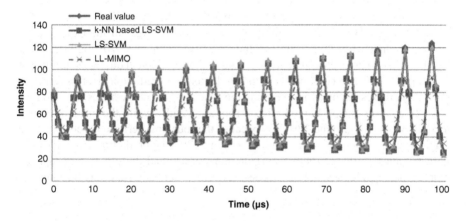

**Fig. 9.6**  Prediction results for chaotic laser

It is easy to see that the predicted values by our proposed $k$-NN based LS-SVM with multi-value integration approach follow the real values quite closely, and it outperforms the traditional LS-SVM and LL-MIMO approaches. The LL-MIMO approach failed to predict and preserve the trend of the time series dataset, especially from $x = 72$ to $x = 88$. Though the traditional LS-SVM approach generated fair prediction results, it consumed a much longer time because of using a large training dataset, which makes it infeasible to do prediction in real time. On average, it took 28.76 s for the traditional LS-SVM approach to execute a 20-step ahead prediction, while the proposed approach only requires 0.51 s.

Figure 9.6 shows a part of the prediction results for time series Chaotic laser. The first 100 predicted data points out of the 381 testing data points are shown in

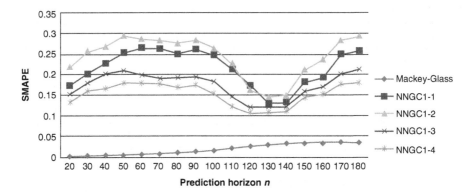

**Fig. 9.7** Performance of the $k$-NN based LS-SVM on long-term time series prediction

the plot. The prediction by our framework again follows the real value very well, while the prediction by LL-MIMO shifts away from the real time series.

We have run more experiments to investigate the prediction ability of the proposed framework when the prediction horizon $n$ is large. In the experiments, $n$ is set to range from 20 to 180, and the results in terms of *SMAPE* are shown in Fig. 9.7. Chaotic laser dataset is not included in this experiment due to the limitation of its length. When prediction horizon $n$ is larger, the prediction error in terms of *SMAPE* increases slightly for predicting the Mackey-Glass time series. For the four time series from NNGC1, prediction errors are steady, while the error measures decrease when $n$ is around 130. It is because the four time series datasets exhibit a roughly periodic pattern with a period of about 130. The prediction error does not increase much as the prediction horizon $n$ increases, and therefore, the proposed approach is capable of handling time series prediction over a very long horizon.

## 9.5 Conclusion

In this chapter, we introduced a $k$-NN based LS-SVM framework with multi-value integration for long-term time series prediction. The proposed approach takes advantage of each of its component, including $k$-NN component, *LS-SVM Regressor* component, *Boundary Constraint* component, and *Multi-Value Integration* component, to generate good prediction efficiently. The usage of the *Boundary Constraint* component is based on the assumption that, for time series which has physical meaning, the values are within certain range and do not change sharply. The experimental results have shown that the proposed framework outperforms the traditional LS-SVM approach and the LL-MIMO approach with lower prediction errors. Also, the proposed approach is capable of handling time series prediction over a very long horizon.

**Table 9.7**  Symbols used in this chapter

| $Tlength$ | Length of the training time series |
|---|---|
| $p_1$ | Input length for $k$-*Nearest Neighbors* component |
| $p_2$ | Input length for *LS-SVM Regressor* component |
| $p$ | $p = MAX(p_1, p_2)$ |
| $d$ | Training time series |
| $D$ | Training dataset |
| $D'$ | Reduced training dataset |
| $TD$ | Testing dataset |
| $n$ | Required prediction horizon |
| $n'$ | Prediction horizon |
| $k$ | Parameter for $k$-*Nearest Neighbors* component |
| $h$ | The number of the values used for integration |
| $e_j$ | Error variable |
| $\gamma$ | Regularization parameter for LS-SVM |
| $\sigma$ | Bandwidth for RBF kernel |
| $UpB$ | Upper bound for *Boundary Constraint* component |
| $LowB$ | Lower bound for *Boundary Constraint* component |
| $E(j)$ | Euclidean distance between the testing instance and the $jth$ training instance |
| $F(j)$ | Euclidean distance between the differential testing instance and the $jth$ differential training instance |
| $Dis(j)$ | Distance measure for $k$-*Nearest Neighbors* component |

# Appendix: Table of Symbols

We list the symbols used in this chapter in Table 9.7.

# References

1. Bontempi, G.: Long term time series prediction with multi-input multi-output local learning. In: 2nd European Symposium on Time Series Prediction pp. 145–154 (2008)
2. Clement, M.P., Hendry, D.F.: Forecasting economic times series. Cambridge University Press, Cambridge (1998)
3. Cristi, R., Tummala, M.: Multirate, multiresolution, recursive kalman filter. Signal Process. **80**(9), 1945–1958 (2000)
4. Crone, S.F.: Artificial neural network & computational intelligence forecasting competition, www.neural-forecasting-competition.com (2010)
5. Farooq, T., Guergachi, A., Krishnan, S.: Chaotic time series prediction using knowledge based green's kernel and least-squares support vector machines. In: IEEE International Conference on Systems, Man and Cybernetics pp. 373–378 (2007)
6. Herrera, L.J., Pomares, H., Rojas, I., Guilln, A., Prieto, A., Valenzuela, O.: Recursive prediction for long term time series forecasting using advanced models. Neurocomputing **70**(16–18), 2870–2880 (2007)
7. Homma, N., Sakai, M., Takai, Y.: Time series prediction of respiratory motion for lung tumor tracking radiation therapy. In: 10th WSEAS international conference on Neural networks pp. 126–131 (2009)

8. Huang, Z., Shyu, M.L.: k-NN based LS-SVM framework for long-term time series prediction. In: 2010 IEEE International Conference on Information Reuse and Integration pp. 69–74 (2010)

9. Huebner, U., Abraham, N.B., Weiss, C.O.: Dimensions and entropies of chaotic intensity pulsations in a single-mode far-infrared $NH_3$ laser. Phys. Rev. A **40**(11), 6354–6365 (1989)

10. Jang, J.S.: Anfis: adaptive-network-based fuzzy inference system. IEEE Trans. Syst. Man Cybern. Syst. Hum. **23**(3), 665–685 (1993)

11. Jiang, T., Wang, S., Wei, R.: Support vector machine with composite kernels for time series prediction. In: 4th international symposium on Neural Networks pp. 350–356 (2007)

12. Jones, R.H.: Exponential smoothing for multivariate time series. J. Roy. Stat. Soc. B **28**(1), 241–251 (1966)

13. Kusiak, A., Zheng, H., Song, Z.: Short-term prediction of wind farm power: A data mining approach. IEEE Trans. Energ. Convers. **24**(1), 125–136 (2009)

14. Lin, K., Lin, Q., Zhou, C., Yao, J.: Time series prediction based on linear regression and SVR. Third International Conference on Natural Computation 1, 688–691 (2007)

15. Liu, P., Yao, J.: Application of least square support vector machine based on particle swarm optimization to chaotic time series prediction. In: IEEE International Conference on Intelligent Computing and Intelligent Systems 4, 458–462 (2009)

16. Lotric, U.: Wavelet based denoising integrated into multilayered perceptron. Neurocomputing **62**, 179–196 (2004)

17. Mackey, M., Glass, L.: Oscillation and chaos in physiological control systems. Science **197**(4300), 287–289 (1977)

18. Maralloo, M., Koushki, A., Lucas, C., Kalhor, A.: Long term electrical load forecasting via a neurofuzzy model. In: 14th International CSI Computer Conference pp. 35–40 (2009)

19. Menezes Jr, J.M.P., Barreto, G.A.: Long-term time series prediction with the narx network: An empirical evaluation. Neurocomputing **71**(16-18), 3335–3343 (2008)

20. Meng, K., Dong, Z., Wong, K.: Self-adaptive radial basis function neural network for short-term electricity price forecasting. IET Gener. Transm. Distrib. **3**(4), 325–335 (2009)

21. Nguyen, H.H., Chan, C.W.: Multiple neural networks for a long term time series forecast. Neural. Comput. Appl. **13**, 90–98 (2004)

22. Pelckmans, K., Suykens, J.A.K., Gestel, T.V., Brabanter, J.D., Lukas, L., Moor, B.D., Vandewalle, J.: LS-SVMlab toolbox user's guide. ESAT-SCD-SISTA Technical Report pp. 1–106 (2003)

23. Puma-Villanueva, W.J., dos Santos, E., Von Zuben, F.: Long-term time series prediction using wrappers for variable selection and clustering for data partition. In: International Joint Conference on Neural Networks pp. 3068–3073 (2007)

24. Renaud, O., Starck, J.L., Murtagh, F.: Wavelet-based combined signal filtering and prediction. IEEE Trans. Syst. Man Cybern. B Cybern. **35**(6), 1241–1251 (2005)

25. Sapankevych, N., Sankar, R.: Time series prediction using support vector machines: a survey. IEEE Comput. Intell. Mag. **4**(2), 24–38 (2009)

26. Sfetsos, A., Siriopoulos, C.: Time series forecasting with a hybrid clustering scheme and pattern recognition. IEEE Trans. Syst. Man Cybern. A Syst. Hum. **34**(3), 399–405 (2004)

27. Soltani, S., Boichu, D., Simard, P., Canu, S.: The long-term memory prediction by multiscale decomposition. Signal Process. **80**(10), 2195–2205 (2000)

28. Sorjamaa, A., Hao, J., Reyhani, N., Ji, Y., Lendasse, A.: Methodology for long-term prediction of time series. Neurocomputing **70**(16-18), 2861–2869 (2007)

29. Suykens, J.A.K., Gestel, T.V., Brabanter, J.D., Moor, B.D., Vandewalle, J.: Least Squares Support Vector Machines. World Scientific, Farrer Road, Singapore (2002)

30. Taieb, S.B., Bontempi, G., Sorjamaa, A., Lendasse, A.: Long-term prediction of time series by combining direct and mimo strategies. In: International Joint Conference on Neural Networks pp. 3054–3061 (2009)

31. Wei, H.L., Billings, S.A.: Long term prediction of non-linear time series using multiresolution wavelet models. Int. J. Contr. **79**(6), 569–580 (2006)

32. Yegnanarayana, B.: Artificial Neural Networks. Prentice-Hall of India Pvt. Ltd, New Delhi (2004)
33. Zhang, G.: Time series forecasting using a hybrid ARIMA and neural network model. Neurocomputing **50**, 159–175 (2003)
34. Zou, H., Yang, Y.: Combining time series models for forecasting. Int. J. Forecast. **20**(1), 69–84 (2004)

# Chapter 10
# Semantic Management of Digital Contents for the Cultural Domain

Carlos M. Cornejo, Iván Ruiz-Rube, and Juan Manuel Dodero

## 10.1 Introduction

The World Wide Web is a space prepared for the exchange of information among networked systems. However, the current web is mainly designed for human consumption, so its contents are not readily understandable by computers. This issue emerges in diverse applications of the Web and becomes especially relevant in the cultural institutions' domain. The aim of Linked Data methods [6] is to extend the published data on the Web so that web analyzing agents are able to process and interpret the semantic metadata of a given domain. All issues related to cooperation, interoperability and accessibility enabled by semantic technologies are growing up to have linked, well represented knowledge domains ready for exploitation by automated agents. By endowing the web with more meaning, you provide solutions to share, process and transfer information among web-based contents and applications from diverse cultural institutions.

To obtain a proper definition of interlinked data, the Semantic Web uses ontologies that represent such information [14]. In the context of digital preservation of Cultural Heritage (CH), the objective is to address a new way of representing, analysing, manipulating and managing different kinds of cultural contents from different media sources [26]. The CIDOC Conceptual Reference Model (CRM)[1] is presented as a set of formal definitions to facilitate the integration, mediation and exchange of heterogeneous CH information [12]. The CIDOC CRM can be used to describe the concepts and relationships contained within information sources

---

[1] CIDOC CRM – http://cidoc.ics.forth.gr/

C.M. Cornejo (✉) · I. Ruiz-Rube · J.M. Dodero
University of Cadiz, Computer Languages and Systems Department, C\Chile. 1,
11002 Cadiz, Spain
e-mail: carlos.cornejo@uca.es; ivan.ruiz@uca.es; juanma.dodero@uca.es

T. Özyer et al. (eds.), *Recent Trends in Information Reuse and Integration*,
DOI 10.1007/978-3-7091-0738-6_10, © Springer-Verlag/Wien 2012

related to cultural issues. It provides a common and extensible semantic framework to formally represent CH information.

The eCultura platform aims to extend over the Web, the knowledge base of cultural institutions, building user communities around it and preparing for its exploitation in several environments. For that, we have extended a content management platform with a set of services and applications related to the cultural domain. Semantic web technologies are used to share information among such services, as well as providing interoperability with external applications. The platform is fed with the CIDOC CRM, extended with other Domain-Specific Ontologies (DSO) that provide an integration platform to manage and access web-based content, applications and services of the CH domain. Semantic technologies have been used to share information amongst integrated applications, as well as to interlink with external applications. We have extended the CH semantic model with other DSO that describe music concepts and relations between people. The Music Ontology[2] links all the information about musical artists and works, while the FOAF Ontology[3] is used for describing persons, their activities and their relations to other people and objects. These ontologies provide a standard vocabulary for extending cultural concepts and also for linking the platform with other application issues.

This chapter is structured as follows. Section 10.2 summarizes works related with semantic content management approaches and systems. Section 10.3 deals with the goals and motivations to build the eCultura solution. Then, we show the method and architecture of our solution, while Sect. 10.4 presents a case study of how eCultura helps in providing an integrative Linked Data platform to host different kinds of CH applications. Finally in Sect. 10.5, we draw some conclusions and provide an outlook to future works.

## 10.2 Semantic Content Management Approaches and Systems

We have organized the related works into four topics. First are ontologies that deal with representation issues of the CH domain. Second are presented some semantic integration approaches and Linked Data platforms. Third are revised semantically-enhanced wiki applications. Finally, several implementations of the data management layer are shown.

### 10.2.1 Ontologies Dealing with Representation of CH

Current efforts to add semantic meaning to resources are related by introducing metadata tags in order to help the computer to understand what it is processing. [16] propose to use a multimedia metadata deployment framework based on

---

[2]Music Ontology – http://musicontology.com/

[3]Friend Of A Friend Ontology – http://www.foaf-project.org/

RDF attributes (RDFa), that includes annotation capabilities for a wide range of multimedia formats.

Within the context of the CH domain, the CIDOC CRM reduces the problem of designing ad-hoc cultural record structures [5] for museum asset management systems [4]. There are other *Specific Domain Ontologies* arising in harmony with the CIDOC CRM, such as the FRBR [11], originally conceived to formally capture the underlying concepts necessary for integrated information systems. Its innovation is under clustering publications and other items around common conceptual notions to support information retrieval.

Ontology extensions bring the opportunity to add more knowledge to a system. For instance, [18] describes how an MPEG-7[4] metadata ontology model is designed to support the description and management of multimedia resources of museums and how to merge them with CIDOC CRM. Other domain-specific extensions are described in [23].

## 10.2.2  Semantic Integration Approaches and Linked Data Platforms

Several approaches over semantic linked-data integration databases have already been built. The outlined idea of interlinking open data on the web, focuses on the existence of large amounts of meaningful interlinked data somehow stored. On one hand, the heterogeneity of data implies that the information held in the database is of a different nature and must be properly mapped. Top-level ontologies (i.e., SUMO [21] or DOLCE [10]) are approached to manage this huge amount of merged raw data. On the other hand, platforms that have already processed and linked the information they hold, offer this information as semantic artifacts, which can be shared amongst users. An example of this can be found under the Metaweb Project[5] which works around the idea of identifying text held in a website and linking it to single entities that are managed internally by the social Metaweb system.

The trend of automatically creating semantic data that are submitted from a website and post-processed by an external agent is faced by Calais.[6] Zemanta[7] also allows the integration with Web 2.0 applications, content management systems (CMS) and web browsers by proposing semantic annotations of the recieved contents.

---

[4]MPEG-7 Ontology – http://metadata.net/mpeg7/

[5]Metaweb Project – http://www.freebase.com/view/freebase/apps/global

[6]OpenCalais – http://www.opencalais.com/

[7]Zemanta – http://www.zemanta.com/

Efforts in converting Wikipedia[8] content into structured knowledge can be found, for instance, on DBpedia and Freebase. These are publicly accessible Linked Data sites that host and manage semantically-enhanced web content [3].

In [2], the extraction of the DBpedia datasets, and how the resulting information is published on the Web for human and machine consumption [7] are described. The DBpedia datasets can be either imported into third party applications or accessed online using a variety of DBpedia interfaces. An alternative to the latter is Freebase [8,9], defined as a collaboratively built, graph-shaped database of structured general human knowledge.

Linked Data systems provide an extension API in order to host specialized applications that exploit the semantic information. For instance, DBpedia provides an API to extract information from Wikipedia to feed its semantic content that helps systems to manage the linked-data information. On the other hand, the Freebase API helps to develop internally managed applications (e.g., Thinkbase [17]) that can exploit the semantic database. It provides its own ACRE[9] hosting platform API to develop specialized applications that exploit the semantics contained by Freebase.

## 10.2.3 *Semantically-Enhanced Wikis*

Web-based CMS used to implement semantic platforms are typically organized as a wiki to manage Linked Data contents and applications. Several wiki implementations have been provided as the basis of semantic content management systems. KiWi[10] is a Java-based semantic wiki engine built on top of the Seam framework.

AceWiki[11] is a type of semantic wiki that uses as knowledge representation language, the controlled natural language ACE[12] as a subset of regular English. In order to help the users to write correct ACE sentences, AceWiki provides a predictive editor which guides them to select concepts and relations within the ontologies.

KnowWE[13] is another Java-based implementation of a semantic wiki based on JSPWiki.[14] Apart from implementing concepts of semantic wikis, such as the definition of classes, taxonomies and user-defined properties, KnowWE is able to represent a reasoner on knowledge that is applied to selected classes of the ontology.

---

[8]Wikipedia – http://www.wikipedia.org/

[9]ACRE – http://www.freebase.com/docs/acre

[10]KiWi Project – http://www.kiwi-project.eu/

[11]AceWiki – http://attempto.ifi.uzh.ch/acewiki/

[12]Attempto Project – http://attempto.ifi.uzh.ch/site/

[13]KnowWE – http://www.is.informatik.uni-wuerzburg.de/en/research/applications/knowwe/

[14]JSPWiki – http://www.jspwiki.org/

A wiki enterprise approach has been made on SMW+.[15] It combines the wiki philosophy of use with enterprise supporting, i.e., authoring, access control restriction and so forth.

### 10.2.4  Semantic Data Store Systems

There are several alternatives for back-end storage. Sesame[16] is an open source RDF framework that supports inference and searching over RDF schema and OWL. It offers an HTTP RESTful API in several languages, e.g., SeRQL[17] and RQL.[18]

Virtuoso is a platform created by OpenLink Software[19] designed to integrate business data and business processes that involve SQL, RDF, XML and web services.

AllegroGraph[20] is a database on its own, so it does not use any relational databases present in the market, but uses simple sentences files (N-triple) for storaging all data, and supports SPARQL and RDFS++.

Oracle Spatial[21] provides an under-license software, based on a graph data model and out-performs results on geographic information systems.

## 10.3  Semantically-Enriched Cultural Content Management

The eCultura approach provides a Linked Data host platform independent of the ontology models and web applications that are to be integrated. We first address our main objectives and the motivation we want to overcome with this paper. Second, we propose our method to create a semantically-enhanced infrastructure. Third, we present our proposal for information modeling and fourth give details of the implementation.

### 10.3.1  Objective

Current Web 2.0 applications provide efficient ways of sharing information, but they do not enable the possibility of engaging human thinking with computing understanding. They do not share any semantic information models about the digital

---

[15]Semantic Enterprise Wiki – http://wiki.ontoprise.com/

[16]Sesame – http://www.openrdf.org/doc/sesame/api/

[17]SeRQL – http://www.openrdf.org/doc/sesame/users/ch06.html

[18]RQL – http://139.91.183.30:9090/RDF/RQL/

[19]OpenLink Software – http://www.openlinksw.com/

[20]AllegroGraph – http://www.franz.com/

[21]Oracle Spatial – http://www.oracle.com/technetwork/database/options/spatial

content they host and manage. For instance, you can add information to a wiki or a social networking application about an event in a place, but this event is not known by other applications (e.g., a time-line renderer, an interactive map, or a learning course) that manage digital contents and could also be interested in the event.

Besides, existing semantic platforms are not easily extensible with standard information models or semantic web ontologies. For instance, DBpedia and Freebase define their own RDF/RDFS schemas that are not standard or specific in the CH domain.

The eCultura platform aims to link data from web-based applications that are relevant to the CH domain. It offers a way of adding knowledge based on standard W3C formats. The architecture allows the integration of external web-based applications of general purpose, such as wikis, blogs and so forth.

In our case, the hosting platform is extended to link their data about common CH topics, then users are enabled to share their experiences across hosted and integrated external applications. Sharing that information, e.g., through a wiki or other kind of input forms, opens the gate of knowledge to a wider community interested in the CH field.

## 10.3.2 Method

eCultura platform aims at integrating the Linked Data that are published by or consumed from external applications. Such applications have their own data models, defined by a semantic information model and conform to W3C standards such as RDF(S), OWL [28] and SPARQL [24].

We show in Fig. 10.1 how the platform works and the required steps to become fully functional. First, it is necessary to add domain-specific knowledge and application-specific information to the repository. The ontology schemas used for defining such models usually require a transformation or extra adaption effort before being integrated in the platform to coexist with other related ontologies [25]. For example, given two ontologies, the task is to find similarities between them and determine which concepts and properties are representing similar notions.

**Fig. 10.1** Platform creation process

Semantic-integration tasks such as mapping discovery, declarative formal representations and reasoning of these mappings are approaches to overcome this issue [22].
These mappings capture the semantic correspondences between concepts held in
schemas or ontologies.

Second, after the platform is fed with knowledge by adding the selected
ontologies, each application may need some initial configuration and enrichment
with semantic information. This semantic enhancement might involve the adaptation
to some ontologies or the definition of how the application should access the
information.

Finally, ontology population involves the process of creating new Linked Data
by the different applications or services integrated through the platform. These can
be information producers, i.e., those capable of generating semantic information
(e.g., a blog) or consumers, i.e., those that can exploit the semantic information,
typically for visualization purposes (e.g., an interactive map, a time-line renderer).
Other services can act as producers and consumers (e.g., a wiki). We discuss below
about such categories of applications.

## 10.3.3   Information Modeling

*Domain-Specific Ontologies* provide a common understanding of a domain (a particular area) in which people and the application system communicate with each
other. Web data sources are independent and communities are free to use their own
vocabulary. Therefore any web source can have its own ontology containing all the
defined terms that has met with the agreements of its community [15]. The use of
domain ontologies is an approach to overcome the issue of interoperability among
systems. Information requests coming from the platform the system are rewritten
locally using these relationships [20] to obtain interontology relationships between
terms or facts of different ontologies.

Furthermore, web applications often have their own data model to represent their
domain concepts, ie. they could publish the stored information according to their
specific *Application Ontologies (AO)*.

Having a Linked Data point-of-view, we could consider a website as a RDF-
compliant metadata container. It provides the ability to view an information source
at the level of its relevant semantic concepts and supports information requests
that are generically independent of the structure or location of the data source. We
represent this fact in Fig. 10.2 by defining an *integration ontology*. It allows having
an abstract organization of the information held on the website as a set of statements.
We have considered some initial attributes associated to the *MetadataContainer* in
order to identify who has created that Linked Data (*author*) and in the time that it
happened (*creationDateTime*).

We address in Fig. 10.3 a way of managing the knowledge base of our platform
by exploiting two scopes of information: terminological statements (TBOX) and
assertion statements (ABOX). In TBOX we make a division between (1) the

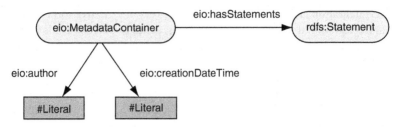

**Fig. 10.2** eCultura integration ontology

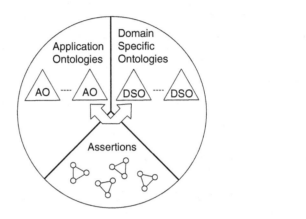

**Fig. 10.3** Composite information modeling

information relative to application concepts which defines how the applications are managed internally (*Application Ontologies*) and (2) the *Domain-Specific Ontologies* (e.g., Music Ontology, CIDOC CRM, FOAF and so forth) that model all the concepts and their inherited properties. The *ABOX* contains all the concept instances or individuals, as well as the relationships between individuals, that are being linked through these DSO.

Because of the independence of the underlying concepts, if a new installation of the platform it is required, we would only need to set up the new applications, not the triple stored information, as it remains unchanged.

### 10.3.4 Architecture

In Fig. 10.4, we show the layered architecture of eCultura, in which we define three levels of abstraction. First, *Metadata Layer* that manages how the triple-stored information is organized and their internal relations. Second the *Service Layer*, that defines how the application should communicate with the repository. Finally, the *Application Layer* where all the applications consume, produce or consume/produce the Linked Data that the platform provides.

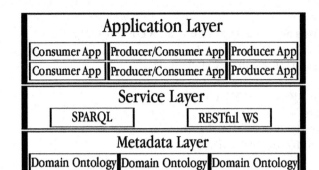

**Fig. 10.4**   eCultura layered architecture

### 10.3.4.1   Metadata Layer

The management of the data sources and how it has to be retrieved is carried out by OntSpace,[22] a Java-based software framework providing an extension of Jena[23] API and the services of a semantic database. In the CH domain, the basis is the CIDOC CRM augmented with other DSO such as FOAF and The Music Ontology.

### 10.3.4.2   Service Layer

We propose to have a generic API that exploits the varied semantic models. Our approach is to use a standard query interface that enable applications to express their needs across diverse data sources, natively stored as RDF. Joseki,[24] is an HTTP server that allows the flow of information with the semantic front-end using SPARQL. This engine supports the SPARQL/Update Protocol and the SPARQL RDF Query language through HTTP binding.

Under our proposal of managing several models of information, we show in Listing 10.1, how Joseki allows the definition of a *composite model*, in which all the information of the platform (TBOX and ABOX) is managed.

**Listing 10.1**   Joseki server configuration excerpt
```
#SPARQL  EndPoint  <#SRV_ecultura>
      joseki:serviceRef     ''ecultura''  ;
      joseki:dataset        <#DS_ecultura> ;
      joseki:processor      joseki:ProcessorSPARQL  ;
```

---

[22]OntSpace – http://code.google.com/p/ont-space/

[23]Jena Framework – http://www.openjena.org/

[24]Joseki Server – http://www.joseki.org/

```
#Dataset
<#DS_ecultura>    rdf:type  ja:RDFDataset  ;
    joseki:poolSize      3;
    ja:defaultGraph  <#MODEL_kb>  ;

#Knowledge Base (RDF Graph)
<#MODEL_kb>    rdf:type  ja:Model;
    ja:subModel  <#MODEL_tbox>;
    ja:subModel  <#MODEL_abox>;

#Terminological Knowledge (TBOX)
<#MODEL_tbox>  rdf:type  ja:Model;
    ja:subModel  <#MODEL_dso>;
    ja:subModel  <#MODEL_ao>;

#Assertional knowledge (ABOX)
<#MODEL_abox>  rdf:type  ja:RDBModel;
    ja:connection   [
        ja:dbType  "MySQL"  ;
        . . .
    ]

#Domain Specific Ontologies
<#MODEL_dso>   rdf:type  ja:Model;
    ja:subModel  <#MODEL_CRM>;
    ja:subModel  <#MODEL_MO>;
    ja:subModel  <#MODEL_FOAF>;

#Application Ontologies
<#MODEL_ao>    rdf:type  ja:Model;
    ja:subModel  <#MODEL_integration>  ;
    ja:subModel  <#MODEL_linkedBlog>  ;
    ja:subModel  <#MODEL_linkedWiki>  ;
```

Understanding the complexity of the information required by each application, for instance, in those cases in which the expressiveness of the query language is not representative enough, we extend the proposal with a second access point through RESTful web services. So, the eCultura platform offers another entry point through RESTful web services using Jersey,[25] which is an implementation of JAX-RS.[26] This also adds capabilities for creating, querying and updating concepts and relationships held in the RDF triple store.

---

[25]Jersey Project – https://jersey.dev.java.net/

[26]JAX-RS – https://jsr311.dev.java.net/nonav/releases/1.1/

### 10.3.4.3  Application Layer

The suite of applications that build up our platform is composed of open-source tools. eCultura architecture, is based on Java/J2EE collaborative portal Liferay,[27] so that each application is integrated as either a portlet, complying with the JSR-286 specification,[28] or wrapped into an *iFrame* Portlet.

## 10.4  Case Studies

Current Web 2.0 applications are amongst the most popular community-oriented systems used to share information between users. For instance, adding semantic features to a wiki or blog opens the challenge of interacting with other information sources. First, we must have a semantic model in order to define and tag the concepts and relationships contained in the information produced and managed by an application. After that, tagged information resources can be queried by other consumer applications or agents. In the following we describe a case study of a producer application (e.g., a wordpress blog), a consumer (e.g., an interactive map) and a producer/consumer (e.g., a wiki).

We require semantically-enhanced versions of existing web applications such as MediaWiki,[29] WordPress.[30] In the wiki case, we used SemanticWiki [19], an already existing extension of Wikimedia semantically enhanced to support OWL annotations. On the other hand, we did not find an appropriate semantic extensions of WordPress, so we have to developed a plug-in component that, after being integrated with the CH ontologies, was named *LinkedBlog*.

### *10.4.1  LinkedBlog*

We have built an approach [27] to ease the task of enriching web content with Linked Data by developing a plug-in for WordPress blog engine, that enables annotating text and video (see Fig. 10.5) with embedded RDFa [1]. We have provided an extended interface to write/annotate texts, videos and video fragments by customizing the open source TinyMCE[31] with RDFa capabilities.

A relevant feature of our proposal lies in the full independence of the ontological model. So, although this software is intended for CH ontologies, the plug-in can work with any ontological model available through an SPARQL-compatible endpoint.

---

[27]Liferay Portal – http://www.liferay.com/

[28]JSR-286 – http://jcp.org/aboutJava/communityprocess/final/jsr286/

[29]MediaWiki – http://www.mediawiki.org/wiki/MediaWiki

[30]WordPress – http://wordpress.org

[31]TinyMCE Editor – http://tinymce.moxiecode.com/

**Fig. 10.5** LinkedBlog video annotation window

The RDF transformation process is performed before saving metadata into the repository. First, the RDF statements are harvested from the RDFa code, using a GRDDL[32] scrapper. Next, the RDF reification mechanism is used to later associate metadata with these RDF statements. Finally, the RDF is contextualized for including application-specific information. In our case, we add some information such as the post author, the publication date and the corresponding post identifier. The RDF final code is according to the application ontology as represented in Fig. 10.6.

## 10.4.2 LinkedWiki

*LinkedWiki* is our proposal for editing and browsing ontologies through wiki pages. This is based on Semantic MediaWiki,[33] which is an extension to MediaWiki that enables RDF annotation and export capabilities, and Semantic Forms,[34] another extension that allows users to add, edit and query data using structured forms.

*LinkedWiki* automatically builds several forms and wiki articles for existing classes and individuals, respectively, using the ontological model accessible through

---

[32]GRDDL Recommendation – http://www.w3.org/TR/grddl/

[33]Semantic MediaWiki – http://semantic-mediawiki.org/

[34]Semantic Forms – http://www.mediawiki.org/wiki/Extension:Semantic_Forms

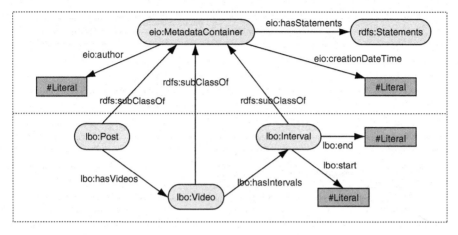

**Fig. 10.6** LinkedBlog ontology

**Fig. 10.7** LinkedWiki resource editing form

eCultura services API. In Fig. 10.7 we show an example of a specific individual wiki form retrieved from the underlying ontology.

*LinkedWiki* is a producer and consumer metadata application, in the sense that when a page is requested, it is properly augmented with semantic metadata and served to the browser. Likewise, when a page is saved, the extracted RDF instances are automatically harvested from the wiki article and stored into the eCultura repository through its services API. These RDF statements are according to the application ontology represented in Fig. 10.8. This way, the semantic wiki database and the eCultura ontology repository keep synchronized.

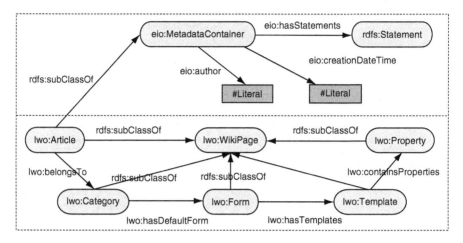

**Fig. 10.8** LinkedWiki ontology

### 10.4.3 LinkedMaps

We have adapted the Mapplet[35] prototype developed by the University of Alcalá [13] as an example of a consumer application. Having all the semantic content previously added to the repository, the semantic map application was conceived to offer users the ability to select, through several parameters, touristic itineraries over a Google Map layer (see Fig. 10.9). The parameters are ontology concepts such as historical characters, places and/or architectural styles of interest. These are positioned on a recommended visit pathway through the city of Alcalá de Henares, calculated by the application from a departure point.

Since the information model is shared, users can have updated information on the recommended sites as provided by the *LinkedWiki* or *LinkedBlog*. Because *LinkedMap* is only a metadata consumer application, an application ontology is not required.

## 10.5 Conclusions and Future Work

In this chapter, we present and discuss the benefits of an exploitable semantic Linked Data platform aimed at managing semantically-enhanced CH assets. Our work was motivated by providing an easy way to interconnect several domains of CH under an integrated, standards-based perspective. We believe that our interontology proposal tackles the problem of interconnecting domains of information through ontologies.

---

[35]Google Mapplets – http://code.google.com/intl/en/apis/maps

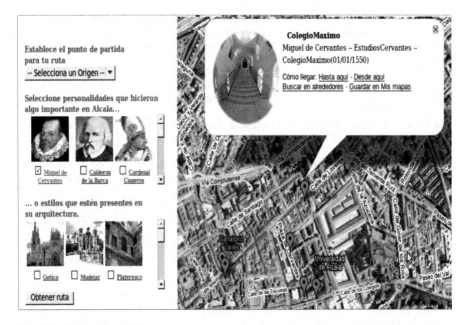

**Fig. 10.9**  LinkedMaps GUI

This chapter has reviewed the current practices for implementing such platforms, and provides an architecture for doing so on the basis of standard ontologies and application integration. Obtaining a higher level of cohesion between the semantic models of different applications is a difficult task, therefore it is necessary to make a comprehensive study of the requirements of each application. This solution is prepared to be extended with new ontologies. According to the arising trends of social network communities, a future review of a social network application should be considered and integrated upon the basis of FOAF ontology. Our goal in this project was to explore the possibilities of reusing interlinked data from externally available applications and adapting them for further use in a cultural context.

**Acknowledgements**  This work has been sponsored by a grant from the ASCETA project (P09-TIC-5230) of the Andalusian Government, Spain.

# References

1. Adida, B., Hausenblas, M.: Rdfa use cases: Scenarios for embedding rdf in html. World Wide Web Consortium, Working Draft WD-xhtml-rdfa-scenarios-20070330 (2007)
2. Auer, S., Bizer, C., Kobilarov, G., Lehmann, J., Cyganiak, R., Ives, Z.: Dbpedia: A nucleus for a web of open data. In: Aberer, K., et al. (eds.) Proc. of (ISWC+ASWC 2007), vol. 4825, chap. 52, pp. 722–735 (2007)
3. Auer, S., Lehmann, J.: What have innsbruck and leipzig in common? extracting semantics from wiki content. In: ESWC. pp. 503–517 (2007)

4. Bekiari, C., Charami, L., Georgis, C., Kritsotaki, A.: Documenting cultural heritage in small museums. In: Annual Conference of CIDOC. Athens, Greece (2008)
5. Bekiari, C., Doerr, M., Constantopoulos, P.: Information patterns for digital cultural repositories [electronic version]. ERCIM News 66, pp. 30–31 (2006)
6. Berners-Lee, T.: World-wide computer. Commun. ACM **40**(2), 57–58 (1997)
7. Bizer, C., Lehmann, J., Kobilarov, G., Auer, S., Becker, C., Cyganiak, R., Hellmann, S.: DBpedia-A crystallization point for the Web of Data. Web Semant. Sci. Serv. Agents World Wide Web **7**(3), 154–165 (2009)
8. Bollacker, K., Cook, R., Tufts, P.: Freebase: a shared database of structured general human knowledge. In: AAAI'07: Proc. of the 22nd national conference on Artificial intelligence. pp. 1962–1963 (2007)
9. Bollacker, K., Evans, C., Paritosh, P., Sturge, T., Taylor, J.: Freebase: a collaboratively created graph database for structuring human knowledge. In: SIGMOD '08: Proc. of the 2008 ACM SIGMOD international conference on Management of data. pp. 1247–1250 (2008)
10. Brodaric, B., Reitsma, F., Qiang, Y.: Skiing with dolce: toward an e-science knowledge infrastructure. In: Proc. of the 2008 conference on Formal Ontology in Information Systems. pp. 208–219 (2008)
11. Doerr, M., Bekiari, C., LeBoeuf, P.: Frbroo, a conceptual model for performing arts. In: Proc. of ICOM-CIDOC Annual Meeting. Museum Benaki Athens, Greece (2008)
12. Doerr, M., Iorizzo, D.: The dream of a global knowledge network—a new approach. J. Comput. Cult. Herit. **1**(1), 1–23 (2008)
13. García-Barriocanal, E., Sicilia, M.A.: On linking cultural spaces and e-tourism: An ontology-based approach. In: The Open Knowlege Society. A Computer Science and Information Systems Manifesto. pp. 694–701 (2008)
14. Gruber, T.R.: A translation approach to portable ontology specifications. Knowl. Acquis. **5**, pp. 199–220 (1993)
15. Hajmoosaei, A., Kareem, S.A.: An approach for mapping of domain-based local ontologies. In: CISIS. pp. 865–870 (2008)
16. Hausenblas, M., Bailer, W., Mayer, H.: Deploying Multimedia Metadata in Cultural Heritage on the Semantic Web. In: First International Workshop on Cultural Heritage on the Semantic Web, collocated with ISWC'07, Busan, South Korea (2007)
17. Hirsch, C., Grundy, J., Hosking, J.: Thinkbase: A visual semantic wiki. In: ISWC'08 (2008)
18. Hunter, J.: Combining the cidoc crm and mpeg-7 to describe multimedia. In Museums, Museums on the Web (2002)
19. Krtzsch, M., Vr, D., Vlkel, M.: Semantic mediawiki. In: Proc. 5th International Semantic Web Conference (ISWC06). pp. 935–942 (2006)
20. Mena, E., Kashyap, V., Illarramendi, A., Sheth, A.: Domain specific ontologies for semantic information brokering on the global information infrastructure (1998)
21. Niles, I., Pease, A.: Towards a standard upper ontology. In: FOIS '01: Proceedings of the international conference on Formal Ontology in Information Systems. pp. 2–9 (2001)
22. Noy, N.F.: Semantic integration: a survey of ontology-based approaches. ACM Sigmod Record **33**(4), 65–70 (2004)
23. Pan, F., Hobbs, J.: Time in owl-s. In: In Proceedings of AAAI Spring Symposium on Semantic Web Services. Stanford University, CA (2004)
24. Pérez, J., Arenas, M., Gutierrez, C.: Semantics and complexity of sparql. ACM Trans. Database Syst. **34**(3), 1–45 (2009)
25. Pinto, H.S., Martins, Jo.aP.: A methodology for ontology integration. In: K-CAP '01: Proceedings of the 1st international conference on Knowledge capture. pp. 131–138 (2001)
26. Ronchi, A.M.: eCulture, Cultural Content in the Digital Age. Springer, Erewhon, NC (2009)
27. Ruiz-Rube, I., Cornejo, C., Dodero, J., Garcia, V.M.: Development Issues on Linked Data Weblog Enrichment. In: Proceedings of the 2010 Metadata and Semantic Research. pp. 235–246 (2010)
28. Saha, G.K.: Web ontology language (owl) and semantic web. Ubiquity **8**(35), 1–1 (2007)

# Chapter 11
# Mediated Data Integration Systems Using Functional Dependencies Embedded in Ontologies

**Abdelghani Bakhtouchi, Chedlia Chakroun, Ladjel Bellatreche, and Yamine Aït-Ameur**

## 11.1 Introduction

In the past few decades, CityplaceEnterprise and Information Integration (EII) became an established business, where commercial and academic tools integrating data from various sources exist. They provide uniform and transparent access to data. The spectacular development of this business is largely due to companies requiring being able to access data located over the Internet and within their Intranets [12, 20].

Data integration problem inputs are a set of distributed, heterogeneous, autonomous, evolving data sources, where each one has its schemes and populations. It outputs a unified description of source schemes via an integrated schema and mapping rules allowing the access to data sources. The construction of data integration systems is a hard task [12]. This difficulty is due mainly to the nature of the manipulated sources: (a) the large number of data sources candidate for integration (b) the lack of explicitation of source semantic, (c) the heterogeneity of sources, (d) the autonomy of sources and (e) the dynamicity of the sources. (a) The *explosion of data sources*: the number of data sources involved in the integration process is increasing. The amount of information generated in the world increases by 30% every year and this rate is bound to accelerate [12], especially in domains such as E-commerce, engineering, environments, etc. Integrating this huge volume of data requires *automatic* solutions. (b) The *lack of explicitation of source semantic*: the semantics of data sources is usually not explicit. Most sources participating in the integration process were designed to satisfy day-to-day applications and

A. Bakhtouchi (✉)
National High School for Computer Science (ESI), Algiers, Algeria
e-mail: a_bakhtouchi@esi.dz

C. Chakroun · L. Bellatreche · Y. Aït-Ameur
LISI/ENSMA – Poitiers University Futuroscope, France
e-mail: chakrouc@ensma.fr; bellatreche@ensma.fr; yamine@ensma.fr

T. Özyer et al. (eds.), *Recent Trends in Information Reuse and Integration*,
DOI 10.1007/978-3-7091-0738-6_11, © Springer-Verlag/Wien 2012

not to be integrated in the future. Often, the small amount of semantic contained in their conceptual models (in the case where each source is a database) is lost, since only their logical models are implemented and used by applications. These models are obtained by translating conceptual models using some fixed rules to satisfy the requirement (normalization) of the target database management systems supporting these applications. The presence of a conceptual model allows designers to express the application requirements and domain knowledge in an intelligible form for a user. Thus, the absence of the conceptual model or any other semantic representation in final databases makes their interpretation and understanding complicated, even for designers who have good knowledge of the application domain. (c) The *heterogeneity* of data sources impacts both the *structure* and the *semantic*. Structural heterogeneity exits because data sources may have different structures and/or different formats to store their data. The autonomy of the sources increases heterogeneity significantly. Indeed, the data sources are designed independently by different designers with different application objectives. Each has a different structural approach. Many studies dealing with structural heterogeneity exist in the contexts of federated databases and multiple databases. Semantic heterogeneity presents a major issue in developing integration systems [19]. It is due to different interpretations of real world objects. Several categories of conflicts occur when integrating sources. [16, 18] suggest the following taxonomy: *naming conflicts*, *scaling conflicts*, *confounding conflicts* and *representation conflicts*. These conflicts may be encountered at schema level and at data level.

1. Naming conflicts: occur when naming schemes of concepts differ significantly. The most frequent case is the presence of synonyms and homonyms. For instance, the status of a person could mean either her family status or her employment status.
2. Scaling conflicts: occur when different reference systems are used to measure a value (for example *price* of a product can be given in dollars or euros).
3. Confounding conflicts: occur when concepts seem to have the same meaning, but differ in reality due to different measuring contexts. For example, the weight of a person depends on the date where it was measured. Among properties describing a data source, we can distinguish two types of properties: context dependent properties (e.g., the weight of a person) and context independent properties (gender of a person).
4. Representation conflicts: arise when two source schemas describe the same concept in different ways. For example in one source a student's name is represented by two elements *FirstName* and *LastName* while in another one it is represented by only one element *Name*.

(d) *Autonomy of sources*: most sources involved in the data integration work are fully autonomous. In other words, the owners of these sources are quitefree to modify their schemas or update their contents without informing the owners of other sources. (e) *Dynamicity of sources*: Since owners of data sources are different; the relationship between the integrated system and its sources is weakly coupled. A source must be able to modify its structure and population without informing the others and without causing abnormalities in integration systems. In such a context asynchronous evolution of the integration system involves schemas and populations [4].

By exploring the existing work and proposals on data integration, we figure out that each one focuses on one step of the chain of construction of data integration systems. This chain has five main steps: (a) source analysis, (b) choice of integration system architecture, (c) definition of global schema, (d) query processing and (e) data consolidation. In the first step, designers of integration systems identify the relevant sources participating in the integration process. This selection may be done manually, when the target sources are well known to the designers (small companies) and automatically by crawling sources (problem known as source selection [23]). After source selection, a fundamental step is also required which consists of analyzing sources to determine their *degree* of heterogeneity and autonomy. Based on this degree, some advanced tools may be considered. The second step consists of choosing the physical architecture of the final data integration system. Three main architectures are possible: materialized, mediator and hybrid (more details can be found in Sect. 11.2). The chosen architecture will have a great impact on query processing. Once sources are selected and the architecture chosen, the designer needs to fix the schema of integrated system (see Sect. 11.2). Finally, the end user may define and execute queries at the top of the integration schema. Data consolidation will happen, especially when mediator architecture is chosen. It occurs when similar objects are described in different data sources which pose the problem of identifying and aligning them [44].

Traditional data integration systems were built manually (e.g. TSIMMIS [11]), where the presence of designers was ubiquitous. With the explosion of the number of data sources, an automatic system becomes highly needed. The main task that requires large amount of automatic effort is the *heterogeneity detection and resolution*. To facilitate the management of heterogeneity, some projects use *metadata* (represented in RDF or RDFS) [5]. Some research efforts propose to use ontologies, where several integration systems were proposed around this hypothesis. We can cite COIN [18], Observer [31], Picsel2 [33], OntoDW [43], etc. In [4], we claim that if the semantic of each source participating in the integration process is explicit (in a priori way), the integration process becomes automatic. This means the used ontology exists before the creation of the sources. This assumption is reasonable since several domain ontologies exist in various areas: medicine, engineering, travel, etc. The explicitation of different concepts used in a database leads to the concept of *ontology-base databases* (OBDB).

Most of existing integration systems suppose that there is a common single identifier for each common concepts between the sources. The assumption facilitates the consolidation of query results but it violates the sources autonomy. Suppose two entities *E1* and *E2* representing the same concept *Person* belonging to two different sources *S1* and *S2*. Each entity has its own identifier (*Social_Security_Number* for *E1* and *Family_Name* for *E2*). Suppose that these two keys are both candidates to identify in a unique way instances of the concept *Person*. If this information (candidate keys) is annotated in ontologies used to integrate sources, the consolidation may be easier. In this chapter, we propose an ontology-based integration approach with mediator architecture. It exploits the presence of ontology referenced by selected sources to explicit their semantic. Instead of annotating different candidate keys of each ontology class, we list its set of functional dependencies defined on its properties. Functional dependencies can be seen as ontological concepts. They can be identified either during the ontology development or by extracting them using work done by [35]. Their presence in ontology determines easily the candidate keys of each class, by running classical algorithms [13]. This information allows the generation of a list of candidate keys for each ontology class. The presence of this list increases the source autonomy, where each one may choose its keys from that list. As consequence, probabilistic methods to conciliate query results are not useful.

The chapter is structured as follows. Section 11.2 presents a state of the art related to data integration problem, where a classification of existing systems is proposed. Section 11.3 proposes the concepts and the formal definitions of ontology and ontology-based databases used by our proposal. The different representations of functional dependencies in ontologies are also given. Section 11.4 describes in details our integration architecture with their main components. Section 11.5 presents a prototype implementing our proposal using OntoDB – an OBDB environment and a real ontology, called EU-Car Rental. Finally Sect. 11.6 concludes the chapter by summarizing the main results and describing possible future work.

## 11.2   State of the Art

A data integration system combines data residing in various sources and provides to users a unified view of these data [29]. It got great attention from academic and business communities, where several prototypes and commercial systems were proposed under some hypothesis. But, numerous open problems remain unsolved. In this section, we propose a classification of integration systems that helps designers and readers to understand each studied system and to analyze it.

### 11.2.1   Data Integration Systems Classification

It is rather difficult to classify the existing integration systems. Most studies classified them using only one criterion. Some studies designate two categories of integrated systems those using Global as View (GaV) approach [11], and those using Local as View (LaV) approach [17,30,33]. Other contributions distinguish between systems using a single ontology, multiple ontologies, and a shared ontology [41]. Some other works focus on the place of data and distinguish mediator and warehouse approaches [39]. We propose below three orthogonal criteria which allow the characterization of the various existing integration systems. These criteria will be described in the next sections.

**Data Representation.** This criterion specifies whether data of local sources are duplicated in a warehouse or remained in local sources and then accessed through a mediator. Two major data integration architectures are proposed in the literature: materialized (warehouse) and virtual (mediator). In the materialized architecture [24, 27] a copy of data recovered from the sources is stored in a single database (called a data warehouse). The data stored in the warehouse are pre-processed in a complex way before storage; it is usually referred to as extract, transform and load (ETL) (Fig. 11.1.a). In this approach, the difficulty is in the data transformation, contrary to the mediation approach, which is oriented especially towards queries rewriting. The data warehouse eliminates several problems of integration, mainly the excessively long server response times, the network clogging, or the sources unavailability. Queries can be also more easily optimized, and the data transformed at the user's discretion. These modifications, even if their usefulness can be proven, will obviously not be reflected on the local sources. A main disadvantage of this approach is that the answers to the queries can frequently be built from outdated data. Data warehouse updating can be expensive, and furthermore copy rights may exist on what certain sources provide.

In the virtual architecture [42], software called a mediator supports a virtual database (without storing data into a database), translates queries into source queries and synthesizes results and returns answers to a user query. A mediator is based on a unified global schema as a synthesis of the source schemas; on this global schema queries are expressed. The most important step in using a mediator is the global schema creation. Contrary to the data warehouse, here the mapping deals with the relationships between the global schema and the local sources. The specification of these correspondences – according to the used method- determines the query reformulation difficulty, as well as the facility of adding or removing of sources to the system. Two main concepts make up this architecture (Fig. 11.1.b): wrappers and mediators. A wrapper wraps an information source and models the source using

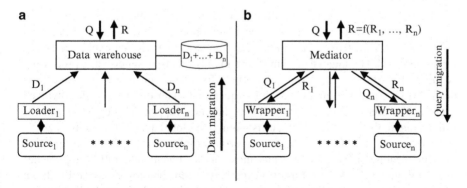

**Fig. 11.1** (a) Materialized architecture, (b) Virtual architecture

a source schema. A mediator maintains a global schema and mappings between the global and source schemas.

A third integration approach mixes fully materialized and fully virtual approaches and combines their advantages [23]. In this approach, part of the data is materialized in a warehouse whereas the rest is not. To reduce the query response time, some frequently sought data are cached. An example of a system adopting this approach is Picsel 3 [34].

**Sense of the Mapping between the Global and Local schemas.** In *Global-as-View* (GaV) systems, the global schema is expressed as a view (a function) over data sources. This approach facilitates the query reformulation by reducing it to a simple execution of views in traditional databases. However, changes in source schema or adding a new data source requires a designer to revise the global schema and the mappings between the global schema and source schemas. Thus, GaV is not scalable for large applications. In the source-centric approach, each data source is expressed with one or more views over the global schema.

*Local-as-View* (LaV) scales better, and is easier to maintain than GaV because the designer creates a global schema independently of source schemas. Then, for a new source schema, the designer has only to give (adjust) a source description that describes source relations as views of the global schema. In order to evaluate a query, a rewriting in terms of the data sources is needed. The rewriting queries using views is a difficult problem in databases [22]. Thus, LaV has low performance when queries are complex.

There is a more general approach in mapping design that generalizes both the GaV and the LaV paradigms. Such an approach, called *Generalized Local-as-View* (GLaV), requires the designer to associate a general query over the source relations to a general query over the global relations. GLaV mappings are more expressive, and are well suited to represent complex relationships in distributed data integration environments. An example of a system adopting this approach is the PIAZZA project [21].

**Mapping automation.** This criterion specifies whether the mapping between the global schema and local schemas is done in a manual, a semi-automatic, or a fully automatic way.

Manual mappings are found in the first generation of integration systems that integrate sources represented by a schema and a population (i.e., each source $S_i$ is defined as: $< Sch_i, Pop_i >$ as in classical databases and without explicit meaning representations. The manual systems focus mainly on query support and processing at the global level, by providing algorithms for identifying relevant sources and decomposing (and optimizing) a global query into sub queries for the involved sources. The construction of the mediators and the wrappers used by these systems is done manually because their focus is mainly on global query processing [10].

To make the data integration process (partially) automatic, explicit representation of data meaning is necessary. Thus most of the recent integration approaches use ontologies [10, 19, 33]. Based on the way where ontologies are utilized, we may discern three different architectures [41]: single ontology methods, multiple ontologies methods, and hybrid methods (see Fig. 11.2). In the single ontology

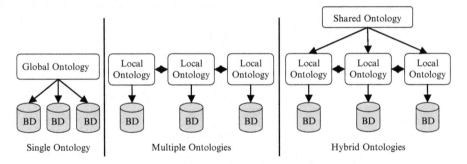

**Fig. 11.2** Different ontology architectures

approach, each source is related to the same global domain ontology (e.g., Lawrence et al. work [28] and Picsel [17] and [19]). As a result, a new source cannot bring new or specific concepts without requiring change in the global ontology. This violates the *source schematic autonomy* requirement (each source can extend its schema independently). In the multiple ontologies approach (e.g., Observer, [31]), each source has its own ontology developed regardless of the other sources. Then, inter-ontology meanings need to be defined. In this case, the definition of the inter-ontology mapping is very difficult as different ontologies may use different aggregation and granularity of the ontology concept [41]. The hybrid approach has been proposed to overcome the drawbacks of single and multiple ontologies approaches. In this approach, each source has its own ontology, but all ontologies are connected by some means to a common shared vocabulary (e.g., KRAFT project [40]).

In all these approaches, ontologies and ontology mappings are defined at integration time. Therefore, they always request a human supervision, and they are only partially automatic.

To enable automatic integration, the semantic mapping shall be defined during the database design. This means that there exists a shared ontology and moreover, each local source shall contain ontological data that refers to the shared ontology. Some systems have already been proposed on that direction such as Picsel2 [33], and COIN [18]. But to remain automatic, these systems do not allow individual data source to add new concepts and properties.

## 11.2.2   Data Consolidation

Data integration has three broad goals: increasing the *completeness, conciseness,* and *correctness* of data that is available to users and applications [12]. *Completeness* measures the amount of data, in terms of both the number of tuples and the number of attributes. *Conciseness* measures the uniqueness of object representations in the integrated data, in terms of both the number of unique objects and the number of

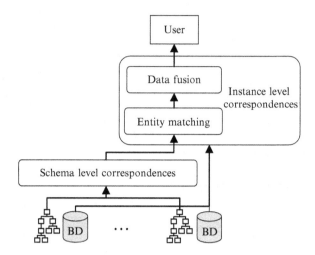

**Fig. 11.3** Tasks in data integration

unique attributes of the objects. Finally, *correctness* measures correctness of data; that is, whether the data conform to the real world.

To meet *conciseness*, and *correctness*, a data integration system needs to perform tow level of tasks: *schema level correspondences* and *instance level correspondences* (see Fig. 11.3).

Schema level correspondences task aims at establishing semantic mappings between contents of disparate data sources. That means identifying tables that represent the same real-world entity type and attributes that represent the same property about some entity type. This task is performed when creating the global schema of the integration system.

The goal of instance level correspondences is to establish reconciliation between instances that represent the same entity in the real world and to produce the correct data of such entity. Some integration systems suppose the existence of a single identifier between the sources for each common concept. This identifier allows the consolidation of the results of a query using relational operations (join and union and their relatives). However, given the autonomy of the sources, this identifier can not exist; How to decide then if two descriptions coming from distinct sources refer or not to the same entity of the real-world (the same person, the same article, the same gene, the same hotel). For that, integration systems perform entity reconciliation methods [7, 32, 36, 44]. The instance level mapping can be treated in tow steps: *Entity matching* and *Data fusion*.

Entity matching (duplicate detection) step aims at resolving heterogeneity at the instance level by detecting records that refer to the same real-world entity. The various approaches to entity matching proposed in the literature can be classified into two broad categories: *rule-based* approaches and *learning-based* approaches [44]. In rule-based approaches, domain experts are required to directly provide decision rules for matching semantically corresponding records. In learning-based

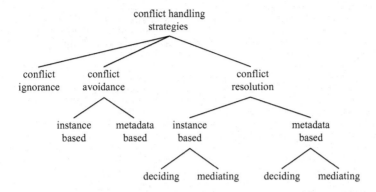

**Fig. 11.4**  A classification of conflict resolution strategies [7]

approaches, domain experts are required to provide sample matching (and non-matching) records, based on which classification techniques are used to learn the entity matching rules.

The objective of data fusion step is to combine records that refer to the same real-world entity by fusing them into a single representation and resolving possible conflicts from different data sources. Schema level correspondences and Entity matching aim at removing redundancy and increasing conciseness of the data, whereas data fusion aims at resolving conflicts from data and increasing correctness of data. There are many different strategies to resolve conflicts adopted by different integration systems [16, 32, 37]. Figure 11.4 classifies existing strategies to resolve data conflicts [7]. In particular, *Conflict ignoring* strategies are not aware of conflicts, perform no resolution, and thus may produce inconsistent results. *Conflict avoiding* strategies are aware of conflicts but do not perform individual resolution for each conflict. Rather, a single decision is made, e.g., preference of a source, and applied to all conflicts. Finally, *conflict resolving* strategies provide the means for individual fusion decisions for each conflict. Such decisions can be *instance-based*, i.e., they deal with the actual conflicting data values, or they can be *metadata-based*, i.e., they choose values based on metadata, such as freshness of data or the reliability of a source. Finally, strategies can be classified by the result they are able to produce: *deciding* strategies choose a preferred value among the existing values, while *mediating* strategies can produce an entirely new value, such as the average of a set of conflicting numbers.

## 11.3   Background

In this section, we present concepts and definitions that facilitate the understanding of our proposal.

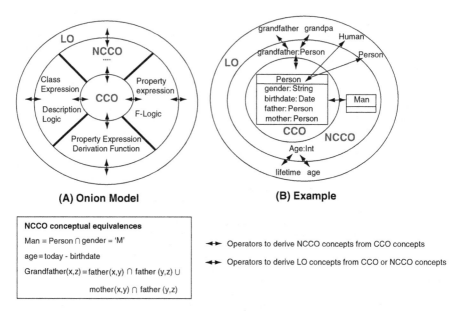

**(A) Onion Model**

**(B) Example**

| NCCO conceptual equivalences |
|---|
| Man ≡ Person ∩ gender = 'M' |
| age ≡ today - birthdate |
| Grandfather(x,z) ≡ father(x,y) ∩ father (y,z) ∪ |
| mother(x,y) ∩ father (y,z) |

↤↦  Operators to derive NCCO concepts from CCO concepts

↤↦  Operators to derive LO concepts from CCO or NCCO concepts

**Fig. 11.5** The onion model

## 11.3.1 Taxonomy and Formal Model for Ontologies

Ontologies can be organized into a layered model, called the Onion Model [26] and shown in Fig. 11.5. The Onion Model is composed of three layers: *Conceptual Canonical Ontologies*, *Non Conceptual Canonical Ontologies* and *Linguistic Ontologies*.

1. Conceptual Canonical Ontologies (CCOs) contain ontologies which describe concepts and not the words of a language and whose definitions do not contain any redundancy. CCOs adopt an approach of structuring of information in term of classes and properties and associate to these classes and properties a single identifiers reusable in various languages. CCOs can be considered as shared conceptual models. They contain the core classes and play the role of a global schema in DB integration architecture.

2. Non Conceptual Canonical Ontologies (NCCOs) contain ontologies which also describe concepts but represent not only primitive concepts (canonical), but also definite concepts (non canonical). i.e. which can be defined from primitive concepts and/or other definite concepts. NCCOs provide mechanisms similar to views in DBs; non canonical concepts can be seen as virtual concepts defined from canonical concepts. These mechanisms may be used to represent mappings between different DBs.

3. Linguistic Ontologies (LOs) are those ontologies whose scope is the representation of the meaning of the words used in a particular universe of discourse,

in a particular language. Beyond the textual definitions, a number of linguistics relationship (synonymous, hyponym, etc) are used to capture, in an approximate and semi-formal way, the relation between the words. LOs may be used to localize similarities between DB schemas, to document existing DBs or to improve the DB-user dialog.

**Formal Model for Ontologies.** Formally, conceptual canonical ontology may be defined as the quadruplet $O :< C, P, Sub, Applic >$ [4], where:

- $C$ is the set of the classes used to describe the concepts of a given domain.
- $P$ is the set of properties used to describe the instances of the $C$ classes.
- *Sub* is the subsumption function defined as $Sub : C \rightarrow 2^C$. For a class $c$ of the ontology, it associates its direct subsumed classes. *Sub* defines a partial order over $C$.
- *Applic* is a function defined as $Applic : C \rightarrow 2^P$. It associates to each class of the ontology, the properties that are applicable for each instance of this class and that may be used, in the database, for describing its instances. Applicable properties are inherited through *is-a* subsumption and partially imported through *case-of* subsumption. Note that for each $c \in C$, only a subset of $Applic(c)$ may be used in any particular database, for describing $c$ instances.

Example 1: we consider the ontology containing the classes (BranchType, Branch, RentalAgreement and Reservation) from the EU-Car Rental ontology of Fig. 12.

- $O :< C, P, Sub, Applic >$, where:
- C = {BranchType, Branch, RentalAgreement, Reservation}
- P = {BranchTypeId, BranchTypeName, BranchId, BranchName, IsOfType, RentalAgreementId, basicPrice, bestPrice, lastModification, onRentInterval, PickUpBranch, DropOffBranch, resrvationDate}
- Sub(BranchType) = ∅
- Sub(Branch) = ∅
- Sub(RentalAgreement) = {Reservation}
- Sub(Reservation) = ∅
- Applic(BranchType) = {BranchTypeName}
- Applic(Branch) = {BranchName, IsOfType}
- Applic(RentalAgreement) = {basicPrice, bestPrice, lastModification, onRentInterval, PickUpBranch, DropOffBranch}
- Applic(Reservation) = {resrvationDate}

## 11.3.2 Ontology-Based Databases

Ontology-based database is a database composed of four parts (Fig. 11.6). Part 1 and 2 are traditional parts available in all DBMSs, namely the data part that contains instance data and meta-base part that contains the system catalog. Part 3 (Ontology)

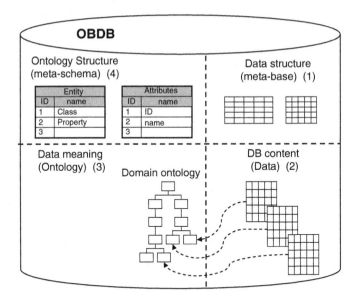

**Fig. 11.6** OntoDB architecture

and part 4 (meta-schema) are specific to OntoDB. The ontology part allows the representation of ontologies in the database. The ontologies supported by OntoDB architecture are all those that can be represented and exchanged as models following Bernstein's terminology [6], i.e., a set of objects accessible from a root object using links between objects. This definition corresponds to most of the ontology models recently developed such as OWL.[1]

The meta-schema part records the ontology model into a reflexive meta-model. For the ontology part, the meta-schema part plays the same role as the one played by the meta-base in traditional databases. Indeed, this part may allow: (1) generic access to the ontology part, support of evolution of the used ontology model, and (2) storage of different ontology models. By means of naming convention, the meta-base part also represents the logical model of the content, and its link with the ontology, thus representing implicitly the conceptual model of data in database relations.

Formally, an OBDB is a quadruplet $< O, I, Sch, Pop >$, where:

- $O$ is an ontology ($O :< C, P, Sub, Applic >$).
- $I$ is the set of instances of the database. The semantics of these instances is described in $O$ by characterizing them by classes and properties values.
- $Pop : C \rightarrow 2^I$ associates to each class its own instances. $Popc$ constitute the population of $c$.

---

- *Sch* : $C \rightarrow 2^P$ associates to each ontology class $c$ of $C$ the properties, which are effectively used to describe the instances of the class $c$. For each class $c$, *Sch(c)* must satisfy: $Sch(c) \subseteq Applic(c)$.

## 11.4  Our Proposal

In this section, we start by incorporating functional dependencies into the ontology formalism then we propose our methodology to integrate ontology-based database sources, where ontologies are enriched by functional dependencies. We describe all components of our mediator.

### 11.4.1  Adding Functional Dependencies to Ontologies

Traditional ontology models do not support functional dependencies in their definition. Data dependencies have been introduced as a general formalism for a large class of database constraints that augments the expressivity of database models [2]. Functional dependencies (FD) compose a particularly interesting data dependency [14] that elegantly models the relationships between attributes of a relation. FDs are used for defining primary keys and in normalization theory avoiding redundant data representation and update anomalies which a database may suffer from. Other important application of FD in database includes query rewriting [22] and query evaluation [1].

A FD noted $R: X \rightarrow Y$, with $X$ and $Y$ being sets of attributes of the relation $R$, means that to each tuple values of the attributes of $X$ corresponds one and only one tuple of attributes of $Y$. If we transpose similar rules to ontology world, we discover that FDs could be very useful to enrich the expressivity of the knowledge representation.

Recently, a couple of studies proposed to enrich ontologies by FDs [8, 9, 35, 37, 38] that we classify into two main categories: (1) *intra-class dependencies* and (2) *inter-class dependencies*.

1. In the first category, we find the work of [3] which is quite similar to those proposed in traditional databases. Each ontology class $c$ may have a set of FD defined on its simple properties $\{p_1, \ldots, p_n\}$. *A simple property is a property with atomic* range. For instance, on the Customer class of the EU-Car Rental ontology presented in the Fig. 11.12 is defined the two FDs; the id property functionally determines the name and the telephone properties (*Customer*: *id* → *name*, *Customer*: *id* → *telephone*) (see Fig. 11.7).
2. The second category involves dependency between classes. Two types arise: FD with a single class in the left part and FD with several classes in the left part. In the first type; we find the work of [35]. The authors define a FD ($C_1 \rightarrow C_2$)

**Fig. 11.7** Example of FDs
intra class

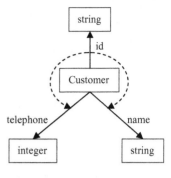

**Fig. 11.8** Example of a
single class left part FDs

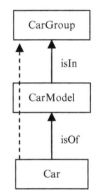

between classes $C_1$ and $C_2$ when each instance of the class $C_1$ determines a single instance of the class $C_2$ (in other word, for each $x$ in $C_1$ corresponds only one $y$ in $C_2$). For example, for an instance of the *Car* class corresponds a single instance of the *CarGroup* class (see Fig. 11.8).

In the second type, we find the work of [8]. The authors define a FD ($R : \{C_1, \ldots, C_n\} \rightarrow C$) between the classes $\{C_1, \ldots, C_n\}$ linked to a root class $R$ by properties (or properties chains) and a class $C$ linked to the root class by a property (or properties chain), if the instances of the $n$ classes determines a single instance of the class $C$. For example, an instance of the *Country* class and an instance of the *BranchType* class determine a single instance of the *RentalDuration* class (see Fig. 11.9).

## 11.4.2 Formalization of Ontologies with FD

Since, FD are a part of ontologies (ontological concept), we propose to extend the traditional formalization of ontologies by considering FD. First of all, we formalize FD. To do so, we use a formalization similar to the one proposed by [8]. A FD

**Fig. 11.9** Example of a multiple classes left part FDs

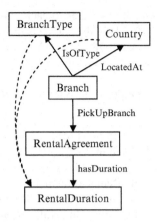

definition *fd*, is composed of the following elements: the left part *LP*, the right part *RP*, and a root class:

$fd = (R, LP, RP)$

This definition can also be expressed as an implication, as in traditional FDs:

*fd* $R : LP \rightarrow RP$

with *R* a class,

$LP = \{\{p_{1,1}, \ldots, p_{1,m1}\}, \ldots, \{p_{n,1}, \ldots, p_{n,mn}\}\}$ a set of paths (properties chains),

$RP = \{p_1, \ldots, p_l\}$ a path (properties chain).

The left part *LP* is a set of *n* paths. A path is composed of a chain of properties, each one being *pi*. The right part is defined by a single path, which is composed of *l* properties. The root class *R* is the starting point of all paths in the left part and the right part, so that a FD expresses relationships among properties of a single instance of the *R* class.

The formal semantics of a FD that the instances of the *n* classes $\{c_1, \ldots, c_n\}$ linked to a root class *R* by the paths $\{\{p_{1,1}, \ldots, p_{1,m1}\}, \ldots, \{p_{n,1}, \ldots, p_{n,mn}\}\}$ determines a single instance of the class *c* linked to the root class by the path $\{p_1, \ldots, p_l\}$ if the instances of the n classes determines a single instance of the class *c*.

A FD can be classified in three classes: (1) basic FD: fd = (R, RP), (2) classic FD: fd = (R, LP, RP), and (3) key FD: fd = (R, LP).

Example 2: let us formalize the FD of Fig. 9.

fd R: $\rightarrow$ RP where:

R = Branch,

LP = {{IsOfType}, {LocatedAt}},

RP = {PickUpBranch, hasDuration}.

Now, we have all the ingredients to extend the initial ontology model proposed in Sect. 11.3.1. As a consequence, it is composed of 5-uplets *O*: <*C, P, Sub, Applic, FD*>, where *FD* is a binary relationship $FD : C \rightarrow (\{2^P \times 2^P \times \cdots \times 2^P\}, 2^P)$

which associates to each class $c$ of $C$, the set of the functional dependencies $(LP, RP)$, where the class $c$ is the root $(fd\ c\colon LP \to RP)$.

A OBDB $< O, I, Sch, Pop >$ with $O :< C, P, Applic, Sub, FD >$ satisfies a FD $(R, LP, RP)$ if $LP \subseteq P$ et $RP \subseteq P$.

Example 3: the ontology of the example 1 becomes $O :< C, P, Sub, Applic, FD >$, where:

– FD(BranchType) = {(BranchTypeId, BranchTypeName)}
– FD(Branch) = {(BranchId, BranchName), (BranchId, IsOfType), (Branch-Name, IsOfType)}
– FD(RentalAgreement) = {(RentalAgreementId, basicPrice), (RentalAgreementId, bestPrice), (RentalAgreementId, lastModification), (RentalAgreementId, onRentInterval), (RentalAgreementId, PickUpBranch), (RentalAgreementId, DropOffBranch)}
– FD(Reservation) = {(RentalAgreementId, resrvationDate)}

**The Impact of FD on Data Consolidation**

In this section, we show the effect of considering functional dependencies in ontologies on data consolidation. In the case where no common identifier is used by various sources, consolidating query results leads to four possible cases: (1) manual consolidation based on the experience and deep knowledge of data sources of designers which is practically impossible in the real life, where a large number of sources is involved. (2) Only sources having common identifiers are taken into consideration to process queries. In this case, mediator propagates query on sources having the common identifiers. This solution compromises the quality of returned results. (3) Query results are consolidated, where some instances overlap which may cause error. (4) Overlapping instances may be discarded using probabilistic reconciliation. To illustrate the role of FD on data consolidation, let us consider the following example.

Example 4: let $S_1$, $S_2$ and $S_3$ be three sources containing the same relation *Customer*. With different properties as follows:

$S_1.Customer(\underline{id}(PK), name, address, telephone)$

$S_2.Customer(\underline{id}(PK), name, telephone)$

$S_3.Customer(\underline{telephone}(PK), name, address)$

On this table, the following FDs are defined:

$fd_1\ Customer : id \to name$

$fd_2\ Customer : id \to address$

$fd_3\ Customer : id \to telephone$

$fd_4\ Customer : telephone \to name$

$fd_5\ Customer : telephone \to address$

This table has two candidate keys: *id* and *telephone*. Suppose that the mediator schema contains a *Customer* relation with the following properties: *id(PK), name, address*. Suppose the following query: "list names and addresses of all customers". The mediator decomposes the query on the three sources. Without FD, we cannot consolidate all sources, since the source 3 has a different identifier. By using $fd_4$ *Customer* : *telephone* $\rightarrow$ *name* and $fd_5$ *Customer* : *telephone* $\rightarrow$ *address*, we notice that the attribute *telephone* is a common candidate key between the three sources. Therefore, a consolidation of the results coming from these three sources becomes possible.

## *11.4.3   Architecture of our Proposal*

Contrary to classical systems which required the existence of all sources before running the integration process, our system integrates the sources on their arrival (on-the-fly integration). The mediator global schema is incrementally enriched by adding a new source.

First, we formalize our proposal before going on to the definition of its architecture. In this formalization we specify the objective, the inputs and the outputs of the proposed approach.

*Objective*: The objective is to allow consolidation of the results coming from different sources in absence of a common identifier.

*Inputs*: We have in input:

1. A shared ontology $O_S$ enriched by functional dependencies defined on each class; hence $O_S$ is represented as a 5-uplets $O_S$ :< $C_S, P_S, Applic_S, Subs_S, FD_S$ >.
2. A set of data sources $S = \{S_1, S_2, \ldots, S_n\}$, each source is an OBDB represented as a 4-uplet $S_i$ :< $O_i, I_i, Sch_i, Pop_i$ >. The ontology $O_i$ references the shared ontology $O_S$.

*Outputs*: We have as output a global schema *Med* which can be defined as a quadruplet:

$$Med :< O, S, Sch, M >$$

1. $O$ is the mediator ontology generated by selection of classes and properties from the shared ontology $O_S$ (Algorithm 3).
2. $S$ is the set of sources. $S = \{S_1, S_2, \ldots, S_n\}$ with $S_i$ defined as a couple $S_i$ :< $OL_i, SchL_i$ >.

   We start with an empty set for $S$ ($S = \emptyset$); by adding a new source $S_i$ having the ontology $O_i$ :< $C_i, P_i, Applic_i, Sub_i, FD_i$ >, we add a source $S_i$ :< $OL_i, SchL_i$ > to $S$. the ontology $OL_i$ is generated from $O_i$. In $OL_i$ we keep only the classes and properties existing in the mediator ontology $O$ ($OL_i = O_i \cap O$). $SchL_i$ is the schema of the classes of $OL_i$.
3. *Sch* is the mediator schema. $Sch : C \rightarrow 2^P$ associates to each mediator ontology class $c$ of $C$ the properties, which are effectively selected to describe the instances of the class $c$ in sources accessible by the mediator.

4. $M$ is the mapping between the classes of mediator ontology $O$ and the classes of source ontologies $OL.M$ : $C \rightarrow 2^{\{C_1 \cup \cdots \cup C_n\}}$ associates to each mediator ontology class $c$ of $C$ the classes of source ontologies in correspondence with the class $c$.

The mapping of the classes of the mediator ontology is initialized at an empty set $(M(c) = \emptyset \forall c \in C)$; by adding a new source $S_i$ having the ontology $O_i$ :< $C_i, P_i, Applic_i, Sub_i, FD_i >$ we update the mapping of these classes $(M(c) = M(c) \cup OL_i.c$ if $c \in C_i)$. Adding a new source and mapping update are performed by algorithm 11.3.

Example 5: Let us consider a mediator ontology containing three classes Customer, Branch and Country and let $S_1$, $S_2$ and $S_3$ be three sources containing the three relations Customer, Branch, and Country with different properties as follows:

1. $S_1$.Customer(idCustomer, CustomerName, telephone, address, BranceName)
2. $S_1$.Branch(BranceName, CountryName)
3. $S_1$.Country(CountryName, carTax)
4. $S_2$.Customer(idCustomer, CustomerName, telephone, BranceName)
5. $S_2$.Branch(BranceName, CountryName)
6. $S_2$.Country(CountryName, carTax)
7. $S_3$.Customer(telephone, CustomerName, address, idBranch)
8. $S_3$.Branch(idBranch, IdCountry)
9. $S_3$.Country(IdCountry, CountryName, carTax)

On the ontology, the following FDs are defined:

1. $fd_1$ Customer : idCustomer $\rightarrow$ CustomerName
2. $fd_2$ Customer : idCustomer $\rightarrow$ telephone
3. $fd_3$ Customer : idCustomer $\rightarrow$ address
4. $fd_4$ Customer : idCustomer $\rightarrow$ BelongsTo
5. $fd_5$ Customer : telephone $\rightarrow$ CustomerName
6. $fd_6$ Customer : telephone $\rightarrow$ address
7. $fd_7$ Customer : telephone $\rightarrow$ BelongsTo
8. $fd_8$ Branch: idBranch $\rightarrow$ BranchName
9. $fd_9$ Branch: idBranch $\rightarrow$ LocatedAt
10. $fd_{10}$ Branch: BranchName $\rightarrow$ LocatedAt
11. $fd_{11}$ Country: idCountry $\rightarrow$ CountryName
12. $fd_{12}$ Country: idCountry $\rightarrow$ carTax
13. $fd_{13}$ Country: CountryName $\rightarrow$ carTax

The component of the mediator $Med$ :< $O, S, Sch, M >$ are the following:
1. The mediator ontology $O$ :< $C, P, Sub, Applic, FD >$: $C = \{$Customer, Branch, Country$\}$

$P = \{$idCustomer, CustomerName, telephone, address, BelongsTo, idBranch, BranceName, BranchAdress, LocatedAt, idCountry, CountryName, carTax$\}$

$Sub(c) = \phi \forall c \in C$

Applic (Customer) $= \{$idCustomer, CustomerName, address, BelongsTo$\}$

Applic(Branch) $= \{$idBranch, BranceName, LocatedAt$\}$

Applic (Country) $= \{$idCountry, CountryName, carTax$\}$

$FD(Customer) = \{fd_1, fd_2, fd_3, fd_4, fd_5, fd_6, fd_7\}$

$FD(Branch) = \{fd_8, fd_9, fd_{10}\}$

$FD(Country) = \{fd_{11}, fd_{12}, fd_{13}\}$

2. $S = \{S_1 < OL_1, SchL_1 >, S_2 < OL_2, SchL_2 >, S_3 < OL_3, SchL_3 >\}$

We assume that $OL_1 = OL_2 = OL_3 = O$

$SchL_1$(Customer) $= \{$idCustomer, CustomerName, telephone, address, BelongsTo$\}$

$SchL_1$(Branch) $= \{$BranceName, LocatedAt$\}$

$SchL_1$(Country) $= \{$CountryName, carTax$\}$

$SchL_2$(Customer) $= \{$idCustomer, CustomerName, telephone, BelongsTo$\}$

$SchL_2$(Branch) $= \{$BranceName, LocatedAt$\}$

$SchL_2$(Country) $= \{$CountryName, carTax$\}$

$SchL_3$(Customer) $= \{$CustomerName, address, telephone, BelongsTo$\}$

$SchL_3$(Branch) $= \{$idBranch, LocatedAt$\}$

$SchL_3$(Country) $= \{$IdCountry, CountryName, carTax$\}$

3. The mediator schema contains the following classes and properties:

Sch(Customer) $= \{$CustomerName, BelongsTo$\}$

Sch(Branch) $= \{$BranceName, BranchAdress, LocatedAt$\}$

Sch(Country) $= \{$CountryName, carTax$\}$

4. The different mappings are:

M(Customer) $= \{OL_1$.Customer, $OL_2$.Customer, $OL_3$.Customer$\}$

M(Branch) $= \{OL_1$.Branch, $OL_2$.Branch, $OL_3$.Branch$\}$

M(Country) $= \{OL_1$.Country, $OL_2$.Country, $OL_3$.Country$\}$

**The proposed architecture.** As shows in Fig. 11.10, our architecture is composed of: (1) a user interface, (2) an OBDB, (3) a query engine and (4) a results consolidator.

1. The user interface allows the user to express his query and is responsible for displaying its results after formatting.
2. The OBDB used adopts the OntoDB model with extensions of the meta-schema (Fig. 11.11). We add to the meta-schema of the OntoDB model, a model of mapping between the mediator ontology and source ontologies. In the ontology part we store the mediator ontology $O$, source ontologies and schemas $< OLi, SchL_i >$, the mapping between the mediator ontology and source ontologies and the functional dependencies between the classes and properties of the mediator ontology.
3. The query engine identifies the concerned sources and rewrites the query written in terms of the mediator ontology into a query written in terms of source

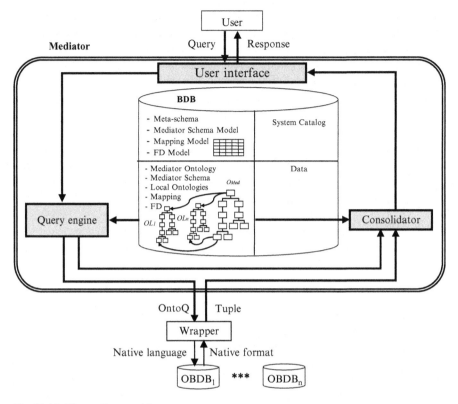

**Fig. 11.10** The mediator architecture

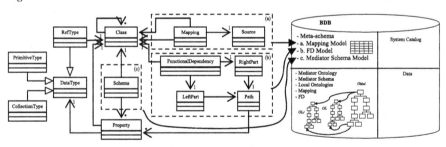

**Fig. 11.11** Extension of OntoDB meta-schema: (**a**) mapping meta-schema, (**b**) FDs meta-schema and (**c**) meta-schema of mediator schema

ontologies using the mapping. It generates then the query consolidation plan. It sends each sub query to the appropriate source and it sends the query consolidation plan to the consolidator.

4. The consolidator recomposes the results returned by the different wrappers after consultation of the query consolidation plan and using functional dependencies stored in the mediator ontology.

### 11.4.4   *Mediator Ontology and Schema Creation*

The global schema of the mediator *Med*: $<O, S, Sch, M>$ contains four components, the first component is the mediator ontology $O$: $<C, P, Applic, Sub, FD>$. This ontology is imported by selection of classes and properties from a shared ontology $O_S$ :$< C_S, P_S, Applic_S, Sub_S, FD_S >$ (Algorithm 11.1). The process is done following these steps:

1. Starting from user requirements, the administrator of the mediator selects classes and properties from the shared ontology.
2. The selected classes and properties are added to the mediator ontology.
3. If an imported class has a super class, its super class is imported also, and so on if its super class has a super class until attaining the first class in the hierarchy.
4. To keep the semantic of complex properties (object properties), the importation of a complex property involves the importation of its range class.
5. Likewise, to keep functional dependencies, the importation of a property appearing in a right part of a FD implies the importation of all the properties of the left part of this FD.

**Schema creation.** The second component of the mediator global schema is the mediator schema *Sch* which is created by selection of classes and properties from the mediator ontology according to users' requirements. The mediator's administrator selects a set of classes among the classes of the mediator ontology. He selects than the properties for each class. To be accessible for users, the selected properties must be valued in at least one source among the integrated sources.

### 11.4.5   *Sources Integration*

We have seen in the previous section the determination of ontology and schema components of the mediator global schema ($O$ and $Sch$), these tow components are determined only once whereas the tow remaining components ($S$ :$< OL_i, SchL_i >$ and $M$) are updated after each integration of a new source (on-the-fly integration (see Sect. 11.4.1)).

The integration of a new source $S_i$ :$< O_i, I_i, Sch_i, Pop_i >$ is run in the following steps (algorithm 11.3):

– First, we create an empty ontology $OL_i$ for the source $S_i$ in $OL$.

---

**Input:** $O_S$ :$< C_S, P_S, Applic_S, Sub_S, FD_S >$: the shared ontology
**Output:** $O$ :$< C, P, Applic, Sub, FD >$: the mediator ontology
**foreach** *class* $c_S \in C_S$ **do**
    **if** $c_S$ *is selected* **then**
        *ImportClass*($c_S$)
    **end**
**end**

**Algorithm 11.1:** Ontology Importation

**Function** *ImportClass(c_S)*;
**begin**
    Add the class $c_S$ to $C$;
    **for** *each property $p_i \in Applic_S(c_S)$* **do**
        **if** *$p_i$ is selected* **then**
            Add the property $p_i$ to $P$
            Add the property $p_i$ to $Applic(c_S)$
            **if** *$p_i$ is an object property* **then**
                | *ImportClass(Range($p_i$))*
            **end**
            $FD_S$ implied properties
            **for** *each $fd_j \in FD_S(c_S)$ and $fd_j : LP_j \rightarrow p_i$* **do**
                **for** *each property $p_k \in LP_j$ and $fd_j : LP_j \rightarrow p_i$* **do**
                    **if** *$p_k \notin P$* **then**
                        Add the property $p_k$ to $P$
                        **if** *$p_k$ is an object property* **then**
                            | *ImportClass(Range($p_k$))*
                        **end**
                    **end**
                **end**
                Add $(LP_j, p_i)$ to $FD(c_S)$
            **end**
        **end**
    **end**
    **if** $c_S \in Sub_S(SuperClass(c_S))$ **then**
        *ImportClass(SuperClass($c_S$))*
        Add the class $c_S$ to $Sub(SuperClass(c_S))$
    **end**
**end**

**Algorithm 11.2:** Function of a class importation

**Input**: $S_i :< O_i, I_i, Sch_i, Pop_i >$ the source $S_i$
**Output**: $Med :< O, S, Sch, M >$: the mediator schema
        $OL_i :< C_{OL_i}, P_{OL_i}, Applic_{OL_i}, Sub_{OL_i}, FD_{OL_i} >$
Add the source $S_i < OL_i, SchL_i >$ to $S$, $OL_i$ is an empty ontology.
**for** *each class $c_j \in C$* **do**
    **if** $c_j \in O_i$ **then**
        Add the class $c_j$ to $C_{OL_i}$
        Add the properties of $c_j$ to $P_{OL_i}$
        Add $Applic_i(c_j)$ to $Applic_{OL_i}(c_j)$
        Add $Sch_i(c_j)$ to $SchL_i(c_j)$
        Add $OL_i.c_j$ to $M(O.c_j)$
    **end**
**end**

**Algorithm 11.3:** Integration of a new source

– Second, we keep to the ontology $OL_i$ from $O_i$ only the classes and properties existing in the mediator ontology.

– Third, we import the schemas of the classes of $OL_i$ from $Sch_i$ to $SchL_i$.

– Finally, we establish the mapping between the classes of the mediator ontology $O$ and the classes of the new source ontology $OL_i$.

**Input**: $Q$ : (SELECT $p_1, \ldots, p_n$ FROM $c$ WHERE $p_1$ op$_1$ $v_1$ AND $\ldots$ AND $p_m$ op$_m$ $v_m$) a
     query on the mediator schema
         $Med :< O, S, Sch, M >$: the mediator schema
**Output**: *Result: list of tuples satisfying the query*
*Concerned Sources identification*
**for** *each source $S_i$ where $OL_i.c \in M(O.c)$* **do**
 | **if** $\{p_1, \ldots, p_n\} \cap SchL_i(OL_i.c) \neq \emptyset$ **then**
 |  | // One property at least is valued in $S_i$ Add $S_i$ to *ConcernedSources*
 | **end**
**end**
*Candidate keys determination*
**for** *each property $p_i$ of $Q$* **do**
 | $LP_i = \emptyset$
 | **for** *each $fd_j \in FD(c)$ and $LP_j \to p_i$* **do**
 |  | Add $LP_j$ to $LP_i$
 | **end**
 | $CandidateKeys = CandidateKeys \cap LP_i$
**end**
$SelectedCandidateKey =$ Choose from *CandidateKeys* a candidate key valued in all
sources of *ConcernedSources*
$Result = \emptyset$
**for** *each source $S_i \in ConcernedSources$* **do**
 | **for** *each property $p_j$ of $c$* **do**
 |  | **if** $p_j$ is valued in $S_i$ **then**
 |  |  | **if** $p_j \in \{p_1, \ldots, p_n\}$ **then**
 |  |  |  | Add $p_j$ to *PropertiesListe$_i$*
 |  |  | **end**
 |  |  | **if** $p_j \in \{p_1, \ldots, p_m\}$ **then**
 |  |  |  | Add $p_j$ op$_j$ $v_j$ to *ConditionsListe$_i$*
 |  |  | **end**
 |  | **end**
 |  | **else**
 |  |  | **if** $p_j \in \{p_1, \ldots, p_n\}$ **then**
 |  |  |  | Add (null AS $p_j$) to *PropertiesListe$_i$*
 |  |  | **end**
 |  | **end**
 | **end**
 | $Result_i =$ Execute of the query (SELECT *PropertiesListe$_i$* FROM $c$ WHERE
 | *ConditionsListe$_i$*) on $S_i$
 | *Results consolidation*
 | *Consolidate(Result, Result$_i$, SelectedCandidateKey)*
**end**

**Algorithm 11.4:** A query processing


## 11.4.6   Querying Sources

The process of querying sources can be decomposed into the execution of simple
queries ($Q$: SELECT FROM $c$ WHERE $p_1$ op$_1$ $v_1$ AND $\ldots$ AND $p_n$ op$_n$ $v_n$). This
process takes as inputs a simple query $Q$ and the mediator schema $Med$: $<O, OL,$
$Sch, M>$. This process runs through four phases (Algorithm 11.4): (1) involved

**Function** *Consolidate(Result$_1$, Result$_2$, Key)*
**begin**
   **for** *each instance I$_i$ of Result$_2$* **do**
      **for** *each instance I$_j$ of Result$_1$* **do**
         **if** *Result$_1$[I$_i$, Key] = Result$_2$[I$_j$, Key]* **then**
           |  *Result$_1$[I$_i$, Key] = Result$_1$[I$_i$, Key] ∪ Result$_2$[I$_j$, Key]*
         **end**
         **else**
           |  Add the instance *Result$_2$[I$_j$]* to *Result$_1$*
         **end**
      **end**
   **end**
**end**

**Algorithm 11.5:** Function of tow results consolidation

sources identification, (2) candidate key determination, (3) interrogation of the sources and (4) the results consolidation.

1. **Involved sources identification**: among all sources we look for the ones in their ontologies a class mapped to the class $c$ and one property of $p_1, \ldots, p_n$ at least is valued in $S_i$. Such sources contain the needed information and will be interrogated in the third phase.
2. **Candidate key determination**: we determine first all candidate keys, then we choose among them, a candidate key valued in all the sources. This candidate key will be used in the results consolidation phase.
3. **Interrogation of the sources**: we rewrite the initial query $Q$ into the fusion of sub queries on the involved sources. Each sub query ($Q_i$: SELECT $p_1, \ldots, p_r$ FROM $c$ WHERE $p_1$ op$_1$ $v_1$ AND $\ldots$ AND $p_r$ op$_r$ $v_r$) on the source $S_i$ contains in the SELECT clause $\{p_1, \ldots, p_r\} \subseteq \{p_1, \ldots, p_n\}$ the properties valued in this source. To obtain the same structure of the results coming from different sources, a not valued property $p_i$ is replaced by the string "null AS $p_i$"; this facilitates the fusion of these results.
4. **Consolidation of the results**: the consolidation of the results coming from different sources is run in an incremental way. The results coming from the second interrogated source are consolidated with those coming from the first one, we thereby obtain the first consolidated results; the results coming from the third interrogated source are consolidated with the first consolidated results; and so on until the consolidation of the results coming from the last interrogated source and the previous consolidated results.

Example 6: on the mediator of the example 5 we execute the query $Q = \Pi$CustomerName, address($\sigma$(Customer)).

**Involved sources identification**: the tree sources are involved.
**Candidate key determination**: from $fd_5$ Customer: telephone $\rightarrow$ CustomerName and $fd_6$ Customer: telephone $\rightarrow$ address the algorithm determines *telephone* as a candidate key.

**Interrogation of the sources**: The query $Q$ will be rewritten on the tree sources as follows:

$$Q_{S_1} = \prod_{CustomerName, address} (\sigma(Customer))$$
$$Q_{S_2} = \prod_{CustomerName} (\sigma(Customer))$$
$$Q_{S_3} = \prod_{CustomerName, address} (\sigma(Customer))$$

**Consolidation of the results**: using the *telephone* property the results coming from the three sources will consolidated.

## 11.5   Validation and Testing of our Architecture

In this section, we propose a validation of our design approach. The database architecture supporting our validation is OntoDB. In this validation we use the following process: (1) make persistency of mediator schema, mapping and FDs, (2) mediator ontology and schema creation and (3) sources integration.

### 11.5.1   Make Persistency of Mediator Schema, Mapping and FDs

The meta-schema of OntoDB contains tow main tables *Entity* and *Attribute* encoding the meta-model level. Entity describes ontological concepts like class, property or data type. Attribute describes attributes related to each ontological concept (name, description, comment . . . ). An extension of the meta-schema of OntoDB is proposed to support the mapping, functional dependencies, and mediator schema. Precisely, three meta-models describing the mapping concepts, functional dependencies concepts, and mediator schema concepts are first proposed (Fig. 11.11). Secondly the three meta-models are instanciated in the meta-schema of OntoDB using OntoQL– an ontology query language proposed to querying OntoDB [25]. The following statements encode this instantiations.

```
CREATE ENTITY #Source( #SourceName STRING,
                       #ConnectionString STRING,
                       #SourceDriver STRING);

CREATE ENTITY #Mapping(#MediatorClass REF (#Class),
                       #SourceClass REF (#Class),
                       #ItsSource REF (#Source));

CREATE ENTITY #Path(#ItsProperties REF (#Property)ARRAY);

CREATE ENTITY #LeftPart(#ItsPaths REF (#Path)ARRAY);

CREATE ENTITY #RightPart(#ItsPath REF (#Path));
```

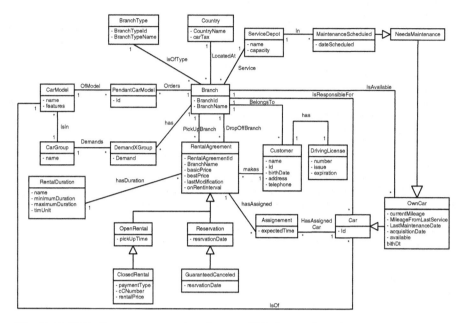

**Fig. 11.12** UML diagram representation of a part of EU-Car rental ontology

CREATE ENTITY #FunctionalDependency(
                          #ItsRootClass REF (#Class),
                          #ItsLeftPart REF (#LeftPart),
                          #ItsRightPart REF (#RightPart),
                          #ItsFunction STRING);

CREATE ENTITY #Schema( #ItsClass REF (#Class),
                          #SchemaProperty REF(#Property));

To illustrate this step, we use the EU-Car Rental ontology [15] represented in
Fig. 11.12. The following OntoQL statements give an example of a creation of the
*Customer* class.

CREATE #Class Customer(
          DESCRIPTOR(#name[en] = 'Customer'
          )
PROPERTIES(idCustomer STRING,
          CustomerName STRING,
          address STRING,
          telephone INT)
          );

## 11.5.2  Mediator Schema Creation

The following OntoQL statements add the CustomerName property to the mediator schema.

```
INSERT INTO #Schema(#ItsClass, #SchemaProperty)
VALUES ( ( SELECT #class.#oid
          FROM #class
          WHERE #class.#name = 'Customer'),
        ( SELECT #Property.#oid
          FROM #Property
          WHERE #Property.#name = 'CustomerName')
        );
```

## 11.5.3  Sources Integration

The following OntoQL statements add the source S, its ontology and give an example of a mapping between two classes.

```
INSERT INTO #Source( #SourceName,
                     #ConnectionString,
                     #SourceDriver)
VALUES ('S1','S1 Connection String','S1 Driver');

INSERT INTO #ontology(#namespace) VALUES ('S1-Ontology');

INSERT INTO #Mapping(#MediatorClass, #SourceClass, #ItsSource)
VALUES (( SELECT #class.#oid
          FROM #class
          WHERE #class.#name = 'Customer'
          and #class.#definedBy = 'Mediator-Ontology'),
        ( SELECT #class.#oid
          FROM #class
          WHERE #class.#name = 'Customer'
          and #class.#definedBy = 'Mediator-Ontology'),
        ( SELECT #Source.#oid
          FROM #Source
          WHERE #Source.#SourceName = 'S1')
        );
```

## 11.6  Conclusion

The need for developing semantic integration systems increases with the evolution of domain ontologies in various systems such as engineering, medicine, etc. The presence of ontologies contributes largely in solving heterogeneity of sources. Some actual integration systems suppose that the manipulated sources have similar keys to ensure data integration which violates the autonomy characteristic of sources. Others use statistical techniques to consolidate data. In sensitive domains, these techniques cannot be used. In this chapter, we propose a rich state of the art of data integration systems, with classification based on three criteria. This classification facilitates the understanding of manipulated systems. Secondly, an integration system of ontology-based database sources is presented in mediator architecture. Its main particularity is that it allows sources to have different keys that ensure a precise consolidation. All components of our system are formalized and the underlying algorithms are presented. Our system is validated using OntoDB – an ontology-based database developed in our laboratory.

Several issues that arise from our work should be explored: the development of semantic cache system to speed up the most frequent queries. A large scale evaluation needs to be conducted to show the real efficiency of our system. Finally, metrics should be defined to measure the quality of our integration system.

# References

1. Abiteboul, S., Duschka, O.: Complexity of answering queries using materialized views. In: Proceedings of the ACM SIGACT-SIGMOD-SIGART Symposium on Principles of Database Systems (PODS), Seattle, WA (1998)
2. Abiteboul, S., CityplaceHull, R., Vianu, V., Wesley, A.: Foundations of Databases Addison Wesley Publishers, ISBN 0-201-53771-0 (1995)
3. Bellatreche, L., Ameur, Y.A., Chakroun, C.: A Design Methodology of Ontology based Database Applications. Logic Journal of the IGPL. **18**(2), (2010)
4. Bellatreche, L., Xuan, D.N., Pierra, G., Dehainsala, H.: Contribution of ontology-based data modeling to automatic integration of electronic catalogues within engineering databases. Comput. Ind. J. **57**(8-9), 711–724 (2006)
5. Berger, S., Bry, F., Furche, T., Linse, B., Schroeder, B.: Beyond XML and RDF: The Versatile Web Query Language Xcerpt. In: Proceedings of WWW'06, pp. 1053–1054 (2006)
6. Bernstein, P.: Applying Model Management to Classical Meta Data Problems. In: Proceedings of the 2003 CIDR Conference (2003)
7. Bleiholder, J., Naumann, F.: Data fusion. ACM Comput. Surv. **41**(1), 1–41 (2008)
8. Calbimonte, J., Porto, F.: Functional Dependencies in OWL A-BOX. XXIV Simpósio Brasileiro de Banco de Dados, 05-09 de Outubro, CityplaceFortaleza, Ceará, Brasil, Anais. (SBBD 2009), pp. 16–30 (2009)
9. Calvanese, D., Giacomo, G., Lenzerini, M.: Identification constraints and functional dependencies in description logics. In: Proceedings of IJCAI, pp. 155–160 (2001)
10. Castano, S., Antonellis, V., Vimercati, S.D.C.: Global Viewing of Heterogeneous Data Sources. IEEE Trans. Knowl. Data Eng. **13**(2), 277–297 (2001)
11. Chawathe, S.S. et al: The TSIMMIS Project: Integration of Heterogeneous Information Sources. in proceedings of the 10th Meeting of the Information Processing Society of Japan, pp. 7–18 (1994)
12. Dong, X., Naumann, F.: Data Fusion – Resolving Data Conflicts for Integration. VLDB '09 (2009)
13. Elmasri, R., Navathe, S.B.: Fundamentals of Database Systems. (3rd edn.), Addison Wesley (2000)

14. Fagin, R.: Functional dependencies in a relational data base and propositional logic. IBM J. Res. Dev. **21**(6), 543–544 (1977)
15. Frías, L., Queralt, A., Olivé, A.: EU-Rent car rentals specification. Technical report, Departament de Llenguatges i Sistemes Informatics (2003)
16. Fuxman, A., Fazli, E., Miller, R.J.: Conquer: Efficient management of inconsistent databases. In: Proceedings of SIGMOD, pp. 155–166, Baltimore, MD (2005)
17. Goasdoué, F., Lattés, V., Rousset, M.C.: The Use of CARIN Language and Algorithms for Information Integration: The PICSEL System. Int. J. Cooper. Inform. Syst. (IJCIS). **9**(4), 383–401 (2000)
18. Goh, C., Bressan, S., Madnick, E., Siegel, M.D.: Context Interchange: New Features and Formalisms for the Intelligent Integration of Information. ACM Trans. Inform. Syst. **17**(3), 270–293 (1999)
19. Hakimpour, F., Geppert, A.: Global Schema Generation Using Formal Ontologies. In: Proceedings of 21th International Conference on Conceptual Modeling (ER'02), pp. 307–321 (2002)
20. Halevy, A.Y., et al.: Entreprise information integration: successes, challenges and controversies. In: Proceedings of the ACM SIGMOD International Conference on Management of Data, pp. 778–787 (2005)
21. Halevy, A.Y., Ives, Z.G., Madhavan, J., Mork, P., Suciu, D., Tatarinov, I.: The Piazza Peer Data Management System. IEEE Transactions on Knowledge and Data Engineering (2003)
22. Hong, J., Liu, W., Bell, D.A., Bai, Q.: Answering Queries Using Views in the Presence of Functional Dependencies. BNCOD, pp. 70–81 (2005)
23. City Hull, R., Zhou, G.: A Framework for Supporting Data Integration Using the Materialized and Virtual Approaches. In: Proceedings of ACM SIGMOD '96, Montreal, Canada (1996)
24. Inmon, B., Wiley, J.: ed. Using the Data Warehouse (1999)
25. Jean, S., Aït-Ameur, Y., Pierra, G.: Querying ontology based databases. The OntoQL proposal. Software Engineering and Knowledge Engineering (SEKE' 06), pp. 166–171 (2006)
26. Jean, S., Pierra, G., Aït-Ameur, Y.: Domain ontologies: a database-oriented analysis. Web Information Systems and Technologies (WEBIST'2006), pp. 341–351 (2006)
27. Kimball, R., Caserta, J., Wiley, J.: Ed. The Data Warehouse ETL Toolkit, Practical Techniques for Extracting, Cleaning, Conforming, and Delivering Data (2004)
28. Lawrence, R., Barker, K.: Integrating relational database schemas using a standardized dictionary. In: Proceedings of the ACM Symposium on Applied Computing (SAC), pp. 225–230 (2001)
29. Lenzerini, M. Data Integration: A Theoretical Perspective. in proceedings of the ACM SIGACT-SIGMOD-SIGART Symposium on Principles of Database Systems (PODS'02), pp. 233–246 (2002)
30. Levy, A.Y., Rajaraman, A., Ordille, J.J.: The World Wide Web as a Collection of Views: Query Processing in the Information Manifold', In: Proceedings of VIEW'1996, pp. 43–55 (1996)
31. Mena, E., Vipul Kashyap, V., Illarramendi, A., Sheth, A.P.: Managing Multiple Information Sources through Ontologies: Relationship between Vocabulary Heterogeneity and Loss of Information. In: Proceedings of Third Workshop on Knowledge Representation Meets Databases (1996)
32. Motro, A., Anokhin, P., Acar, A.C.: Utility-based resolution of data inconsistencies. In: Proceedings of IQIS Workshop, pp. 35–43, CityplaceParis, country-regionFrance (2004)
33. Reynaud, C., Giraldo, G.: An Application of the Mediator Approach to Services over the Web. Special track Data Integration in Engineering, Concurrent Engineering (CE'2003), pp. 209–216 (2003)
34. Reynaud, C., Safar, B.: Construction automatique d'adaptateurs guidée par une ontologie pour l'intégration de sources et de données XML. Technique et Science Informatiques (TSI). **28**, 199–228 (2009)
35. Romero, O., Calvanese, D., Abello, A., Rodriguez-Muro, M.: Discovering Functional Dependencies for Multidimensional Design. ACM 12th Int. Workshop on Data Warehousing and OLAP (2009)

36. Saïs, F., Pernelle, N., Rousset, M.C.: Réconciliation de références : une approche adaptée aux grands volumes de données. Colloque sur l'Optimisation et les Systémes d'Information (COSI), pp. 521–532 (2007)
37. Schallehn, E., Sattler, K.U., Saake, G.: Efficient similarity-based operations for data integration. Data Knowl. Eng. **48**(3), 361–387 (2004)
38. Toman, D., Weddell, G.E.: On keys and functional dependencies as first-class citizens in description logics. J. Automat. Reas. **40**(2–3), 117–132 (2008)
39. Ullman, J.D.: Information Integration Using Logical Views. In: Proceedings of the International Conference on Database Theory (ICDT), Lecture Notes in Computer Science, 1186: 19–40 (1997)
40. Visser, P.R.S., Beer, M., Bench-Capon, T., Diaz, B.M., Shave, M.J.R.: Resolving Ontological Heterogeneity in the KRAFT Project. In: Proceedings of DEXA'99, pp. 668–677 (1999)
41. Wache, H., Vögele, T., Visser, U., Stuckenschmidt, H., Schuster, G., Neumann, H., Hübner, S.: Ontology-based Integration of Information – A Survey of Existing Approaches, In: Proceedings of the International Workshop on Ontologies and Information Sharing, pp. 108–117 (2001)
42. Wiederhold, G.: Mediators in the architecture of future information systems. IEEE Comput. **25**(3), 38–49 (1992)
43. Xuan, D.N., Bellatreche, L., Pierra, G.: OntoDaWa, un système d'intégration à base ontologique de sources de données autonomes et évolutives. Ingénierie des Systèmes d'Information. **13**(2), 97–125 (2008)
44. Zhao, H., Ram, S.: Entity matching across heterogeneous data sources: An approach based on constrained cascade generalization. Data Knowl. Eng. archive (ACM). **66**(3), 368–381 (2008)

# Chapter 12
# Algorithms for Spatial Data Integration

**Mark McKenney**

## 12.1 Introduction

The traditional basic units of spatial representation are the *point, line, and region* [5,15]. Points represent collections of points in space, lines are used to represent one dimensional features and networks, such as roads and rivers, and regions represent areas, such as states or countries. These spatial primitives form the foundation of spatial storage techniques; however, spatial visualization typically occurs in the form of *maps*. Maps are a familiar concept that incorporate spatial, thematic, and even temporal data into an intuitive representation that inherently manages large amounts of data in a visual form: for example, adjacency of regions, connectivity of road networks, and the relationship between thematic data and spatial and geographical phenomena are easily identified in maps. Despite the utility of maps, maps are represented, stored, and managed as collections of their component points, lines, and regions in spatial systems. For example, a map of the United States may show the fifty states as a single map, but in current storage approaches, fifty separate regions representing the states will be stored, not a single map. Therefore, the information inherent in a map is not represented and must be explicitly computed from a collection of spatial primitives.

The use of maps as primarily a visualization tool in spatial systems leads to two problems: (1) maps are difficult to reuse in forming new maps or in computing operations over maps because they are not a data item in themselves, but a visualization of many separate data items, and (2) information that is naturally encoded into maps, such as adjacency of items, must be computed on demand because no relationships are maintained among the separate constituents of the map since they are independent data items. We propose a new concept of maps in spatial

M. McKenney (✉)
Department of Computer Science, Texas State University, 601 University Drive,
San Marcos, TX 78666-4684, USA
e-mail: mckenney@txstate.edu

T. Özyer et al. (eds.), *Recent Trends in Information Reuse and Integration*,
DOI 10.1007/978-3-7091-0738-6_12, © Springer-Verlag/Wien 2012

systems in which maps are individual data items, not conglomerates of separate data items. This new concept of maps provides three benefits: (1) visual information and spatial relationships inherent to map structure are automatically maintained, (2) map reuse is facilitated because operations to manipulate, extract information from, and combine maps as individual data items are well documented [6, 16, 17], and (3) algorithms for maps can be expressed over single map data items instead of collections of spatial primitives, which allows for much more efficient algorithm implementations enabling map management on a large scale. However, the creation of map data from region data is problematic.

The *map construction problem* (MCP) is the problem of creating a single map data item from a collection of individual regions. The MCP is essentially a spatial integration problem in which separate spatial components and their associated thematic information must be preserved in the context of a more complex data item representing an entire map. In this paper, we consider maps that contain only region structures. Given a collection of $m$ regions, a mechanism is required to construct a single data item that represents the map defined by the collection of regions. From a spatial and geometric perspective, mechanisms exist to combine separate region items into a single partition of space [2]; however, maintaining the associated thematic data with portions of regions that overlap in a map is not addressed. Currently, a map must be computed incrementally by repeatedly adding a single region to a growing map; effectively, this causes regions to be processed multiple times once they are added to the growing result map. For $m$ input regions, this results in $O((mn \lg mn)^2 + k)$ complexity, where $n$ indicates the average size of a region in terms of its representation and $k$ is the number of intersections among boundary line segments. This quadratic behavior is observed in commercial applications such as geographic information systems, and typically occurs when data layers, each consisting of multiple component regions, are overlaid to form a map [10]. We propose a new integration mechanism that can achieve a map from a collection of $m$ regions in $O((n + k)(\lg n + m + \lg m^2))$ time and $O(n + k + m^2)$ space for $n$ total boundary line segments. Our approach considers the entire input as a single collection of *labeled* region boundary elements. The labels allow the linking of thematic data to the regions in the result map, and by defining label representation thoughtfully, they can be leveraged to preserve all original thematic information from the original component regions in the resulting map, and can be processed efficiently. This, coupled with the fact that our approach processes each input region only once leads to an efficient solution.

Although the algorithmic solution to the MCP presented in this paper is efficient in terms of both space and time complexity, it does not address the underlying characteristic of spatial and geographic data that such data tends to be very large. For example, one data set used for testing represents the counties of ten states in middle of the US and is roughly 155 Mb. A second data set is roughly 610 Mb, creating a maximal memory usage of over one gigabyte since our first solution to the MCP must keep the input regions, output map, and various data structures in memory throughout the algorithm; thus, the first algorithm requires computers with significant memory resources for larger data sets. Therefore, we present an external memory version of the algorithm that accepts a memory bound which limits the algorithm's memory usage. The external memory version of the algorithm is also efficient, and allows the proposed algorithm to be run on standard desktops and laptops. The experimental results in Sect. 12.6 show the performance of our algorithms over small to large data sets on laptop, desktop, and server hardware.

This paper is structured as follows. Sect. 12.2 describes existing work relating to spatial data type definitions and plane sweep algorithms for region processing. Implementation considerations of the plane sweep algorithm and notations used in this paper are given in Sect. 12.3. We describe our algorithm for integrating regions into maps in Sect. 12.4, and provide an external memory version of the algorithm in Sect. 12.5. We report performance results of our implementations of the algorithms in Sect. 12.6. Finally, in Sect. 12.7, we draw some conclusions.

## 12.2   Related Work

The traditional spatial data type of complex regions [15] describes a region as a collection of *faces* such that faces must be disjoint or meet at a finite set of points. A face can contain zero or more *holes*, describing area that is not part of a face, but surrounded by it. For example, Italy contains multiple faces, its mainland and islands, and a hole that does not belong to Italy where Vatican City lies. From an implementation perspective, regions are typically represented by a set of straight line segments indicating the boundary of the region (Fig. 12.1).

In this paper, we refer to maps as defined in the model of spatial partitions [6, 13]. We choose this model because it is mature, and a full algebra is defined around it, including operations and predicates. In short, a spatial partition defines a collection of regions such that regions either meet along boundaries, or are pairwise disjoint. Each region is assigned a unique *label*, which models thematic information. Regions are not allowed to overlap. If two regions $r$ and $s$ do overlap, their intersection is computed and three regions are actually inserted into the map, the intersection of $r$ and $s$, and the differences $r - s$ and $s - r$. The label of the intersecting portions of the regions will contain the labels of both regions; thus, three *region primitives*, each with unique labels, represent the original two regions. Figure 12.1 depicts and example.

The efficient combination of both the geometric and thematic portions of regions into a map is an open problem that we address in this paper. Current solutions,

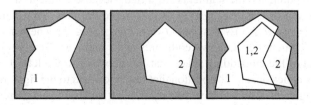

**Fig. 12.1**  Two labeled regions, and the map formed when they are overlaid

as observed through experimentation with commercial GIS applications, depict a quadratically scaling running time based on input size [10]. Our proposed solution to this problem was first described in [12]; we improve that solution in this paper so that the resulting algorithm labels overlapping segments correctly. Furthermore, this paper addresses the known problem that spatial data tends to be large, requiring external memory algorithms to implement spatial operations in general.

The combination of the geometric portions of regions in 2-dimensional space into a map is achieved through line segment intersection algorithms. In this paper, we employ a version of the *plane sweep algorithm* for computing overlays of regions and maps [2, 10]. Given two input regions, their overlay consists of both regions overlaid on each other such that boundary line segments intersect only at end points. The plane sweep algorithm proceeds by sweeping an imaginary line over a pair of regions or maps, $R$ and $W$. Each time the line encounters a new line segment on the boundary of one of the input geometries, that segment is added to a list of segments actively being considered, known as the *active list*. Each time the line moves past a segment, that segment is removed from the active list. The active list stores the line segments in the order in which they intersect the sweep line; thus, if the sweep line is traveling in the $x$ direction of a euclidean plane, the segments in the active list are sorted based on the $y$ value of their intersection point with the imaginary sweep line. Because segments in the active list are sorted, the segments in the active list adjacent to a segment $s$ from region $R$ that is newly inserted into the active list contains the information necessary to determine whether $s$ lies on the interior, boundary, or exterior of $W$. This knowledge is deduced because each line segment carries two identifiers, one indicating the region that lies above the segment, and the other indicating the region that lies below the segment. If no region lies above or below a segment, then a default label indicating the exterior of the region is assigned. The plane sweep algorithm *processes* one segment in each iteration of the algorithm. Processing a segment involves discovering if it intersects any other line segments, adding it to the active list, and determining which region interiors lie above and below the segment by looking at its immediate neighbors in the active list.

## 12.3 Implementation Aspects of the Plane Sweep Algorithm

In implementation, the sweep line does not move continuously in a sweep line algorithm, but instead progresses based on segment endpoints. Therefore, each segment is actually represented twice, once for its beginning end point, and once for its ending end point, and the segments are sorted. Because segments are represented twice, the notion of a *halfsegment* is used to represent them. We define the type *halfsegment* $= \{(s, d, l, r)|s \in segment, d \in bool, l, r \in \mathbb{Z}\}$ where a segment is a straight line segment between two endpoints and $l$ and $r$ are *labels* corresponding to the portion of the embedding space that lies above or to the left of the line that the halfsegment lies on, and the portion that lies below and to the right, respectively. Thus, a halfsegment is said to have two *sides*, a left and right side corresponding to each label. As a matter of notation, we refer to $l$ as the *above label* of $h$, and $r$ as the

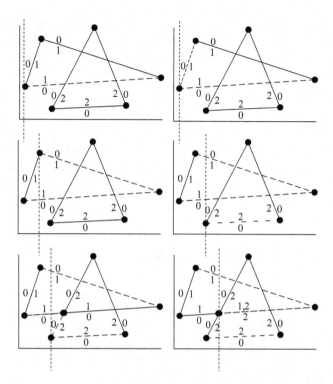

**Fig. 12.2** A partial example sequence of a sweep line algorithm integrating two triangles into a map. Segments are shown with labels indicating where triangle interiors lie. Note that only labels to the left of the sweep line are finalized. The label 0 indicates the exterior of both triangles

*below label* of $h$. For a halfsegment $h = (s, d, l, r)$, if $d$ is true (false), the smaller (greater) endpoint of $s$ is the *dominating point* of $h$, and $h$ is called a *left (right) halfsegment*. Hence, each segment $s$ is mapped to two halfsegments $(s, true, l, r)$ and $(s, false, l, r)$. Let $dp$ be a function which yields the dominating point of a halfsegment and $len$ be a function that returns the length of a halfsegment. For two distinct halfsegments $h_1 = (s_1, d_1, l_1, r_1)$ and $h_2 = (s_2, d_2, l_2, r_2)$ with a common endpoint $p$, let $\alpha$ be the enclosed angle such that $0° < \alpha \leq 180°$. Let a predicate $rot(h_1, h_2)$ be true if, and only if, $h_1$ can be rotated around $p$ through $\alpha$ to overlap $h_2$ in counterclockwise direction. We define a total order on halfsegments as:

$$h_1 < h_2 \Leftrightarrow$$
$$dp(h_1) < dp(h_2) \vee$$
$$(dp(h_1) = dp(h_2) \wedge ((\neg d_1 \wedge d_2) \vee$$
$$(d_1 = d_2 \wedge rot(h_1, h_2)) \vee$$
$$(d_1 = d_2 \wedge collinear(s_1, s_2) \wedge len(s_1) < len(s_2))))$$

Figure 12.3 depicts an example sequence of the sweep line algorithm. In Fig. 12.3, part of the sweep line algorithm is shown in which two triangles are combined into a map. Line segments are shown with their labels on each side.

The sweep line is dotted, and the segments currently in the active list are shown dashed. The sweep line visits halfsegments in halfsegment order. The labels of all halfsegments behind the sweep line are finalized, anything in front of the sweep line must still be processed. Note the final step depicted: when the halfsegment is added to the sweep line, it is clear (based on the labels of the segment being added and the labels of the segment below it in the sweep line) that the interiors of both triangles intersect above the segment. This intersection of triangle interiors is indicated in the labels of the segment. Therefore, the sweep line progression provides the ability to compute the necessary geometric information to integrate regions into maps. Furthermore, the opportunity to integrate thematic information exists, but the complexity of handling thematic information can be dramatic since an arbitrary number of regions may intersect.

## 12.4  Integrating Regions Into a Map

In this section, we describe our proposed algorithm that integrates regions and their associated thematic values into a map form that maintains all spatial and thematic data from the input regions. The input to our algorithm is a set of regions. We assign each region a label in the form of a unique integer identifier, known as the *region identifier*, that will be used to link each region with its thematic data. We then prepare the regions for input to the algorithm.

### 12.4.1  Preparing the Input

As mentioned previously, the plane sweep algorithm can detect line segment intersections, and determine whether the currently processed line segment lies in the interior, boundary, or exterior of other regions. The input to the plane sweep algorithm is a set of halfsegments sorted in halfsegment order. Our version of the algorithm requires that each halfsegment carry two labels, indicating the identifiers of all regions that respectively lie immediately above and below the halfsegment. Initially, the halfsegments forming an individual region will each have the region identifier for that region on one side, and the region exterior label on the other. Some data representations will only store segments, or halfsegments that do not carry labels; in this case, the halfsegments can be generated and labels assigned to the appropriate sides of the halfsegments in $O(n \lg n)$ time for each input region of $n$ halfsegments on average [11]. Once the halfsegments for all input regions are prepared, they must be sorted into a single list. For $m$ regions containing $n$ halfsegments on average, sorting takes $O(mn \lg mn)$ time.

## 12.4.2    Spatial Processing of a Halfsegment

Once the algorithm input is computed, the plane sweep portion of the algorithm commences. We assume a sweep line that travels from left to right across the euclidean plane in the $x$ direction. The plane sweep portion of the algorithm serves two main functions: (1) detect and remedy line segments that intersect at points other than end points, and (2) merge the labels of the intersecting portions of regions. These functions are managed simultaneously within the algorithm. Recall that the plane sweep algorithm proceeds by processing a single halfsegment at a time. At each step of the algorithm in which a left halfsegment $h$ is being processed, $h$ is inserted into the active list, and the immediate neighbors of $h$ in the active list are computed. These neighbors are $a$ and $b$, the neighbor halfsegment that lies above $h$ and below $h$ in the active list, respectively. The intersection points between $h$ and $a$, and $h$ and $b$ are computed, the segment $s \in \{a, b\}$ that forms the least intersection point $p$ with $h$ in halfsegment order is chosen. $h$ and $s$ are removed from the active list, split according to $p$, and the resulting segments that are less than $h$ in halfsegment order are inserted back into the active list. This step is the traditional computation of line segment intersections performed by the sweep line.

## 12.4.3    Thematic Processing of a Halfsegment

Once the spatial portion of the processing of halfsegment $h$ is complete, the second function of the algorithm proceeds. The goal of the second part of the algorithm is to maintain thematic information associated with regions in the presence of multiple overlapping regions. Recall that if regions overlap in the spatial partition map model, then the overlapping portions of the regions carry the labels of all the overlapping regions. We proceed with this discussion using examples. Consider Fig. 12.3a in which two regions overlap. The region's have identifiers which we assume are reflected in the halfsegment's labels. The segments in the figure are labeled. For figures where segments are labeled, we use the notation $1_l$ and $1_r$ to indicate the left and right halfsegments corresponding to segment 1, respectively. The plane sweep begins and processes halfsegments $1_l, 2_l, 2_r,$ and $3_l$, all of which belong to the same region and do not intersect any other halfsegments. When the plane sweep algorithm begins to process $5_l$, the active list contains $1_l$ and $3_l$; $1_l$ is the neighbor below $5_l$ and $3_l$ is the neighbor above $5_l$ in the active list. It follows from halfsegment ordering that the above label of the halfsegment below $5_l$ in the active list will indicate which region $5_l$ lies inside, or if it lies in the exterior of all other regions. Therefore, $5_l$ lies in the interior of region 1 because the interior of region 1 lies above $1_l$, and this will be reflected in its above label. Because the interior of region 2 lies above $5_l$, it follows that the interiors of both region 1 and region 2 lie above $5_l$. From the spatial partition definition, we follow the convention that overlapping portions of regions receive the labels of all regions involved in the overlap; therefore a new

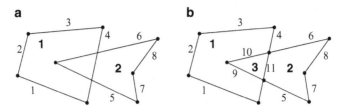

**Fig. 12.3** Two regions superimposed (**a**), and the result of their integration into a map (**b**). Segments are labeled

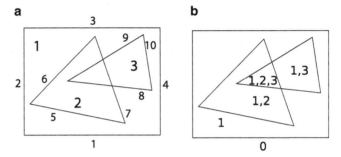

**Fig. 12.4** Three overlapping regions. (**a**) has labeled segments. (**b**) shows the region labels indicating which region interiors lie in each region primitive. 0 indicates the exterior of all regions

label is assigned to the overlapping portion, and the labels of $5_l$ must be changed to reflect the new topology (label management details will be discussed below). When the plane sweep algorithm processes $4_l$, it will only have one neighbor in the active list: $5_l$. An intersection is detected between the two, and the segments are split. Because no segments exist in the active list below $4_l$, it cannot lie in the interior of another region, and its labels remain unchanged. If the overlapping portion of the region 1 and 2 receives the identifier 3, then Fig. 12.3b indicates the result of the algorithm. Because the label 3 indicates the overlapping portion of regions 1 and 2, some record keeping must exist to indicate that region identifier 3 is equivalent to the pair of region identifiers 1 and 2.

Managing the labels of a pair of overlapping regions, as in the example above, is relatively straightforward; however, difficulty arises in cases where multiple regions overlap. For example, in Fig. 12.4 three regions are shown, all of which overlap other regions. Again, computing the line segment intersections is straightforward, but managing region identifiers becomes much more challenging. For instance, when segment $8_l$ is processed, the halfsegment below it in the active list is $5_l$. In order for us to correctly label the above label of $8_l$ as the overlap of regions 1, 2, and 3, the above label of $5_l$ must indicate that its above label is the overlap of regions 1 and 2, or we must proceed further down the active list than the halfsegment immediately below the one currently being processed. The time complexity bounds of the plane sweep algorithm rely on the fact that only the immediate neighbors of a segment in the active list are examined when processing the current segment, so we cannot look beyond the immediate neighbors without causing an $O(n^2)$ time complexity. Therefore, we must devise a mechanism such that when a halfsegment $h$ is processed, all regions whose interiors intersect $h$ can be identified by examining only the halfsegment below $h$ in the active list. In essence, the labels of $h$ must be merged with the labels of the halfsegment below $h$ in the active list to reflect the overlapping regions; we denote this *label merging*.

A successful label merge of a halfsegment results in the halfsegment's labels indicating the identifiers of all regions that lie immediately above and below the halfsegment. Therefore, the label of a halfsegment must be a set of region identifiers indicating all regions whose interiors lie on either side of the halfsegment. For example, when halfsegment $8_l$ in Fig. 12.4a is processed, it must indicate that the interiors of regions 1 and 2 lie immediately below the halfsegment, and the interiors of regions 1, 2, and 3 lie immediately above it. The original above and below labels of halfsegment $8_l$ are 3 and 0, respectively, where 0 indicates the exterior of the region. When $8_l$ is processed, we can determine the identifiers of all regions that must be involved in labels of $8_l$ by looking at the above label of the halfsegment below it in the active list, $5_l$. Because $5_l$ is less than $8_l$ in halfsegment order, it will be processed before $8_l$ and its labels will be known when $8_l$ is processed. The above label of $5_l$ must indicate that the interiors of regions 1 and 2 lie above it. Therefore, the identifiers for regions 1 and 2 must be merged into $8_l$s labels, indicating that $8_l$ lies within regions 1 and 2. When $9_l$ is processed, the halfsegment below it in the active list, $8_l$, indicates that region identifiers 1, 2, and 3 must be merged with the labels of $9_l$. Because the labels of $9_l$ indicate that it bounds region 3, the identifier for region 3 will not be added to the above label of $9_l$ (i.e., the interior of region 3 does not extend above $9_l$). Therefore, the above and below labels of $9_l$ must indicate the region identifiers 1 and 2, and 1, 2, and 3, respectively. Formally, a label merge is represented as the following:

**Definition 1.** let $h = (p, q, A, B)$ be a halfsegment and $h_b = (p_b, q_b, A_b, B_b)$ be the halfsegment below $h$ in the active list when $h$ is being processed by the plane sweep algorithm. Because $h$ is not processed, it will have the identifier of the input region it bounds as one of its labels, and the other will be the identifier of the exterior. A *label merge* is then:
$$A = A \cup (A_b - B)$$
$$B = B \cup (A_b)$$
These formulas follow directly from the definition of regions and the fact that a boundary always separates the interior of a region from the exterior of a region.

A single exception to the above definition exists when a halfsegment is being processed, and a spatially equivalent halfsegment (i.e., the halfsegments have identical endpoints and are both left or right halfsegments) is already present in the active list. Such a situation is especially common when integrating maps consisting of political divisions, which typically contain many regions that are adjacent and share boundary segments. Because the new halfsegment does not lie completely above the spatially equivalent halfsegment in the active list, it does not lie in the interior of the region primitive bounded by the halfsegment in the active list. Therefore, the new halfsegment cannot obtain the labels of all region interiors that lie below the halfsegment by examining only the labels of regions that lie above the halfsegment in the active list. Instead, the labels of all region interiors that lie *above* the new halfsegment can be determined from the labels of region interiors that lie above the corresponding halfsegment in the active list, and the region interiors that lie *below* the halfsegment must be determined by the region interiors that lie below the corresponding halfsegment in the active list. Using the original label merge technique will cause a halfsegment $s$ to appear to be lying on the interior of regions who's interiors do not actually include $s$. Formally, a label merge for two spatially identical halfsegments is defined as follows.

**Definition 2.** let $h = (p, q, A, B)$ be a halfsegment and $h_b = (p_b, q_b, A_b, B_b)$ be a halfsegment with identical end points to $h$ that is in the active list when $h$ is being processed by the plane sweep algorithm. Because $h$ is not processed, it will have the identifier of the input region it bounds as one of its labels, and the other will be the identifier of the exterior. A *label merge* for spatially identical halfsegments is then:
$A = A \cup (A_b - B)$
$B = B \cup (B_b)$
These formulas follow directly from the definition of regions and the fact that a boundary always separates the interior of a region from the exterior of a region.

Figure 12.5 depicts an example of merging region labels over three boundary spatially identical segments, each from a different region. If the standard label merge is used, the label for region 1 will be incorrectly propagated above the segment

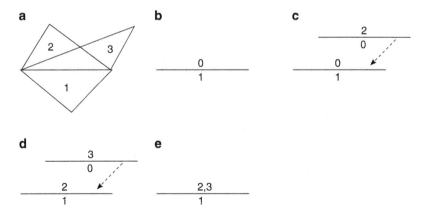

**Fig. 12.5** A possible sequence of encountering spatially identical segments with different labels in the plane sweep portion of the algorithm. **a** shows the three triangular regions that are to be integrated into a map. The remainder of the figures show the sequence in which spatially identical halfsegments are integrated into a map, and the resulting labeling. By using the label merge formula for overlapping segments, the correct labeling will be computed

shown in Fig. 12.5b when the segment from region 2 is added. As the sequence of figures indicates, using the label merge for overlapping segments ensures the correct labeling scheme propagates through the resulting map.

When a halfsegment is being processed, it will always have exactly one identifier in one of its labels, and the exterior identifier in the other label. The halfsegment below it in the active list may have many region identifiers in its above and below labels. It follows from the definition of halfsegment merging that computing the below label of the current halfsegment $h$ involves removing a single region identifier from the above label of the halfsegment below $h$ in the active list. Furthermore, removing the below label of $h$ from the above label of the halfsegment below $h$ in the active list and inserting the above label of $h$ also involves removing and adding single region identifiers. Therefore, the label merges can be expressed as three separate operations in which a single region identifier is added to, or subtracted from, a set of region identifiers. When integrating $m$ regions, the size of the set of region identifiers on the halfsegment below the halfsegment being processed is bounded by $m$; therefore, efficient insertion and removal operations are required. We assume these sets of region identifiers are implemented as a height bounded binary search tree that supports removal and insertion in $O(\lg m)$ time.

If many regions overlap in a map, then a label may contain many region identifiers. Storing a list of region identifiers for each segment forming the boundary of a map is inefficient in terms of space, especially when a map must be stored on disk; therefore, we maintain a mapping that defines a relationship between *label identifiers* and labels. A label identifier is a unique identifier that corresponds to a label. Because a label may contain many region identifiers, it is bounded by the number of regions being integrated ($m^2$ for $m$ regions). A label identifier can be implemented as a single integer, which is much more storage efficient and fits with the definition of halfsegments which specifies that labels are single integers, and not sets of integers. Again, a height bounded binary search tree is a good candidate for implementing this mapping and maintaining $O(\lg m^2)$ time complexity for adding, removing, and looking up elements. $O(\lg m^2)$ behavior is the worst case in which all possible combinations of overlapping regions exist, which is rare in practice. A reverse mapping from labels to label identifiers is also required and can be implemented efficiently using a trie, providing insertion and lookup in $O(m)$ time for $m$ input regions [3, 7].

When a halfsegment $h$ is being processed by the sweep line algorithm, the halfsegment below $h$ is computed and its label is found. That label is copied twice, and a label merge is performed with the label for the region above $h$ and below $h$ respectively with each copy. The result of the label merges are labels indicating all regions whose interiors lie immediately above and below $h$. Because many halfsegments may contain identical labels, and these labels may contain up to $m$ region identifiers for $m$ input regions, representing each label with a label identifier can drastically reduce the memory requirements of the algorithm in situations when labels contain many region identifiers. Therefore, a trie associating labels with label identifiers is queried to see if the computed label is already assigned to another halfsegment and already has an associated label identifier. If the label has not yet been encountered, then a new label identifier is created for the label, and the trie is updated. A height balanced binary tree is also updated to reflect the mapping from the label identifier to its corresponding label. Therefore, for each halfsegment visited by the plane sweep algorithm, a trie must be queried and possibly modified twice, and a height balanced binary tree must be queried and possibly modified twice, leading to a time complexity of $O(m + \lg m^2)$ for $m$ regions for each halfsegment visited by the plane sweep. The optimal plane sweep has a time complexity of $O(n \lg n + k)$ for $n$ segments with $k$ intersections [4]; however, the version of plane sweep that is used more typically in practice (and the version we use for implementation) has complexity $O(n \lg n + k \lg n) \Leftrightarrow O((n + k) \lg n)$ where $n + k$ segments are visited and the $\lg n$ term reflects operations on data structures necessary for each segment visited by the algorithm. Therefore, the overall time complexity of the algorithm is $O((n + k)(\lg n + m + \lg m^2))$. The space complexity of the algorithm depends on the number of unique labels created during the integration. In the worst case, every region will overlap every other region in such a way that $m^2$ labels will exist for $m$ regions. Because each unique label is stored, and every segment from regions must be stored, the space complexity is $O(m^2 + n + k)$ for $m$ regions containing $n$ line segments with $k$ intersections. In practice, the number of unique labels generated is typically much less then $m^2$; furthermore, $m$ is typically much smaller than $n$, especially in geographic data sets, so a larger space or time complexity on the $m$ term is acceptable.

## 12.5 External Memory Algorithm

At this point in the paper, the proposed algorithm is described as an in-memory algorithm. The input to our algorithm is an unsorted list of labeled halfsegments. This list of halfsegments is generated by simply appending lists of halfsegments that define the boundaries of all input regions; therefore, the first step of the algorithm is to sort the list in halfsegment order to prepare it as input to the plane sweep algorithm. However, spatial and geographic data tend to be very large. For example, one of the data sets we use for testing contains the counties of ten states in the US: Arizona, Utah, Colorado, Kansas, Oklahoma, Texas, Missouri, Arkansas, and Louisiana. This input list is approximately 155 Mb, which is well inside the bounds of most modern computers, but is a relatively small data set. A second data set used for testing consists of all counties or county equivalents for all US states and territories, which requires approximately 610 Mb to represent. When testing the in memory algorithm with this data set, performance degraded rapidly since the virtual memory system was having to perform a large amount of memory page swapping to disk. Because many geographic applications that would utilize the proposed algorithm are run on laptop and desktop scale computers, the in-memory algorithm is not sufficient due the potentially large data sizes of spatial data; therefore, we extend our algorithm as an external memory algorithm in this section.

Our proposed algorithm was designed with structural simplicity in mind in order to ease implementation; this structural simplicity also lends the algorithm to a straightforward conversion to an external memory algorithm. The basic steps in the algorithm are to sort the input, run a plane sweep algorithm in which labels are merged and maintained, and store the output. The first step of sorting the input can be implemented with an external merge sort algorithm. In an external merge sort algorithm, the maximum amount of memory that the sort uses can be arbitrarily assigned, making it straightforward to adapt to any reasonably powerful computing hardware. Furthermore, with good choices for buffer sizes, the algorithm is very efficient.

The usage of the plane sweep algorithm in an external memory setting has been considered in the realm of databases where the plane sweep algorithm can be used to perform spatial joins over relations containing regions [1, 9]. Because the plane sweep algorithm reads its input from a sorted list of halfsegments, it is clear that the entire list need not be kept in memory. Therefore, we implement a buffered disk reader that reads a chunk of halfsegments into memory at a time for the plane sweep to take as input. Organizing the plane sweep active list as an external memory structure is less clear; however, the size of the plane sweep active list tends to follow the *square root rule* [8, 14], meaning the size of the active list does not, in practice, exceed the square root of the number of input halfsegments. Because of this observation, creating an external memory active list is not necessary. Indeed, in our experimental runs of the algorithm , the active list rarely exceeds a few megabytes in size even when the input data is dense.

Finally, the resulting map that is output from the algorithm must be written to disk as it is produced. However, the halfsegments produced by the plane sweep are not necessarily produced in the correct order. The version of the plane sweep algorithm that does produce its output in sorted order is, in practice, complex and difficult to implement [4]. The original plane sweep design is slightly less efficient, but significantly more straightforward to implement, and does not produce a sorted output. Most plane sweep implementations, including ours, follow the original design, and so a sorting step is required for the output. Again, we use a buffered file interface for quick file writing, and an external memory sorting algorithm.

## 12.6   Experimental Results

We have implemented the region integration algorithm and the external memory algorithm described in this paper, as well as the traditional $O((nm \lg nm)^2 + k)$ algorithm, in C++. Although the proposed algorithm is clearly faster in terms of algorithmic complexity, a practical comparison verifies the utility of the new algorithm. Furthermore, the need for the external memory algorithm becomes apparent through the analysis of the in-memory algorithm. We ran the new and traditional algorithms over geographic data sets consisting of the counties of Florida, Alabama, and Texas. The input counties were taken from freely available US Census

**Table 12.1** Running times for the new algorithm and the traditional algorithm for constructing maps consisting of the counties of Florida, Arizona, and Texas. The number of halfsegments for each input collection of regions and the resulting map is shown. Each time is the average of three runs of the respective algorithm for the given data

| State | Halfsegments input | Halfsegments output | New algorithm (s) | Traditional algorithm (s) |
|-------|---------|--------|------------|------------|
| AL | 302,602 | 165,202 | 12.291 | 80.295 |
| FL | 413,742 | 226,870 | 16.891 | 149.319 |
| TX | 1,211,768 | 664,580 | 60.954 | 1095.028 |

data sets, and the counties for each state were integrated into a single map of the state. The algorithms were run on a desktop computer with a 2.8 GHz processor and 2 GB of RAM. Table 12.1 shows the running times of the algorithms, and the sizes of the resulting maps. Note that the resulting maps contain fewer halfsegments than the input since many counties share boundary segments. Our new algorithm is clearly more efficient than the traditional method, and scales much better for larger input. Testing of larger data sets was not carried out due to the large performance gap already present between the proposed algorithm and the traditional algorithm with the existing data sets.

The running time of the proposed algorithm jumped drastically between the Florida and Texas data sets. Although an increase in execution was expected, the degree of change was surprising. Further analysis showed that the jump was due to an increase in memory page faults when running the Texas data set. The data set was large enough that the computer on which the experiments we executed filled its physical memory. Therefore, clear limit of usefulness for the proposed algorithm exists since a large amount of memory is required for large data sets. The fact that memory usage was an issue when only integrating the counties of a single state, albeit a large state, was disappointing, and led to the development of the external memory version of the algorithm.

Table 12.2 lists the results of running the external memory version of the proposed algorithm on the original three data sets, and two additional data sets. The first additional data set consists of the counties of a subset of the states in the US: Arizona, Utah, Colorado, New Mexico, Texas, Oklahoma,Kansas, Missouri, Arkansas, and Louisiana. The second additional data set consists of the counties or equivalent constructs in all states and territories belonging to the US. Each input consisted of labeled, unsorted halfsegments. The external memory algorithm first sorts the input data set, then performs the data integration, and finally sorts the result. The algorithm was executed on a laptop with a 2.4 Ghz processor and 2 GB of ran, the same desktop computer on which the in-memory algorithm was run, and a server with a 2.3 Ghz processor and 8 Gb of ram. The external memory algorithm ran slightly faster than the internal memory algorithm on the smallest two data sets due to better memory management and the use of a different sort algorithm in memory; furthermore, the smaller data sets fit entirely in memory and did not require writes to disk. The data set consisting of the counties of Texas benefited significantly from

**Table 12.2** Running times for the external memory version of the algorithm. The subset of states data set consists of the counties of Arizona, Utah, Colorado, New Mexico, Texas, Oklahoma, Kansas, Missouri, Arkansas, and Louisiana. The all states data set consists of the counties and county equivalents of all US states and territories. All data is from the U.S. Census Tiger data set

| Data set | Halfsegments input | Halfsegments output | Laptop (s) | Desktop (s) | Server (s) |
|---|---|---|---|---|---|
| AL | 302,602 | 165,202 | 16.3 | 1.9 | 7.3 |
| FL | 413,742 | 226,870 | 18.8 | 14.9 | 8.8 |
| TX | 1,211,768 | 664,580 | 53.1 | 42.5 | 23.1 |
| Subset of states | 3,885,270 | 20,208,92 | 177.3 | 143.6 | 74.2 |
| All states | 15,251,180 | 7,716,312 | 778 | 632.4 | 300.6 |

**Table 12.3** The size in bytes of each of the data sets used in the analysis of the proposed algorithm

| Data set | Input data size (bytes) |
|---|---|
| AL | 12,104,120 |
| FL | 16,549,720 |
| TX | 48,470,760 |
| Subset of states | 155,410,840 |
| All states | 610,047,240 |

the external memory algorithm. The performance of the external memory algorithm held even for the larger data sets. The "all states" data set was terminated by the operating system when run under the internal memory algorithm on the laptop due to memory limits; however, it ran to completion in about 13 min using the external memory version of the algorithm.

Table 12.3 shows the sizes of the data sets used in our experiments. During peak memory usage, the in-memory algorithm essentially duplicates the data set by storing both the input regions, and a nearly completed map consisting of the integrated input regions. Thus, the largest data set required at least 1.2 GB of memory to run. The external memory algorithm was given a 200 MB memory limit, and so all data sets ran without stressing the memory limits of the computers. The 200 MB limit can be adjusted upwards for better performance on even larger data sets, or to take advantage of more powerful hardware.

## 12.7 Conclusion

In this paper, we have introduced the map construction problem as a data integration problem requiring the handling of both geometric and thematic aspects of spatial data. We proposed a new algorithm that solves the map construction problem for $m$ regions containing $n$ line segments on average in $O((n+k)(\lg n + m + \lg m^2))$ time complexity and $O(m^2 + n + k)$ space complexity. We then extended this algorithm into an external memory algorithm capable of efficiently integrating extremely large data sets. These algorithms create new opportunities for the use of maps as reusable data items in spatial and geographical systems, rather than simply visualizations.

Future work in this area will focus on investigating further improvements of execution time through the use of parallelism. Although parallelism in the plane sweep paradigm has been investigated, the variable length nature of labels does not fit well into the known paradigms.

# References

1. Arge, L., Procopiuc, O., Ramaswamy, S., Suel, T., Vitter, J.S.: Scalable sweeping-based spatial join. In: VLDB '98: Proceedings of the 24rd International Conference on Very Large Data Bases. pp. 570–581. Morgan Kaufmann Publishers Inc., San Francisco, CA, USA (1998)
2. Bentley, J., Ottmann, T.: Algorithms for Reporting and Counting Geometric Intersections. IEEE Trans. Comput. $C$-$28$(9), 643–647 (1979)
3. de la Briandais, R.: File Searching Using Variable Length Keys. In: Proceedings AFIPS Western Joint Computer Conference (1959)
4. Chazelle, B., Edelsbrunner, H.: An Optimal Algorithm for Intersecting Line Segments in the Plane. J. ACM. $39$(1), 1–54 (1992)
5. Egenhofer, M.J., Herring, J.: Categorizing Binary Topological Relations Between Regions, Lines, and Points in Geographic Databases. Technical report, National Center for Geographic Information and Analysis, University of California, Santa Barbara (1990)
6. Erwig, M., Schneider, M.: Formalization of Advanced Map Operations. In: 9th Int. Symp. on Spatial Data Handling. pp. 8a.3–17 (2000)
7. Fredkin, E.: Trie memory. Comm. ACM. $3$(9), 490–499 (1960)
8. Güting, R.H., Schilling, W.: A practical divide-and-conquer algorithm for the rectangle intersection problem. Inform. Sci. $42$(2), 95–112 (1987)
9. Kriegel, H., Brinkhoff, T., Schneider, R.: Combination of Spatial Access Methods and Computational Geometry in Geographic Database Systems. In: SSD '91: Proceedings of the Second International Symposium on Advances in Spatial Databases. pp. 5–21. Springer, London, UK (1991)
10. McKenney, M.: Map Algebra: A Data Model and Implementation of Spatial Partitions for Use in Spatial Databases and Geographic Information Systems. Ph.D. thesis, University of Florida (2008)
11. McKenney, M.: Region extraction and verification for spatial and spatio-temporal databases. In: SSDBM. pp. 598–607. Lecture Notes in Computer Science, Springer (2009)
12. McKenney, M.: Geometric and Thematic Integration of Spatial Data into Maps. In: Information Reuse and Integration (2010)
13. McKenney, M., Schneider, M.: Topological Relationships Between Map Geometries. In: Advances in Databases: Concepts, Systems and Applications, 13th International Conference on Database Systems for Advanced Applications (2007)
14. Ottmann, T., Wood, D.: Space-economical plane-sweep algorithms. Comput. Vis. Graph. Image Process. $34$(1), 35–51 (1986)
15. Schneider, M., Behr, T.: Topological Relationships between Complex Spatial Objects. ACM Trans. Database Syst. (TODS). $31$(1), 39–81 (2006)
16. Scholl, M., Voisard, A.: Thematic Map Modeling. In: SSD '90: Proceedings of the first symposium on Design and implementation of large spatial databases. pp. 167–190. Springer, New York, USA (1990)
17. Tomlin, C.D.: Geographic Information Systems and Cartographic Modelling. Prentice-Hall (1990)

# Chapter 13
# Integrating Geographic and Meteorological Data for Storm Surge Planning

**Jairo Pava, Fausto Fleites, Shu-Ching Chen, and Keqi Zhang**

## 13.1 Introduction

According to the National Oceanic and Atmospheric Administration (NOAA), hurricanes are the most destructive natural hazards to threaten the United States East and Gulf coasts. With approximately 90% of hurricane-related deaths, the greatest threat to human life and property by a hurricane is storm surge. Hurricane Katrina, which made landfall in southeast Louisiana on August 29, 2005, is a prime example of the potential danger of storm surge. Many of the 1,500 deaths caused by Hurricane Katrina may be directly or indirectly attributed to storm surge. Salt water flooding is also a major cause of damage to buildings and infrastructure [13, 19]. According to AIR Worldwide, a provider of risk modeling software and consulting services, the property damage from Hurricane Katrina's storm surge has an estimated value of $44 billion [1].

Weather projections are typically updated every six hours, but to help residents prepare for evacuation, the frequency of updates is increased to every three hours as a storm approaches landfall. Coastal residents are warned and advised on evacuations through television, Internet, radio, and newspapers. Residents, however, have been unable to relate the two-dimensional evacuation maps broadcast over media outlets to their three-dimensional life experiences. As a result, local governments have struggled to avoid under and over-evacuation of coastal areas. Hurricane Floyd, for instance, triggered the second largest hurricane-related evacuation to date.

J. Pava (✉) · F. Fleites · S.-C. Chen
Distributed Multimedia Information Systems Laboratory, Florida International University, Miami, FL 33199, USA
e-mail: jpava001@cs.fiu.edu; fflei001@cs.fiu.edu; chens@cs.fiu.edu

K. Zhang
International Hurricane Research Center, Florida International University, Miami, FL 33199, USA
e-mail: zhangk@fiu.edu

T. Özyer et al. (eds.), *Recent Trends in Information Reuse and Integration*,
DOI 10.1007/978-3-7091-0738-6_13, © Springer-Verlag/Wien 2012

As Hurricane Floyd approached the United States, 2.6 million coastal residents from five states started to evacuate. The mass evacuation resulted in heavy traffic and obstructed or hampered the evacuation efforts of many who really needed to evacuate. Evacuees from Charleston, SC, for instance, had an average travel time of nine hours. When the National Hurricane Study Program conducted its assessment of Hurricane Floyd, most coastal residents agreed that if they could have done anything differently, they would have begun evacuating much earlier [12].

To help residents make timely evacuation decisions, NOAA provides many services that provide free access to up-to-date weather data. NOAA's public Simple Object Access Protocol (SOAP) service, for instance, may be used to retrieve meteorological data including wind, precipitation, and cloud cover [17] over the Internet. The SOAP service provides the public, government agencies, and commercial enterprises with data from the National Digital Forecast Database (NDFD) in extensible markup language (XML) format. The database contains data for up-to-date climate outlook probabilities, convective outlook hazard probabilities, probabilistic tropical cyclone surface wind speeds, and other weather elements including temperature, sky cover, and precipitation. Furthermore, as low pressure weather systems, such as hurricanes, approach a coastal area, storm surge models are used to estimate the potential storm surge at the projected landfall area. One such type of model is the Coastal and Estuarine Storm Tide (CEST) model developed by the International Hurricane Research Center (IHRC) [36, 40]. The model takes into consideration the expected tide at landfall and atmospheric pressure and wind of the weather system. It also takes into consideration major coastal topographic features such as coastal ridges and barrier islands.

Additionally, there exist many tools that enable the rapid and accurate creation of virtual city models. Light Detection and Ranging (LiDAR) systems are used to collect topographical measurements [5]. LiDAR data are collected with aircraft-mounted lasers and stored as XYZ data points in text files where X is longitude, Y is latitude, and Z is the elevation at that point. The NOAA Coastal Services Center provides free public access to its repository of LiDAR data sets [18]. The laser-scanned data include topographical measurements as well as non-ground objects such as cars, buildings, and vegetation. Digital Line Graphics (DLG) from the United States Geological Survey (USGS) are used to gather the coordinates of all public roads [28]. DLG data are digital vector representations of cartographic information. The data are publicly available for download from the official United States Geological Survey (USGS) website. Since the road data captured in DLG comes from scanned cartography, it does not contain data regarding road widths, names of streets, or traffic direction. However, the precise geo-coordinates of the DLG files allow us to identify public roads with a high degree of accuracy.

To address the challenge of integrating information and knowledge for enhancing timely evacuation decision-making, we propose an extension of the three-dimensional storm surge simulation system presented in [23]. Three-dimensional is defined by a computer-generated virtual environment that may be interacted with and viewed within the spatial dimensions of depth, height, and width. The system presented in [23] integrates CEST, LiDAR, and DLG data to produce storm surge simulations of easily recognizable community areas. However, the simulations are based only on projected storm surge height and do not consider other meteorological forecasts. Hence, the simulations may lack accuracy and lose credibility among coastal residents. We extend the system in [23] by integrating NDFD data to accurately reflect ocean wave height, precipitation, thunderstorm, and wind forecasts. With more accurate simulations, coastal residents will be better informed to make potentially life-saving evacuation decisions.

The main contributions of the storm surge simulation system are as follows: (1) it extends the system in [23] to enhance storm surge simulations with meteorological forecasts including wind, rain, and ocean wave heights among others; (2) it presents a novel methodology to accurately visualize real-time and historical hurricane storm surge scenarios in a three-dimensional environment; and (3) it provides a web-based integration of two and three-dimensional storm surge visualizations for coastal residents and emergency planners.

The rest of this chapter is organized as follows. The next section discusses work that is related to our simulation system. Section 13.3 provides details of our system's architecture. Section 13.4 describes how our system procedurally generates digital terrain models (DTMs) and builds footprints from LiDAR measurements and roads from DLG data. Section 13.5 describes how storm surge and meteorological data are used to animate the visualization. Section 13.6 describes a web-based interface for coastal residents and emergency planners to view storm surge visualizations. Section 13.7 presents concluding remarks and acknowledgements.

## 13.2   Related Work

The work presented by [3] and [30] is the most similar to our proposed system. The work shows how various data sources, ranging from computational simulations of storm surge to satellite imagery, may be used to create interactive visualizations of hurricanes. A visualization of hurricane Katrina is presented as a case study. The work explains how heterogeneous data are integrated into a unifying data format and rendered using the Amira visualization framework [2]. However, since the Amira framework does not have any Level-of-Detail functionality, it is unable to efficiently handle large LiDAR data sets to construct detailed simulations. Furthermore, it is unclear how accessible their proposed system is to the general public.

To encourage hurricane preparedness, the work in [16] proposes a PC-based instructional program that educates children on the effects of hurricane winds on residential structures. Children configure hurricane wind speeds and select from various structure types to witness, through realistic graphics and sound, the degree of damage based on their choices. The study claims that the program significantly captured the interest of children and effectively taught concepts on hurricane preparedness. Our approach may benefit from the work in [16] by incorporating field testing techniques to evaluate the effectiveness of preparing coastal residents to evacuate.

The VTP is used in [7] to construct virtual models of large cities and regions. The models are used to visualize, in three-dimensional form, alternative futures. A case study is discussed whereby GIS data for buildings, terrain, rivers, and roads are integrated to model the city of Hangzhou, China. The virtual model of the city constructed with the VTP was utilized to aid city planners in their efforts to identify areas to place strategic public investments in transportation, infrastructure, and civic buildings. The approach presented in this chapter differs from [7] in that virtual models of the cities to be visualized under storm surge conditions are programmatically constructed from LiDAR and DLG data. While the approach presented in this chapter requires significantly less effort to construct virtual city models, it could benefit from the methods presented in [7] to evaluate the system's ability to operate in an urban planning environment.

The work in [15] and [38] presents a storm surge visualization system which displays the effect of a rising storm surge on public roads and traffic. However, it is highly scripted, and the individual components used to animate the storm surge, wind, traffic, and rain have to be manually configured upon every initialization. The work in [23] extends the work in [38] by automating the integration of terrain and storm surge data to rapidly create storm surge simulations. However, the simulations are based on projections from the CEST storm surge model while other weather phenomena such as wind, precipitation, or ocean wave height are not based on any official forecasts. Furthermore, the web-based interface proposed by [23] greatly limits the ability of residents to interact with the storm surge simulation system by only providing videos of pre-selected locations. Our proposed storm surge simulation system will extend the work in [23] by automatically integrating up-to-date weather forecasts from the NDFD. These weather forecasts will include weather phenomena such as wind, precipitation, and ocean wave height to enhance the accuracy of the simulation. Additionally, a new interface to allow users to request simulations by address is introduced to enable residents to view videos of their own home or businesses under projected storm surge conditions.

## 13.3 System Architecture

The storm surge simulation system proposed largely extends the open-source Virtual Terrain Project (VTP) visualization system [32]. The VTP is a set of tools used to construct and render three-dimensional visualizations of geographic data. The VTP's two main tools, VTBuilder and Enviro, are used to construct and render virtual terrains. VTBuilder provides a graphical user interface to construct terrains, man-made structures, and roads. Enviro provides an interactive runtime environment using OpenGL [22]. The VTP was designed to run well on personal computers with relatively low hardware configurations. As of the latest April 2010 release, the minimum PC requirements to run the VTP include a computer running Windows ME or above, a graphics rendering card, and at least 128MB of RAM.

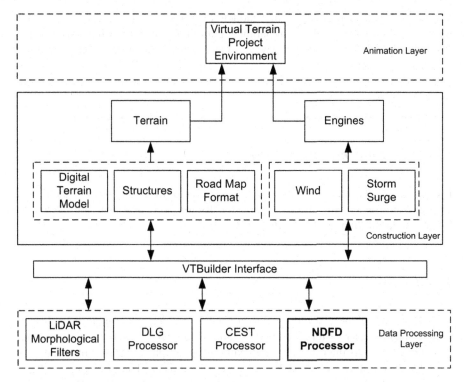

**Fig. 13.1** System architecture

Enviro's extensibility through engines is the main reason we chose to work with the VTP. An engine in the VTP is a block of code that is executed once per frame in the animation. Typically, a simulation runs at 30 frames per second. Engines allow us to efficiently implement animations for storm surge, wind, rain, traffic, vegetation, lightning, and other weather elements using OpenGL. OpenGL is a cross-platform API for writing 2D and 3D applications. Moreover, VTBuilder's capability to import GIS data, like LiDAR, expedites our processing of terrain data to three-dimensional interactive digital elevation models. Therefore, with the VTP we can combine the power of VTBuilder and Enviro to integrate GIS terrain data to real-time storm surge and meteorological data in a three-dimensional environment.

Figure 13.1 presents an extended version of the system architecture demonstrated in [23]. In the data processing layer, LiDAR, DLG, NDFD, and CEST data are converted into formats that are compatible with the VTBuilder. The NDFD processor, which is presented in Fig. 13.1 with a bold outline, extends the previous architecture by providing the facilities to parse NDFD SOAP responses and extract meteorological forecasts including wind, cloud cover, precipitation, wave height, thunderstorm, and weather advisories. After processing all data, terrains and engines are constructed and configured in the construction layer. LiDAR and DLG data are used to construct Digital Terrain Models (DTM), buildings, and roads which are then integrated to create virtual models of cities. CEST and NDFD data are used to configure the graphics animation engines in the VTP for wind, storm surge, vegetation, rain, traffic, and ocean waves. Once a city's models are created and the engines have been configured, they are integrated in the VTP's Enviro. Enviro is able to render the cities and engines, and it provides an interactive navigation of the digital terrain.

## 13.4 Terrain Data Integration

LiDAR and DLG data are used to construct three-dimensional terrain, buildings, and roads in our system. In this section we discuss how these data are integrated into a standard format usable by the VTP.

### 13.4.1 LiDAR

LiDAR is one of the common methods used to collect topographic data. It is a remote sensing laser system mounted on aircraft to record elevation measurements from large areas. As the LiDAR system gathers topographical measurements, it uses a Global Positioning System (GPS) to store data as XYZ data points in text files. X is longitude, Y is latitude, and Z is the elevation at that point. LiDAR systems cannot automatically differentiate measurements collected from natural elevation, vegetation, or man-made structures. Therefore, to be usable in three-dimensional visualizations, LiDAR data must be categorized to separate ground and non-ground measurements.

### 13.4.2 Digital Terrain Model

The first step in creating a three-dimensional storm surge simulation environment is to create a virtual terrain based on topographical LiDAR measurements. LiDAR measurements, however, include man-made structures such as vehicles, light posts, or buildings and vegetation. To construct an accurate DTM from LiDAR data, measurements from ground and non-ground features have to be separated. The progressive morphological filter proposed by [37, 42] provides a methodology to remove non-ground measurements from LiDAR data. The filter is able to systematically remove non-ground measurements without prior knowledge of the size and elevation of the terrain. It is based on dilation and erosion principals of set theory. Furthermore, the morphological filter compensates for gaps in topographical measurements due to irregularly spaced LiDAR data points using nearest neighbor approximation. The filter's output is in the same XYZ format used to store LiDAR data. It is then converted to the Binary Terrain (BT) [4] format with VTBuilder. A BT file is an elevation grid consisting of elevation values at specified geo-coordinates. BT files are used by Enviro to render three-dimensional representations of DTMs.

Using this approach, DTMs of nine cities in the State of Florida have been constructed. They are Pensacola, Jacksonville, Tampa, St. Petersburg, West Palm Beach, Ft. Lauderdale, Miami Beach, Key Biscayne, and Key West.

### 13.4.3   Automatic Building Construction

By extracting elevation measurements from LiDAR data using the progressive morphological filter, we have implicitly identified non-ground measurements of man-made structures such as buildings. Therefore, we use the framework in [39, 41] to automatically construct virtual building models from LiDAR measurements. A region-growing algorithm based on plane-fitting techniques is used to identify building footprints which are outlines of building shapes from a top-down view. The heights of the buildings are identified by averaging the elevation differences between building measurements and the DTM constructed from non-ground measurements. Results from [39] demonstrate that the framework is able to accurately identify building footprints and heights well with a 12% average omission and commission rate of error.

The building footprint coordinates are automatically converted to the Virtual Terrain Structure (VTST) [33] file format. VTST is based on the XML format. It is readable by the VTP and allows for the systematic production of 3D building models. VTST is built upon the definition for the OpenGIS Geography Markup Language (GML) Implementation Specification, version 2.1.2 [21]. It is the standard for encoding geographic data in XML.

Figure 13.2 presents a sample building footprint in the VTST format. Building footprint points are specified in counter-clockwise order using the VTP's own two-dimensional coordinate system which corresponds to the geo-coordinates in the LiDAR data. The building's floor height is specified in meters, and a building may have more than one floor with respect to the story count attribute. The footprint vertex with the lowest elevation becomes the height of the base of the building. Furthermore, buildings may be systematically textured by adding the Edge element with the appropriate attributes for material and color.

Using LiDAR data to construct virtual building models offers advantages to using 3D model creation software as proposed in [38]. Although automatically applying photo-realistic textures to virtual building models still presents a significant challenge, the VTST file format provides some capability to apply generic textures. This facility is useful when constructing buildings that will not be the focal point of the storm surge simulations. Another significant challenge that is presented when

```
<StructureCollection
 xmlns:gml=''http://www.opengis.net/gml''>
<Building>
  <Level  FloorHeight=''5.0''
    StoryCount=''10''>
      <Footprint>
          <gml:coordinates>
              587213.716,2850851.91
              587150.444,2850849.74
              587150.623,2850826.46
              587180.351,2850830.25
          </gml:coordinates>
      </Footprint>
      <Edge  Material=''Siding''
        Color=''ffffff''>
          <EdgeElement  Type=''Wal''/>
      </Edge>
  </Level>
</Building>
</StructureCollection>
```

**Fig. 13.2** VTST structure file

using LiDAR is that the building footprint shape may not be entirely accurate since LiDAR is captured from directly above the building. However, the ability to use geo-referenced LiDAR data points allows us to systematically place the buildings models at their exact geo-coordinates on the DTM and with accurate distance between neighboring buildings and roads. Furthermore, we can be much more certain of the building heights from LiDAR data as opposed to visual estimation. Therefore the relative storm surge height in the simulation is also precise.

### 13.4.4 Road Distribution

The USGS provides public access to its DLG data through its GeoCommunity website [28]. DLG data are digital vector representations of cartographic information in raster form. Geographic coordinates and traffic lane width of all public roads in the United States are available in DLG data. With DLG data we are able to systematically place roads on DTMs. This is done by converting the DLG data into Road Map Format (RMF) [25], a binary file format that stores the extents of the DLG dataset, road coordinates, and intersections. Enviro uses the RMF files format to render the roads over the DTM. Since road data in DLGs come from scanned cartography, we cannot accurately identify specific road widths, street names, or traffic direction. However, publicly available road map data [10] allow us to manually append this information.

## 13.5   Weather Integration

NOAA provides a Simple Object Access Protocol (SOAP) service to access the National Digital Forecast Database (NDFD). The service provides the capability to request NDFD data over the Internet and receive the information back in an XML format called the Digital Weather Markup Language (DWML) [6]. NOAA exploits XML to facilitate the ease with which software developers may integrate NDFD data into software applications and therefore enhance the value of weather information by encouraging the development of value-added products. The NDFD is updated once every hour with the latest weather forecasts, watches and warnings, and current meteorological observations. The NOAA SOAP service offers more unrestricted access to weather data than other online weather service providers [34, 35].

NDFD data is integrated into the proposed storm surge simulation system to accurately introduce meteorological elements not available from storm surge projections. For instance, wind forecasts from the NDFD allow us to introduce accurate visualizations of wind speed, direction, and gusts in the simulation. Lightning, cloud cover, and other meteorological elements may also be accurately integrated. More importantly, the NDFD database contains ocean wave height projections which enable the simulation system to improve the accuracy of the storm surge height.

A server-side Java application was developed using the Java Net package [14] to establish an HTTP URL Connection and request data from the NOAA SOAP service. Two SOAP messages are defined using XML and sent to the SOAP service to request specific weather NDFD data. Figure 13.3 presents a subset of the first message which makes a request for all of the NDFD grid points within a rectangular subgrid as defined by points at the lower left and upper right corners of a rectangle. The default resolution of NDFD data is 5 km; however, to reduce the amount of grid points requested, a lower resolution, in this case 10 km, is selected. The response is received in XML format and is parsed to extract a collection of grid points within the rectangular grid requested. A second request, presented in Fig. 13.4, is then dispatched to retrieve weather data from the NDFD. In this request, the exact longitude and latitude in the NDFD database are specified. A time-series request is made to capture all weather data from the specified start and end times. As shown

```
<SOAP-ENV: Envelope
SOAP-ENV: encodingStyle = ''http://schemas.xmlsoap.org/soap/encoding/''>
   <SOAP-ENV: Body>
     <ns5689 : LatLonListSubgrid>
        <lowerLeftLatitude >25.77</lowerLeftLatitude>
        <lowerLeftLongitude >−80.13</lowerLeftLongitude>
        <upperRightLatitude >25.78</upperRightLatitude>
        <upperRightLongitude >−80.12</upperRightLongitude>
        <resolution >10.0</resolution>
     </ns5689 : LatLonListSubgrid>
   </SOAP-ENV: Body>
</SOAP-ENV: Envelope>
```

**Fig. 13.3** NDFD request 1

```
<SOAP–ENV: Body>
  <ns3591:NDFDgen ns3591 = ' 'DWMLgen' ' >
    <latitude >25.775</latitude >
    <longitude >−80.125</longitude >
    <product>time−series </product>
    <startTime >2010−10−31T00:00:00 </startTime>
    <endTime>2010−10−31T23:59:59 </endTime>
    <weatherParameters>
      <wdir_r >1</wdir_r >
      <wspd_r >1</wspd_r >
      <wgust>1</wgust>
      <sky_r >1</sky_r >
      <waveh>1</waveh>
      <pxtotsvrtstm >1</pxtotsvrtstm >
    </weatherParameters>
  </ns3591:NDFDgen>
</SOAP–ENV: Body>
```

**Fig. 13.4** NDFD request 2

**Table 13.1** Weather information integrated from the NDFD

| Weather element | NDFD code | Unit of measurement | Used by engine |
|---|---|---|---|
| Wind speed | wspd | Knots | Wind module |
| Wind gusts | wgust | Knots | Wind module |
| Wind direction | wdir | Degrees | Wind module |
| Cloud cover | sky | Percent | Cloud |
| Precipitation | qpf | Inches | Rain |
| Wave height | waveh | Feet | Ocean wave |
| Thunderstorm | ptotsvrtstm | Percent | Cloud |
| Advisories | wwa | Plain text | Web interface |

in Table 13.1, the following weather parameters are requested: wind direction, wind speed, wind gust, cloud cover, precipitation, wave height, and warning notifications. Wind measurements are measured in knots, cloud cover as a percent, precipitation in inches, wave height in feet, and warning notifications from a set of plain text keywords such as Flood, Flash Flood, Lakeshore Flood, and many more [6].

Figure 13.5 presents a high-level diagram of the storm surge simulation system. Storm surge and meteorological data are gathered and stored periodically via web agents. Storm surge projections from the CEST storm surge model are retrieved from the Internet as the latest data become available every three to six hours as described in [23]. Wind, cloud cover, ocean wave height, and weather warnings are gathered from the NDFD via NOAA's SOAP service once every hour. Data from the agents are stored in a shared repository available to all engines in the system.

As shown in the minimal class diagram in Fig. 13.6, the VTP provides an abstract engine class, vtEngine, which coordinates the behavior of every animation within the visualization. The eval() function within the vtEngine class is invoked once

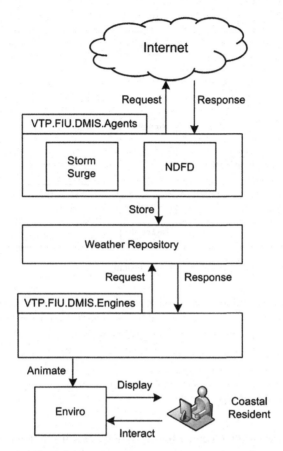

**Fig. 13.5** Diagram for simulation system

per frame in the simulation and must be implemented by every class that inherits from vtEngine with code, often written in OpenGL, that renders some change to the animation. On average, a simulation will render 30 frames per second. All of the engines that inherit or are otherwise associated with the vtEngine class are described in the following sections.

### 13.5.1   Time Engine

The engines in the storm surge simulation system are synchronized by the time engine which is implemented by the Time class. Upon initialization of the simulation system, the engine keeps track of the time in days, hours, minutes, and seconds. The time engine may be configured to use the system clock time of the computer the simulation is running on or may be configured to initialize with specific dates for simulation of historical or projected storms. The time engine may also be configured

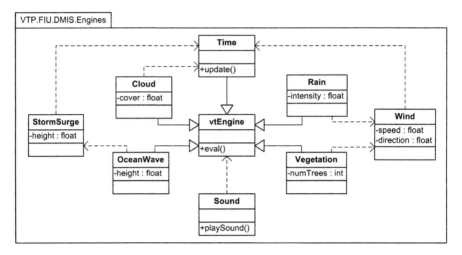

**Fig. 13.6** Minimal class diagram for simulation system

to elapse at an accelerated rate. Upon every invocation of its eval() function, the time engine updates the current time and uses event-driven techniques to notify each engine of the time change. Each engine then makes a request to the localized weather repository to retrieve weather data at a specific time.

### 13.5.2 Wind Module

The wind module is implemented by the Wind class. It maintains track of wind speeds, gusts, and direction for the current time based on the location being simulated under storm surge conditions. Wind speeds and gusts are measured in knots and wind direction in degrees clock-wise from true north. It does not inherit from the vtEngine class since it is only updated when the time engine notifies it of a time change. Upon notification of a time change, the wind engine retrieves wind conditions for a specific time from the weather repository and calculates an average speed, gust, and direction. On its own, the wind module does not provide any visual effect in the simulation. However, it is used by the vegetation, rain, ocean wave, cloud, and sound engines to dictate each of their behavior with respect to wind.

### 13.5.3 Vegetation and Rain Engines

The Vegetation class implements the vegetation engine which manages tree placement and movement induced by wind in the simulation. An implementation of the vertex weighting technique [26, 27] is used to animate each tree. Upon invocation of its eval() function, the vegetation engine rotates, bends, and ruptures individual tree branches based on wind speed, gust, and direction data from the wind module.

Animating trees based on wind provides users with an aid to visually perceive wind data. It also provides ambient details that add to the experience of the visualization.

Similarly, the rain engine is implemented by the Rain class which animates rain droplets based on wind data from the wind module and precipitation data from the weather repository. The quantity of rain droplets rendered is directly proportional to the forecast rain fall from the NDFD. Likewise, the direction and speed with which the rain droplets fall is dependent upon wind direction, gusts, and velocity.

### 13.5.4   *Storm Surge Module and Ocean Wave Engine*

Implemented by the StormSurge class, the storm surge module retrieves storm surge projection data from the weather repository and calculates an average of all the storm surge measurements within a one mile radius of the currently simulated location. The storm surge module need not inherit from the vtEngine class since it shall be updated when it receives a time notification from the time engine. The ocean wave engine invokes the storm surge module once per frame to render the varying height and waves of the ocean over time.

The virtual ocean is animated by the OceanWave class using OpenGL. Its rhythmic, wave-like motion is modeled using Fournier's model for ocean waves [9]. Fournier's model takes into account the depth of the ocean to model breaking waves on the shore. It is also used to animate waves at deeper ocean depths by modeling the movement of individual water particles in elliptical stationary orbits around their rest position $(x_0, y_0, z_0)$. The particle's motion is given by the following parametric equations in $x_0$ for a given $t$ and a constant $z_0$:

$$x = x_0 + r sin(kx_0 - \omega t)$$

$$z = z_0 - r cos(kx_0 - \omega t)$$

The horizontal plane is defined on the $XZ$ plane. The height of a wave is defined on the $Z$ axis. Note that $z_0$, the initial wave height, is retrieved by the ocean wave engine from the weather repository. The equations above delineate the curve generated by a point, or water particle, a distance $r$ from the center of a circle with radius $\frac{1}{k}$. The circle rolls over a line at distance $\frac{1}{k}$ under the $X$ axis [9]. For $t = 0$ and $z_0 = 0$, the equations are

$$x = -\frac{\alpha}{k} - r sin(\alpha)$$

$$z = -r cos(\alpha)$$

where $\alpha = -kx_0$, the height of the wave is $H = 2r$, the wavelength is $L = \frac{2\pi}{k}$, the period is $T = \frac{2\pi}{\omega}$, the phase speed is $c = \frac{L}{T} = \frac{\omega}{k}$, and the phase is $\phi = kx_0 - \omega t -$ assuming a phase of 0 for $x_0$. In deep ocean, the period and wavelength are related by $L = \frac{gT^2}{2\pi}$ [9].

The ocean floor surface is accounted for by including the depth $h$ at the point $(x_0, y_0, z_0)$ in a cumulative way. That is, the phase delay introduced by the depth effect is carried over from deep ocean to the shore. The wave number $k$ (the reciprocal of the wavelength) is a function of $h$, $h$ is a function of $x_0$, and the phase is given by the equation:

$$\phi = -\omega t + \int_0^{x_0} k(x)dx \text{ where } k(x) = \frac{k_\infty}{\sqrt{tanh(k_\infty h(x))}}$$

The term $k_\infty$ is the wave number at deep ocean (infinite depth) and is calculated by $k_\infty = ktanh(kh)$. An approximation to the above equation is given by [9]:

$$\phi = -\omega t + \sum_0^{x_0} \frac{k_\infty}{\sqrt{tanh(k_\infty h(x))}} \Delta x$$

The values for the angular frequency $\omega$ and the wave number $k$ are received as parameters. The time $t$ is computed from the succession of frames the ocean engine goes through after the simulation starts. The wave trains begin their motion at deep ocean, a depth marked by an arbitrary large number MAXDEPTH, and sweep the terrain towards the shore; through this motion, the cumulative term in the above equation gathers information from the bottom and correspondingly alters the shape of the wave. Near the shore, the term $h(x)$ simulates an increasing slope from MAXDEPTH to zero depth, causing a reduction in the wave length and producing "small" waves that break at the shore. Figure 13.7 shows the water animation using the model described.

**Fig. 13.7** Ocean wave simulation

## 13.5.5   Clouds and Lightning

To increase the realism and similarity of the visualizations in the simulation to actual storm surge conditions during a hurricane, we have integrated virtual clouds and lightning based on weather forecasts from the NDFD. To maintain low computational overhead, clouds are animated using the billboard modeling technique. With this technique, three-dimensional cloud models are simulated using a combination of translucent two-dimensional images. The cloud engine uses percent cloud cover data stored in the weather repository to calculate how many clouds it shall render and animate. Clouds are positioned at varying altitudes, may overlap with each other, and move across the sky based on wind conditions provided by the wind module. As show in Fig. 13.8, clouds affect brightness, length of

**Fig. 13.8**  Percent cloud cover affects lighting conditions

visibility, and shadows caused by buildings, vegetation, and the clouds themselves. Similarly, severe thunderstorm forecasts stored in the weather repository are used by the cloud engine to identify the frequency and intensity with which it renders lightning in the simulation. Lightning is rendered as a sequence of bright lines which grow like an inverse tree from the sky to the ground.

## 13.6  Public Simulation Access

The work presented in [23] proposes a web-based interface that provides public access to videos of storm surge simulations based on our system. It is based on the Google Maps API [11] and uses HTML, JSP, Ajax, JavaScript, and Flash web technologies. Figures 13.9 and 13.10 present a screen capture of the web interface. Users may query location-specific storm surge projections by submitting an address or zip code. Once submitted, the map centers on the submitted address, a color-coded graphic of storm surge heights is superimposed over the map, and a pre-recorded video of the storm surge video is played. The web-based interface for the storm surge simulation system plays a crucial part in disseminating information to the public. Doing so with video enables anyone with a personal computer and an Internet connection to view the simulations despite the performance capabilities of their hardware configurations. However, pre-recording videos introduces a setback to the goal of our system because the video may not necessarily focus on areas of interest to the users, such as their homes or businesses.

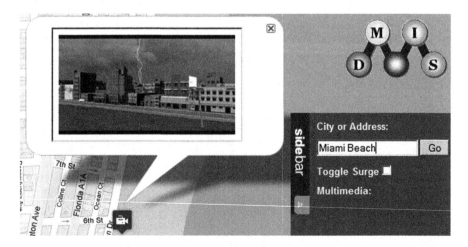

**Fig. 13.9** Users may query videos of storm surge via web interace

**Fig. 13.10** Color-coded graphics indicate projected heights of storm surge

A methodology to allow users to request videos was therefore designed. Geographic coordinates of the user-submitted address or zip code are retrieved through the Google Maps API. An internal MySQL database is queried for any videos within a quarter mile radius of the coordinates that have been recorded within the last storm surge projection updated. Typically, projections are updated every three to six hours. If a video is found, then it is presented to the user in a Flash-based video player [8]. Otherwise, the storm surge system creates and records a simulation. The open-source PROJ.4 library [24] is used to automatically convert the geographic coordinates of the address or zip code to the Universal Transverse Mercator coordinate system [29], which is used by the VTP. The storm surge simulation system is then configured to automatically load the latest storm surge projection and NDFD data and begin recording a 30-second video from the UTM coordinate. The video is recorded from a height of 30 feet and tilt of 50° toward the natural ground. The video rotates westward at 12° per second so that a 360° view of the area is recorded in 30 s. Video is recorded using the VTP's native screen capture utility. Sound is recorded using OpenAL [20]. The video is integrated with the sound and compressed with VirtualDub [31], the open-source video processing utility. The MySQL database is updated with a link to the recorded video which is then presented to the user. On average, the time it takes between receiving a request, recording the video, and displaying to the user is approximately one minute.

## 13.7   Conclusions and Future Work

In this chapter we presented a storm surge simulation system designed to help local governments in the State of Florida overcome difficulties in persuading residents to adhere to official evacuation notices. The system integrates LiDAR, DLG, NDFD, and storm surge projection data to construct accurate virtual models of cities in the State of Florida under projected storm surge conditions. LiDAR data are used to extract building footprints and topography measurements and construct digital terrain models. DLG are used to place public roads on the terrain models. A virtual ocean is positioned over the digital terrain model using official storm surge projections. The ocean waves, clouds, lightning, wind, and vegetation in the simulation are animated based on weather data from the National Digital Forecast Database (NDFD). A web-based interface to the storm surge simulation system was also presented. The interface uses the Google Maps API to enable residents access to the latest storm surge projections. Users may also request and view videos of the storm surge simulation in the location of their homes or businesses. By integrating LiDAR, DLG, NDFD, and storm surge projection data, we have created a public relations system that enables residents in the State of Florida to make potentially life-saving decisions.

Future work calls for a study to evaluate the system's ability to have an impact on coastal resident evacuation. Investigating how the community reacts to the storm surge visualizations and evaluating what is the most useful information to coastal residents and public decision makers is crucial towards further development of the system. Furthermore, as computer users become more use to and expectant of interactive web applications, our system must be extended to provide interactive visualizations of storm surge. In addition to videos, the system shall provide an interactive interface whereby a user may directly manipulate the depth of storm surge and orientation of point of view. Finally, more extensive LiDAR data sets available from NOAA will be used to create storm surge visualizations for the entire State of Florida.

**Acknowledgements** This project was supported in part by a grant from NOAA. Any opinions, findings and conclusions or recommendations expressed in this material are those of the author(s) and do not necessarily reflect those of NOAA.

## References

1. AIR Worldwide (Aug. 2010). http://www.air-worldwide.com/newsandeventsitem.aspx?id=12648
2. Amira (Apr. 2010). http://www.amira.com/
3. Benger, W., Venkataraman, S., Long, A., Allen, G., Beck, S.D., Brodowicz, M., Maclaren, J., Seidel, E.: Visualizing katrina – merging computer simulations with observations. Springer Verlags Lecture Notes in Computer Science Series. pp. 340–350. Springer, Berlin (2007)

4. BT Format (Apr. 2010). http://vterrain.org/implementation/formats/bt.html
5. Campbell, J.B.: Introduction to remote sensing, 3rd edn. Guilford Press, New York. (2002)
6. Digital weather markup language (Sep. 2010). http://www.weather.gov/forecasts/xml/dwmlgen/schema/dwml.xsd
7. Flaxman, M.: Using the virtual terrain project to plan real cities: alternative futures for hangzhou, china. In: ACM SIGGRAPH, pp. 340–350. San Antonio, TX, USA (2002)
8. Flowplayer (Apr. 2010). http://flowplayer.org/
9. Fournier, A., Reeves, W.T.: A simple model of ocean waves. In: SIGGRAPH, pp. 75–84 (1986)
10. Google Maps (Apr. 2010). http://maps.google.com
11. Google Maps API (Apr. 2010). http://code.google.com/apis/maps
12. Hurricane floyd impacts (Aug. 2010). http://www.erh.noaa.gov/mhx/floyd/impacts.php
13. IHRC (Apr. 2010). http://www.ihrc.fiu.edu/about_us/hurricane_hazards
14. Java net package (Sep. 2010). http://download.oracle.com/javase/1.4.2/docs/api/java/net/package-summary.html
15. Li, Y., Chatterjee, K., Chen, S.-C., Zhang, K.: A 3-d traffic animation system with storm surge response. In: ISM '09: Proceedings of the 2009 11th IEEE International Symposium on Multimedia, pp. 257–262. IEEE Computer Society, Washington, DC, USA (2009)
16. Luo, J., Makwana, A.P., Liao, D., Kincaid, J.P.: Hurricane! – a simulation-based program for science education. In: WSC '08: Proceedings of the 40th Conference on Winter Simulation, pp. 2543–2548. Winter Simulation Conference (2008)
17. NOAA SOAP service (Aug. 2010). http://mi.nws.noaa.gov/xml/
18. NOAA (Apr. 2010a). Coastal services center. http://www.csc.noaa.gov/
19. NOAA (Apr. 2010b). http://www.nhc.noaa.gov/haw2/english/storm_surge.shtml
20. OpenAL (Apr. 2010). http://connect.creativelabs.com
21. OpenGIS (Apr. 2010). http://www.opengeospatial.org
22. OpenGL (Apr. 2010). http://www.opengl.org
23. Pava, J., Fleites, F., Ruan, F., Chatterjee, K., Chen, S.-C., Zhang, K.: A three-dimensional geographic and storm surge data integration system for evacuation planning. In: 11th IEEE International Conference on Information Reuse and Integration, pp. 181–188. IEEE SMC Society, Las Vegas, NV, USA (2010)
24. Proj4 (Feb. 2010). http://trac.osgeo.org/proj/
25. RMF (Apr. 2010). http://vterrain.org/doc/roads.html
26. Saleem, K., Chen, S.-C., Zhang, K.: Animating tree branch breaking and flying effects for a 3d interactive visualization system for hurricanes and storm surge flooding. In: ISMW '07: Proceedings of the Ninth IEEE International Symposium on Multimedia Workshops, pp. 335–341. IEEE Computer Society, Washington, DC, USA (2007)
27. Singh, P., Zhao, N., Chen, S.C., Zhang, K.: Tree animation for a 3d interactive visualization system for hurricane impacts. In: ICME 2005, pp. 598–601, IEEE International Conference on Multimedia and Expo. (2005)
28. United States Geological Survey (Apr. 2010). http://data.geocomm.com
29. Universal transverse mercator coordinate system (Apr. 2010). http://egsc.usgs.gov/isb/pubs/factsheets/fs07701.html
30. Venkataraman, S., Benger, W., Long, A., Jeong, B., Renambot, L.: Visualizing hurricane katrina: large data management, rendering and display challenges. In: GRAPHITE, pp. 209–212 (2006)
31. Virtual dub (Apr. 2010). http://www.virtualdub.org/
32. Virtual Terrain Project (Apr. 2010). http://vterrain.org
33. VTST Format (Apr. 2010). http://vterrain.org/implementation/formats/vtst.html
34. Weather underground (Sep. 2010). http://www.wunderground.com/
35. Weather.com XML data feed (Sep. 2010). http://www.weather.com/services/xmloap.html
36. Xiao, C., Zhang, K., Shen, J.: A three-dimensional coastal and estuarine storm tide model. J. Coast. Res. p. 20 (2006)

37. Zhang, K., Chen, S.-C., Whitman, D., ling Shyu, M., Yan, J., Zhang, C., Member, S.: A progressive morphological filter for removing nonground measurements from airborne lidar data. IEEE Trans. Geosci. Rem. Sens. **41**, 872–882 (2003)
38. Zhang, K., Chen, S.-C., Singh, P., Saleem, K., Zhao, N.: A 3d visualization system for hurricane storm-surge flooding. IEEE Comput. Graph. Appl. **26**(1), 18–25 (2006a)
39. Zhang, K., Yan, J., Chen, S.-C.: Automatic construction of building footprints from airborne lidar data. IEEE Trans. Geosci. Rem. Sens. **44**(9), 2523–2533 (2006b)
40. Zhang, K., Xiao, C., Shen, J.: Comparison of the cest and slosh models for storm surge flooding. J. Coast. Res. **24**, 489–499 (2008)
41. Zhang, K., Yan, J., Chen, S.-C.: Automatic 3d building representation from airborne lidar measurements. In: Joint Urban Remote Sensing Event, pp. 483–487. IEEE Computer Society, Shanghai, China (2009a)
42. Zhang, K., Yan, J., Chen, S.-C.: A framework for automated construction of building models from airborne lidar measurements. In: Shen, J., Toth, C. (eds.) Topographic Laser Ranging and Scanning: Principles and Processing, pp. 511–534. CRC Press, Boca Raton (2009b)

# Chapter 14
# A Supervised Machine Learning Approach of Extracting and Ranking Published Papers Describing Coexpression Relationships among Genes

**Richa Tiwari, Chengcui Zhang, Thamar Solorio, and Wei-Bang Chen**

## 14.1 Introduction

With the advent of technology and Internet, information exchange and storage has increased tremendously. There is a huge amount of information available in almost all fields on the Internet. The technological development has led biomedical research scientists to publish and share their findings and results online. PubMed is one such source where people can find a large amount of data and publications [16]. According to a recent fact sheet produced by MEDLINE a component of PubMed, it consists of almost 712,615 indexed citations by the end of 2009 (http://www. nlm.nih.gov/pubs/factsheets/medline.html) [13]. Figure 14.1 shows some of the statistics of the trend of citations in the past five years in MEDLINE. We can clearly see that there is 6% increase in the indexed citations from 2008 to 2009 in MEDLINE/PubMed alone. Such a vast source of information being readily available is a great resource for new research and hypotheses. However, this also leads to some of the problems related with handling massive amount of data, such as data management, storage, extraction of precise information and appropriate information retrieval.

It is often times essential to obtain very precise and relevant information from published papers without wasting too much time in reading the whole paper. As research is progressively increasing in the biomedical discipline more and more papers are being published. Information extraction from these papers has become a very challenging task for both computer scientists and computational biologists. A computer scientist can help biomedical researchers in managing this data by building semantic information extraction and retrieval tools that can suit their needs.

R. Tiwari (✉) · C. Zhang · T. Solorio · W.-B. Chen
Department of Computer and Information Sciences, The University of Alabama at Birmingham,
115A Campbell Hall, 1300 University Boulevard, Birmingham, Alabama 35294
e-mail: rtiwari@cis.uab.edu; zhang@cis.uab.edu; solorio@cis.uab.edu; wbc0522@cis.uab.edu

T. Özyer et al. (eds.), *Recent Trends in Information Reuse and Integration*,
DOI 10.1007/978-3-7091-0738-6_14, © Springer-Verlag/Wien 2012

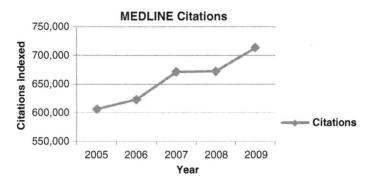

**Fig. 14.1** Graph showing the increase in the rate of citations being indexed in MEDLINE in past 5 years

One piece of such precise information that is needed by biomedical researchers is Gene-Gene Relationships, often referred to as Protein-Protein Interactions (PPI). These relationships can lead to the discovery of new hypotheses. Several kinds of interactions exist between and among genes and one such relationship is called "Co-expression" relationship. If two genes are expressed together, they are said to be coexpressed. This is a very important relationship and property among genes. If an unknown gene is expressed together with a known gene, we can easily assume that there exists some functional relationship between them. And we can then determine some of the properties of this unknown gene based on the known gene. These coexpressed genes can share the same pathway and can lead to several other functionalities. Also, the extracted relationships can be further used to build up networks that can describe complex biological pathways.

Often this coexpression relationship can be determined by various biological experiments such as microarray experiments and immune staining. There are publicly available gene expression repositories based on the results of microarray experiments that can give ranked lists of coexpressed genes [7]. The papers that are published with the results of these microarray experiments are very rich in information and hold a lot of other useful details related to the experiments that were performed and the by-products of this coexpression. Abstracts of these papers are readily available on PubMed, but they do not have extended information as what a full length papers can provide us with. Full length papers can help others to replicate those experiments, make more hypothesis based on the information present in the paper, etc. Often, just information extraction from the paper is not enough and we need to present this information in a ranked list. This ranked list of papers or documents can help the user to select the critical information from the top of the list more easily. For example, when we perform a search in an information retrieval system, we assume that the top few documents/web pages are the most relevant to our query. Similarly it is also essential to provide researchers with the papers that have the coexpression relationships as the main content of the paper. To provide the researchers a ranked list of papers based on the information content, it is essential to have a good extraction model first. Once we have a good information extraction model, we can provide a better retrieval results based on it.

To the best of our knowledge, there is no existing tool that can extract predicates that talk about coexpression relationships between and among genes from published articles based on text analysis and information extraction techniques. A work presented by [23] is the first step towards extracting such relationship from published literature. Our goal in this chapter is to present a framework that can retrieve papers that talk about gene-gene coexpression relationships, by using machine learning approach to first extract the predicates describing these relationships and then using our scoring scheme to rank the retrieved positive papers. Good retrieval results can be achieved based on a good extraction model in the back-end of it and hence we propose a retrieval framework that is based on our sentence classification model. As our classification model is especially trained to extract coexpression sentences irrespective of whichever ways they have been expressed in the paper, our retrieval system which is based on it will be able to rank the papers better even if the query terms use only one of the possible ways to this concept. We have used Dynamic Conditional Random Fields (DCRFs), a graphical probabilistic model, trained on these predicates [21]. Afterwards, we present our results by testing this model on full-length biomedical papers collected from PubMed Central, a free digital archive of biomedical and life sciences journal literature of U.S. National Institute of Health (NIH). For the information retrieval part, we have created our own scoring scheme to rank the papers for 5 different query genes and their coexpression. We then compared our ranking with Google and PubMed for the same set of papers and query terms using a well known rank comparison metric, Mean Average Precision (MAP) and a modified version of MAP that we created to agree with our needs [24]. The ranking comparison results show that our model performed much better than both of these search engines in retrieving the papers that talk about coexpression of genes.

The experiments performed in this work can be divided into two main parts. The first part includes the information extraction part, where we train a DCRFs model that can classify sentences as positive or negative based on whether they talk about gene-gene coexpression or not. In the second part we present the evaluation results of the proposed information retrieval system that ranks a paper according its relevance to the query gene.

The rest of the chapter is arranged as follows. Section 14.2 discusses the related work in this field. Section 14.3 introduces the methodology of our framework, and we present and analyze our results in Sects. 14.4 and 14.5.

## 14.2   Related Work

Information extraction from text is the process of automatically extracting structured data from unstructured data, usually text written in a natural language like English. There has been a lot of work done in the area of information extraction in the field of biomedical sciences. Noticeable conferences such as, Knowledge Discovery and Data Mining (KDD) challenge held in 2002 involved extraction of information from full-length papers in the field of biomedical research [6]. Another major conference in this field is BioNLP, which encourages research in natural language processing of biological text.

Aaron M. Cohen and William R. Hersh give an up-to-date survey of work done in biomedical text mining field [11]. Some effort in this area incorporates extraction of information such as named-entities, different medical terms, relationships etc from either the abstracts of the published literature or the full-length papers themselves. In the past, researchers have used various natural language processing techniques such as hand-written rules based on linguistic knowledge, to extract information from text [22]. However, these approaches are very expensive and require a lot of time and effort from domain experts, which involves writing all the rules manually. Hence, more statistical approaches such as machine learning have been introduced to aid the Information Extraction task. One of the early works done in this field was using Naïve Bayes probabilistic model to extract information about protein localization pattern from MEDLINE abstracts [8]. Similarly, Hidden Markov Models, Neural Networks and similar stochastic classification algorithms have been successfully used in extracting information from text [5]. Few other works towards extracting interactions between biomedical entities includes use of machine learning classification algorithms like Support Vector Machine (SVM) and Neural Nets. Probabilistic graphical models like Hidden Markov models (HMM) and Conditional Random Fields (CRFs) have proven to be quite successful in classifying and extracting relationships between biomedical entities [18, 19]. CRFs are undirected graphical models developed by Lafferty et al. in 2001, and have been applied to various text mining tasks such as table extraction and named entity extraction in biomedical text [12, 25]. [4] used linear chain CRFs to extract gene-disease relationships from biomedical literature. DCRFs are an extension of CRFs and have been shown to outperform CRFs in natural language chunking tasks [21].

Gene-gene coexpression relationships extraction is a subtask of relationship extraction, but differs in the way that this relationship is expressed in several ways in natural language in actual text. This is a widely researched area in computational biology and both manual and automatic work is being done in this area. There are existing databases that contain manually curated 12,000 Medline articles for protein-protein interactions (PPI) [17]. Work done towards automatic extraction of PPI using natural language processing tools (NLP) includes Miyao et al. in which they evaluated the contribution of natural language parser to protein-protein interactions (PPI) [14]. They showed that the appropriate natural language parsers can help researchers extract information such as PPI from unstructured data such as published literature. Another work in this direction proposes to use a rich NLP feature vector set and Support Vector Machine classification algorithm to extract PPI from sentences [15]. Another approach towards extracting PPI from papers is the co-occurrence counts of words, that is, if the two entities occur often or have a textual pattern of occurrence in a paper, then we can predict if a relationship among them exists or not by using statistical measures like pointwise mutual information, chi square, etc. A work by Mooney et al. combines this co-occurrence statistics of words approach as well as the information extraction approaches that use classification algorithms to extract sentences and presents an IR system for PPI from Medline [3]. SUISEKI, is another framework developed for detecting PPI from published text by Valencia et al. in 2002 [2]. Generally there is a pattern of expressing a relationship between two entities, for instance the relationship word, usually a verb, will occur in between the entities in the text and hence can be extracted if we know the entity boundaries. This tool uses a set of predefined rules or frames along with the other statistical and linguistic tools to extract PPI from literature. The rule based extraction makes the process non expandable and time consuming to prepare. It is not manageable to collect all the possible rules for sentences and thus the retrieval system will only be able to retrieve limited number of documents. Coexpression relationships are not necessarily written in this format, certain challenges can occur in classifying those using fixed patterns. For example, there are several different words that can express this interaction among genes. Some of the commonly used words are – "Coexpress", "expressed together", "both were expressed", "up-regulated", etc.

In this approach we have provided a complete framework for extracting and retrieving papers about coexpression relationship among genes. But, as our classification and indexing scheme is independent of any specific retrieval algorithm it can be used with any technique. Also, as our work focuses on the retrieval of papers based on not just the query keywords but semantic meaning of those keywords, it provides a better result than PubMed and Google in the case of papers talking about coexpression relationship among genes. This can be seen from the comparison results later in Sect. 14.4 of this chapter.

## 14.3 The Proposed Approach

Automatic relationship extraction from unstructured machine-readable text is a complex and challenging task. Following recent successes in using CRFs for relationship extraction, we have explored their use in our application. Relationship extraction is comprised of annotating the unstructured text with the entities involved and the relationship between those entities. In our case, entities are the genes and the relationship terms are the ones that express coexpression relationships. As illustrated in Fig. 14.2, our framework for the sentence classification task can be divided into five steps: (1) Data collection, (2) Pre-processing, (3) Feature extraction and class assignment, (4) DCRFs classification model training, and (5) Feature and class analysis.

We also prepared a scoring scheme and performed rank comparison with Google and PubMed. Figure 14.3 represents our information retrieval and rank comparison model in detail.

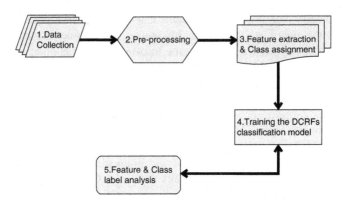

**Fig. 14.2** Block diagram of our proposed sentence classification framework

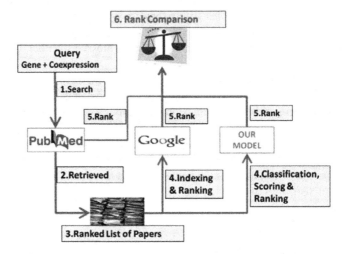

**Fig. 14.3** Block diagram showing our paper ranking and comparison scheme

## 14.3.1  Data Collection

For the sentence classification purposes, we collected 500 full-length papers from PubMed and manually divided them into positive (285) and negative papers (215) depending upon whether a paper contains predicates about some genes being coexpressed and not just the coexpression process in general. A paper that contains positive sentences has both gene names as well the terms defining the coexpression relationships in them. This collection helped us to prepare a model that was later used for testing the papers in our retrieval process. Figure 14.4 gives some example sentences that help understand the difference between the positive sentences (sentences that talk about gene-gene coexpression) and negative sentences (sentences that talk about just coexpression in general or do not talk

**Positive Sentences**
Functionally , coexpression of Kir2.1 and PSD-93 had no
discernible effect upon channel kinetics but resulted in cell
surface Kir2.1 clustering and suppression of channel
internalization.
The coexpression of the c-Myb DBD dominant negative
mutant protein was effective in blocking the up-regulation of
all three promoters.

**Negative Sentences**
We further identified the partial co-expression relationship
between genes: gene profiles may simultaneously rise and
fall in a sub-range of the time course rather than the overall
time course.
One important goal of analyzing gene expression data is to
discover co-regulated genes.

**Fig. 14.4** Examples of positive and negative sentences in literature

about coexpression at all). Overall, we have 15,010 sentences in our repository to
experiment with.

For the second part of our experiments, we collected a different set of papers
from PubMed using five different query genes. The five query genes were – "Bcl-2
coexpression", "ErbB coexpression", "IL coexpression", "Myc coexpression", and
"p53 coexpression". In total we collected 500 papers by taking the top 100 papers
for each query. These papers were tested against the DCRFs model created in the
first part, scored and ranked.

## 14.3.2   Pre-processing

The collected papers are all in PDF format and were converted into text using
the PDFbox java library (http://incubator.apache.org/pdfbox/index.html) [1]. Once
these papers were converted into text, we deleted the sentences or parts from the
papers that we believe do not provide any useful information about coexpression
relationships. These parts include – everything before the abstract section of the
paper and everything after the acknowledgment section of the paper. As mentioned
earlier, relationship extraction includes tagging the entities and the relationship
among them. In our case these entities are gene names. Hence, we assume that
the sentences that do not contain the desired entities (gene names) usually would
not correspond to any useful information about the relationship. We consequently
deleted all the sentences that do not contain gene names. To determine if any gene
name is mentioned in a given sentence, we tagged each word in a sentence using the
GENIA tagger, a freely available tool that has been trained using HMMs to identify
gene names in any given text GENIA Tagger 3.0, (http://www-tsujii.is.s.u-tokyo.ac.

jp/GENIA/tagger) [9]. The sentences that are not tagged by the GENIA tagger as containing gene names are discarded from the succeeding steps.

### 14.3.3  Feature Extraction and Class Assignment

This step is performed for the task of building a classification model. Feature extraction is a very important task in machine learning techniques. It is important to provide a good set of features to any machine learning algorithm, if we want the model to perform well. However, there is no fixed set of features that can be used to improve the performance of any model, as they are very task specific.

The basic set of features is extracted using the GENIA tagger. GENIA parses the sentences and tokenizes them into words. It then assigns features to each word. These features include, the root form of the word, e.g., the word *coexpression* has the root *coexpress*. It also performs grammatical analysis of each sentence and extracts the part of speech tag feature for each word. Part of speech (POS) feature of a word describes the grammatical role of the word in the sentence, e.g., the word "books" has a POS tag as *noun* in the sentence "Books are made of paper ink and glue". But the same word is tagged as *verb* for the sentence "book your flight soon". Another set of features that GENIA tagger extracts is often referred to as the *Chunk tags*. Chunk tags are the tags for the constituent phrases in a sentence. A constituent phrase is a group of words that functions as a single unit in a sentence. Therefore, all the words in a noun phrase will get the chunk tag NP (Noun Phrase) as a feature. It is worth noting that a biomedical entity term will be assigned a chunk tag as "B-NP" by the GENIA tagger. Finally as mentioned earlier, GENIA tagger also assigns biomedical entity tags to the entities, such as "B-protein" and "B-RNA" and these tags act as additional features of those words. The final basic feature set can be divided into two categories:

Local:

- The word ($w_i$, where $w$ indicates the word and $i$ indicates the position of the word) itself and the root of the word.
- POS tag of the word and its chunk tag, i.e., the tag at the phrase level
- Biomedical named-entity tag.

Contextual:

- POS and chunk tags of the words $w_{i-3}, w_{i-2}, w_{i-1}, w_{i+1}, w_{i+2}, w_{i+3}$.
- Biomedically named entity tags of the words $w_{i-3}, w_{i-2}, w_{i-1}, w_{i+1}, w_{i+2}, w_{i+3}$.

We train a DCRFs model on the class labels assigned to words in a way as follows:

- RE class labels assigned to the words that express the coexpression relationships.
- GE class labels assigned to the gene name words that are involved in the coexpression.
- Chunk tags assigned to the rest of the words in the sentence.

**Table 14.1** Example of class labeled words in a sentence

| Words | Class tags | Meaning |
|---|---|---|
| We | B-NP | Noun phrase |
| Conclude | B-VP | Verb phrase |
| That | B-SBAR | Subordinating conjunction |
| coexpression | **RE** | **Relationship word** |
| of | B-PP | Propositional phrase |
| LacZ | **GE** | **Gene name** |
| and | B-NP | Noun phrase |
| M71 | **GE** | **Gene name** |
| Occur | B-VP | Verb phrase |
| Frequently | B-ADVP | Adverb phrase |
| When | B-ADVP | Adverb phrase |
| The | B-NP | Noun phrase |
| Two | B-NP | Noun phrase |
| Gene | B-NP | Noun phrase |
| Be | B-VP | Verb phrase |
| Present | B-ADJP | Adjective phrase |
| In | B-PP | Propositional phrase |
| Cis | B-NP | Noun phrase |

Overall, we have 12 class labels including the GE, RE and all the chunk tags, out of which we are interested in classification of RE and GE labeled words by our model. These are the basic or the original set of features and classes that we started our experiment with, and later on we refined the feature set by analyzing the results in Steps 4 and 5. Table 14.1, shows an example of labeled sentence with each word labeled according to their phrase tag/chunk tags and GE and RE tags.

## 14.3.4 Training the DCRFs Model

We perform supervised learning using DCRFs, a probabilistic graphical model that can be effectively used for sequence classification problems. It is a generative model that relaxes the Markov assumption of HMMs regarding the input and output sequence. DCRFs combine the concept of CRFs (conditional probability distribution which allows rich feature set) and Dynamic Bayesian Network (DBN). DCRFs calculate the conditional probability distribution $p(y/x)$ of output label sequences $y$ given a particular observation sequence $x$, rather than finding the joint probabilities of both label and observation sequence [18]. Equation 14.1 shows how to calculate conditional probabilities of output labels given the input observations using DCRFs.

$$p(y/x) = \frac{1}{z(x)} \prod_t \prod_{c \in C} \exp\left(\sum_k \lambda_k f_k(y_{t,c}, x, t)\right). \qquad (14.1)$$

where $f_k(y_{(t,c)}, x, t)$ is referred to as the transition feature function which is equivalent to transition probabilities in HMM. $\lambda_k$ is the learned parameter vector in the model. $Z(x)$ represents the normalization function, and $K$ is the number of feature functions. $t$ is the time step (state) and $C$ being the set of clique indices. A clique in an undirected graph G is a set of vertices $V$ such that for every two vertices in $V$, there is an edge connecting them.

In our framework we have used a GRaphical Model for Mallet toolkit (GRMM) which has a java implementation of DCRFs [20]. Once we have the training data ready, we run Mallet for DCRFs algorithm and train a classification model on the classes present in the training set. This model learns the conditional probability values for the occurrence of each class given any input sequence. Subsequently when a testing data is given to it, the model will assign class labels to each of its input words based on the learned parameters.

Once the input files are tagged with their true class labels and features, we divide the data into training and testing sets. We also create a small subset of tagged data referred to as the development set, which consists of 4,109 sentences (1,333 positives ones and 2,776 negative ones). A development set is a set of testing data that is used for parameter tuning and feature assessment and is different from the main testing file. This set is prepared separately from the main testing set such that while parameter tuning our model does not get biased towards our main testing file. This development set was used as the testing data set in the next step to select the appropriate set of features and class labels. In other words, the next step is to determine which features to use for predicting the class labels.

### 14.3.5 Feature and Class Selection

In this section we present our process of feature selection and class label analysis. We use our development set as the testing data for this process and performed several experiments with different feature combinations. DCRFs are good for representing complex interactions between class labels. In our framework, this will refer to the interaction between class labels for genes that are actually involved in coexpression relationships and the labels for coexpression relationship terms. This leads us to experiment with various sets of appropriate class labels because gene name extraction in itself is a big area of research and we are more interested in finding the coexpression terms. Finally we come up with a set of features and class labels that have the strongest correlation with the classification variable, i.e., the feature and class label combination that gives the best results for tagging RE labels. We experimented with different combinations of features and classes, shown in Fig. 14.5. Each of these experiments and their results are explained in more details in Sect. 14.4. The list of experiments performed includes:

1. Experiment with the original feature set, which includes all the features mentioned in Sect. 14.3.3.

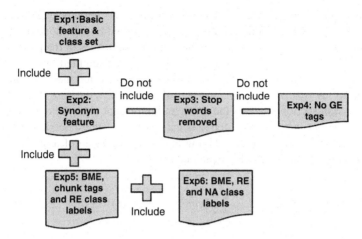

**Fig. 14.5** The process of experimenting with different combinations of features and class labels

2. Experiment with trigger words gazetteer with synonym features.
3. Experiment with removing stop words after assigning contextual features to the data set.
4. Experiment with learning different sets of class labels, which include – 1) no GE class labels, such that all the gene names get their chunk tags as their class labels, and 2) all the biomedical entity words get the biomedical entity (BME) tags assigned by the GENIA tagger as their class labels. So, no GE class labels but three other class labels – B-protein, B-DNA, and B-RNA.
5. Experiment with different combinations of settings from Experiments 3 and 4.

Experiment with just a few class labels including B-protein, B-DNA, B-RNA, RE and NA, where NA class label was given to all of the irrelevant words in the sentence (the ones that were given their chunk tags as their class labels in earlier experiments).

Once we decide on an appropriate set of features and class labels, we perform training and testing on our larger data set. The comparison results are presented in the next section.

### 14.3.6   Scoring Scheme

Once we have a DCRFs model trained to classify sentences talking about coexpression, we use that model to test the second set of 500 papers that we collected using 5 query genes from PubMed. The end product of this testing are the class labels generated for each sentence in the testing papers, which distinguish between a positive sentence and a negative sentence. We propose a scoring scheme to score these test papers in this section. We score a paper based on three main criteria: (1)

Number of positive sentences tagged by our model in the paper, (2) Location of those sentences, and (3) Number of query genes tagged by our model in the paper. We implemented a location based weighting of the sentences in the paper, with the *Abstract, Results and Discussion* sections of the paper weighted twice as much as the rest of the paper. We made this assumption because often the important results are stated in the *Abstract* section, and it is almost always the case that the *Results* and *Discussion* section contains the result and important observations of the paper. Therefore, if a coexpression sentence occurs in one of these sections of a paper, we can assume that this paper is talking about coexpression of a gene. This zone weighted scoring scheme can be explained by 14.2.

$$ws = \sum_{i=1}^{4} \{count_s\,(i) \times w\,(i)\}. \tag{14.2}$$

In this equation, $ws$ is the total weighted score of a paper, $i$ is the location or zone and we have 4 different zones. $count_s(i)$ is the total number of sentences at location $i$. $w(i)$ is the weight of each particular zone $i$, which can be either 1 or 2 depending upon whether the sentence was found at *Abstract* and *Result & Discussion* sections ($w(i) = 2$) or *Introduction* and *Material & Method* sections ($w(i) = 1$).

To calculate the number of query genes tagged by our model, we first prepare a manually built dictionary of all the gene synonyms and the family names of the query genes from PubMed and then perform a dictionary matching from the tagged list of genes by our model to count the number of occurrences of query genes in the whole paper.

Once we collect all three counts, i.e., the number of positive sentences, their location based weights, and the number of occurrences of query genes, we then normalize the whole score by the number of pages in the paper. Hence, the final score assigned to each paper can be calculated as in 14.3.

$$Score = \frac{tcount + ws + gs}{pages}. \tag{14.3}$$

Where score indicates the total score assigned to a paper which is the sum of the number of positive sentences (*tcount*), the zone weighted score (*ws*), and (*gs*) the total number of query gene occurrences, and this sum divided by the total number of *pages* in the paper. This division helps us to normalize the scores relative to the length of the paper. After we have the score of each paper, we rank them with the highest scoring paper getting the top rank.

## 14.4  Results

In the following two subsections, we present results of our sentence classification experiment and information retrieval experiments, respectively.

**Fig. 14.6** Comparison of the results obtained by the baseline method and that of the model trained with the original feature set

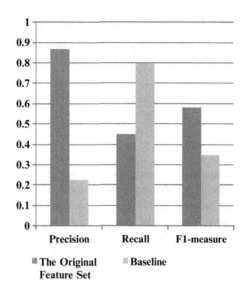

## 14.4.1   Sentence Classification

In this section we present the experimental results of our model in detecting the coexpression predicates from text using DCRFs. We started with 500 full length papers in our experiment and performed pre-filtering on it. After down sampling we were left with approximately 15,010 sentences all together. As we used a supervised learning approach, each word in each of these sentences was manually tagged as the ground truth for learning. Out of the total 15,010 sentences we have around 3,130 positive sentences which are only 20% of the total data set. Once the sentences are tagged appropriately, we divide the whole data set into training and testing files and perform training and testing on them using DCRFs. We present our results in the form of Precision, Recall and F-measure. F-measure is the weighted harmonic mean of the both Precision (P) and Recall (R). Figure 14.6 shows a comparison between the results of our **baseline model** and that of model trained with the **original set of features**. In the baseline experiment we perform the basic term matching from training to testing file. All the terms in the training files that correspond to coexpression terms were collected in a training corpus, and then each testing file was matched against it. If any word in the testing file matches some word in the training corpus, then the file is tagged as positive. This can also be considered as dictionary matching where the dictionary is built each time from a set of training files. The experiment with the original set of features involves experimenting with the set of features mentioned in Sect. 14.3.3.

We can see that although the baseline approach has a higher Recall score, it has a very poor Precision. This is because the baseline approach is solely based on keyword matching, and the contextual information is missing. One common

**Table 14.2** Experimental results with different combinations of features and class labels

|                              | Precision | Recall | F1   |
|------------------------------|-----------|--------|------|
| Trigger words                | 0.81      | 0.51   | 0.62 |
| No stop words                | 0.81      | 0.49   | 0.61 |
| No GE labels                 | 0.81      | 0.58   | 0.67 |
| BME tags as labels           | 0.82      | 0.56   | 0.66 |
| No stop words & BME tags      | 0.80      | 0.50   | 0.61 |
| BME, RE and NA class labels  | **0.81**  | **0.63** | **0.71** |

example is the word "express" along with some other word like "together" or "both". This word, if used alone without a helping word like "both" or "together," does not indicate anything about coexpression relationships, but in the baseline approach every "expression" word will be tagged as positive in a test file. This will result in lots of false positives and lead to low Precision. Our model, trained with just the original set of features without feature selection, is almost 23% better than the baseline model in terms of F-measure.

As mentioned in Sect. 14.3.5, we performed some experiments to come up with the best set of features and class labels. Table 14.2 shows the tenfold cross validation result of each of those 6 experiments, which include different combinations of features and classes.

The first experiment involves experimenting with **Trigger words** and the results are presented in the second row of the table. Trigger words are the words that most commonly describe the variable that we want to classify. A gazetteer of these words can be created and a model can be tuned on them. We created a list of approximately 200 trigger words by including words like "coexpress" and their inflected forms. Whenever a word in a file matches some word in the trigger word gazetteer, it is assigned a feature referred to as "synonym feature," i.e., tagged with the word "coexpress". This improved the Recall of our system by 6%, which means that our model was able to identify more of the positive sentences. Since we achieved a significant improvement in the performance by adding this feature (as shown in Table 14.1), we included it for the rest of our experiments.

Stop words are the commonly occurring words like articles and prepositions that do not contain any useful information regarding the semantic content of the text. It is a common practice in natural language processing to filter out these words from the text as a preprocessing step. In the second trial we removed all the stop words from the training files but kept the contextual features for each remaining word. The result of this **No stop words** trial is presented in row 3 of Table 14.1. Although this approach performed better than the original feature set experiment, we did not achieve any significant improvement in this experiment (as shown in Table 14.1). This is probably because once these words were removed we also lost their syntactic tags/labels. And as we know DCRFs also learn the relationship between labels, thus losing those labels within a sentence may throw off the model and lead to poorer performance.

As aforementioned, DCRFs help in capturing complex relationship between labels and are useful in chunking task. In our task of relationship extraction, we want to identify the words describing coexpression as well as genes, and the performance of our DCRFs model is influenced by the prediction accuracy of GE as well as RE tags in a sentence. Therefore, we decided to further experiment with the class labels too. In the third experiment, mentioned as **No GE labels** in Table 14.1, we replaced all the GE class labels for gene names with their chunk tags "B-NP" assigned by GENIA tagger. In this case we made no distinction between the gene names and the other regular words in terms of the way their class labels are assigned. This leads to the least guessing work in model training and the prediction of class labels for non RE terms as there is a near-deterministic relationship between the input feature and the class label. This approach shows the highest improvement on our dataset, a 9% increase from the experiment with the original set of features, but does not give any specific information about the genes that are coexpressed. Consequently we did not include this in our final experiment.

Thinking along the same lines as before, we replaced all the GE class labels from the gene name words with the biomedical named-entity tags (not the same as chunk tags) that GENIA tagger assigns to those words. We also assigned those class labels to all the words that are tagged as biomedical entities by GENIA. Hence not just the genes involved in the coexpression have those labels, but all of the gene names, protein names, and RNAs are assigned those labels. Though, our model now had three additional class labels to learn (B-protein, B-DNA and B-RNA), it still performed almost as well as the **No GE labels** experiment (see Table 14.1, row 5 **BME tags as labels**). However, with this approach, we not only get the coexpression relationship words extracted but also the genes that are involved in the relationships. Recall of our system is also significantly higher, almost 11% higher than that of the experiment with the original set of features.

In the 5th experiment, we combined experiments 2 and 4, i.e., all the biomedical entities got their GENIA assigned tags as the class labels and all the stop words were removed from the sentences. However, this combination did not obtain any performance gain probably for the same reasons as stated earlier for the second experiment.

In the last experiment, we decreased the number of classes from 12 to only 5 by assigning just one class label to all the words that are not useful for us and were given their chunk tags as the class labels. We kept the three biomedical entity name tags as the class labels (B-protein, B-DNA, and B-RNA), the RE class labels and all the other words got an NA class label. We see a high improvement in our Recall (7%) and F-measure (10%) values from experiment 4 above. This was due to the fact that in this case our model had fewer classes to learn and more instances. The noise produced by too many class labels in the form of chunk tags was reduced in this experiment.

Finally by analyzing all of these experiments on the development test set and training set, we came up with the following set of features and class labels for training our model. These include:

**Table 14.3** Comparison results with the baseline approach and another machine learning approach

|                  | Precision | Recall | F1-measure |
|------------------|-----------|--------|------------|
| DCRFs            | 0.81      | 0.63   | 0.71       |
| SVM              | 0.68      | 0.75   | 0.67       |
| Bayesian network | 0.45      | 0.66   | 0.53       |
| Naïve bayes      | 0.38      | 0.71   | 0.49       |
| Baseline         | 0.23      | 0.80   | 0.35       |

1. Local and contextual features as mentioned in Sect. 14.3.3.
2. Synonym features as used in experiment 1.
3. Class label RE for words that express coexpression relationships.
4. Class labels B-protein, B-DNA and B-RNA for tagging biomedical entities.
5. Class labels NA for all the other words in the sentence.

Although the Precision of the original feature set model is higher its Recall value is not as good as the others. There is always a trade-off between the Precision and Recall values, but it is essential to keep both of them as high as possible. We can see that the Recall values of our experiments with feature tuning, when compared with the original baseline feature set, have increased significantly more, almost 28% increase than the decrease in the Precision values of only 6.8%.

We also compared the performance of DCRFs with the baseline approach, Support Vector Machine (SVM), Bayes Net and Naïve Bayes, few of the well known classification algorithms. We used the Weka implementation of these algorithms and tested them with the same final set of features and class labels as was used for DCRFs [10]. Table 14.3 shows the comparison result of our DCRFs model with the baseline approach, Bayes Net, SVM, and Naïve Bayes algorithms. The Baseline results shown in Table 14.3 are the same as mentioned in Fig. 14.6 and the row representing the DCRFs result is the result of the best combination of features and class labels mentioned in Table 14.2 (combination of classes BME, RE and NA).

## 14.4.2 Ranking Comparison

In this section we show the comparison results of our model's ranking with those of Google's and PubMed's ranking. As mentioned in Sect. 14.3.1 of data collection, we downloaded a different set of papers from PubMed for five different query genes and kept the ranking given by PubMed to each of those papers. These papers acted as our repository for testing against our model as well as for indexing and ranking for Google. We gave the same set of 100 papers for each query to Google's custom search engine to index and rank. These papers were also tested against our model created in the first part of our experiment and later scored and ranked with our scoring scheme mentioned in Sect. 14.3.6. To test the ranks of all the three search results, we first needed to manually know the ground truth of these papers.

**Table 14.4**  Ground truth of all five query genes

|  |  | Gene names | | | | |
|---|---|---|---|---|---|---|
|  |  | Bcl2 | ErbB | IL | Myc | p53 |
| Category | (Relevant) 1 | 52 | 71 | 46 | 53 | 55 |
|  | (Not-main) 2 | 15 | 17 | 21 | 23 | 18 |
|  | (Irrelevant) 3 | 30 | 11 | 33 | 23 | 26 |

Therefore, we collected the ground truth of these papers and divided them into three categories – Relevant, Not-main and Irrelevant. Table 14.4, shows the ground truth for all the papers for each query set, which are the different gene names. The relevant category in the table is referred to as category 1 and contains the total number of papers that have one of their main results as the coexpression of the query gene with other genes. Not-main category is category 2 and contains the total number of papers that do mention the coexpression of the query genes but not necessarily as their main result. Finally, Irrelevant category is the 3rd category with the papers that do not talk about coexpression of the query genes but may contain the coexpression word or the query gene names in them. Each column of Table 14.4 represents the ground truth results for each different query gene.

Once we have the ground truth, we can confidently say that in an ideal retrieval system which of the papers should come in the top retrieval result list and which ones should come at the bottom.

To compare the ranking results of our model with Google and PubMed, we used a well know evaluation metric known as Mean Average Precision (MAP) [21]. It calculates the mean of the Average Precisions (AveP) of the system on each of the query term. AveP is a measure, which computes the relevancy of the document at each retrieval step. It gives better results for a system that has relevant documents ranked at a higher rank. Equation 14.4 gives the formula for computing AveP for each retrieval result.

$$AveP = \frac{\sum_{r=1}^{N} (P(r) \times rel(r))}{\text{\# of relevant documents}} \qquad (14.4)$$

where $r$ is the rank, $N$ the number of documents retrieved, $rel()$ a binary function on the relevance of a given rank, and $P(r)$ the precision at a given cut-off rank $r$, given by 14.5 below:

$$P(r) = \frac{|\{\text{relevant retrieved documents}\}|}{r} \qquad (14.5)$$

Finally, if we have a set of queries, the MAP is the sum of all the average precisions (*AveP*) for each query divided by the total number of queries. This is given by Equation 6 below:

$$MAP = \frac{1}{Q} \left( \sum_{q=1}^{n} AveP \right) \qquad (14.6)$$

where $Q$ is the set of queries from $q = 1$ to $n$ and *AveP* is the Average Precision for each of the query term.

As we have three different categories, we want to compare the ranking results for each of the different categories individually. Hence, for each category, the other two categories were considered as totally irrelevant documents while calculating *AveP*. In this way, if a category 2 paper is wrongfully retrieved and placed in category 1, it will be treated the same as if a category 3 paper is wrongfully included in category 1. Although, this is a simple approach, it does not take into account any inter-category relationship among categories 1, 2 and 3. It should be noted that categories 1 and 2 are much more related to the query term as they both include the papers that talk about coexpression of the query gene, even though in category 2 it is not the main subject of the paper. Category 3 is the most irrelevant category in our case because it contains the papers that either talk about coexpression of some other genes or have the query terms in them but not really the coexpression of the query gene.

Therefore, to identify with this inter-category relationship, we prepare a modified AveP that can better evaluate the quality of ranking results. This can be done by changing the value of *rel*() function from just being binary to having different weights for different category misplacement in the retrieval results. The *rel*() function in the original *AveP* equation gives 0 weight to all misplaced papers and 1 to correctly placed ones. In the Modified Average Precision (*MAveP*) the *rel*() is no longer a binary function, but has 4 different values depending upon which category a retrieved paper is placed in and which category it truly belongs to. Table 14.5 shows the different values we assign to the *rel*() function. The first column and the first row of the matrix are the categories of the paper returned by the ground truth and those returned by the search system and rest of values in the matrix are the values assigned to *rel( )*. We can see from the table that the weight assigned to a mismatch between categories 1 and 2 is larger than that between categories 1 and 3. And the weight assigned to a mismatch between categories 2 and 3 is slightly larger than that between categories 1 and 3 but smaller than that between categories 1 and 2. If a paper is placed in its correct category, the *rel*() function will return a weight 1. The smaller the weight, the less related are two categories.

We compared the ranking results in each category among our model, PubMed and Google using both AveP and MAveP. Due to space constraint we are not showing the AveP and MAveP scores for each query gene, but it was seen that for most of the query terms and categories our model outperformed both Google and PubMed except for *IL coexpression* query term in which Google outperformed our model.

| | | Ground truth categories | | |
|---|---|---|---|---|
| | | 1 | 2 | 3 |
| Predicted categories | 1 | 1 | 0.75 | 0.25 |
| | 2 | 0.75 | 1 | 0.5 |
| | 3 | 0.25 | 0.5 | 1 |

**Table 14.5** Different values of rel() function for MAveP

**Table 14.6** MAP comparison results for the three search systems

| Category | PubMed | Google | Our model |
|---|---|---|---|
| 1 | 0.36 | 0.65 | 0.7 |
| 2 | 0.05 | 0.09 | 0.14 |
| 3 | 0.09 | 0.24 | 0.31 |

**Table 14.7** Modified-MAP comparison results for the three search systems

| Category | PubMed | Google | Our model |
|---|---|---|---|
| 1 | 0.47 | 0.76 | 0.78 |
| 2 | 0.13 | 0.2 | 0.31 |
| 3 | 0.15 | 0.32 | 0.38 |

We present the MAP and Modified-MAP (Mean of *MAveP*) comparison results, which are the average performance of the three search systems on all the five query terms, in Tables 14.6 and 14.7. It is quite clear from the values that although Google is close to our model in categorization, still our model has a better retrieval result. This high MAP score of our model is attributed to our model's capability to better distinguish between the positive and non-positive papers. The proposed model just does not search only the paper for the occurrence of query terms, but looks for the semantic meaning of the query terms in the paper.

Our DCRFs model has been trained for tagging all the sentences that talk about coexpression of genes even though those sentences do not contain the word "coexpression" itself. And hence when a user searches for the coexpression of a particular gene with the other by providing query term as "gene name + coexpression", while other search engines try to find these two words in papers, our search engine looks for all possible ways it can be written and extracts the results that are more semantically related to the query.

## 14.5   Conclusion

In chapter we present a framework for extracting and ranking papers containing predicates that state coexpression relationships among genes. We trained a DCRFs model based on semantic analysis of text to classify papers that talk about gene-gene coexpression relationships. Later, we devised a scoring scheme to score and rank papers for five different query terms and compared them against Google and PubMed.

There has been a lot of work done towards relationship extraction from biomedical literature, but the work presented here accomplishes more than just information extraction. In this work, we not only present a model that can extract a specialized relationship from these papers written in natural language but also a retrieval system that can be trained on this model to index and rank papers according to the importance of information present in them with regards to this specialized relationship, i.e. coexpression. Our complete framework helps in building a unified information resource that can help a researcher to get all the information needed regarding coexpression of any given gene with ease. In addition, our classification model has been trained on different ways the coexpression relationships can be expressed in literature, and so while retrieving the papers it not only retrieves the words present in the query but also all possible synonyms of those words too. This work can be extended to other domains that involve classification and ranking of unstructured data. One of the important steps in any classification algorithm is the meaningful feature extraction and accuracy of the classifiers depends upon the type of features used. Also, one commonality among all the relationship classification tasks is that we need to identify the entities, the keywords, and the positive contextual words. In this work we have emphasized the use of contextual features such as the words that occur before and after the given word and also their part of speech. This helps in giving more contextual knowledge to the model. Hence, this work can be extended and most of the features used in this work can be applied directly in any text classification task. Also, the used of Conditional Random Fields help in classification of sequential input data such as sentences and have proven to perform better than Bayes Net, SVMs and Naïve Bayes as shown in our work too.

We also use dictionary matching for the query gene to get all possible synonyms and family names of that gene. Consequently, in a way we are also performing a query expansion during the retrieval. This helps us to further present a more specialized and detailed result to the user. We can see the effect of this specialization by the improvement in the ranking result that we obtain as compared to Google and PubMed.

Some of the future ideas towards further improvement in this work can be to investigate different grammatical parsing methods, like dependency parsing and use the parse trees generated as additional features to train this model. We believe that this can improve the classification results. Another direction to explore is semi-supervised learning, as it requires less human effort of labeling the ground truth needed for training a model, and may still give sufficiently high accuracy.

This work can be considered as a step towards the semantic based information retrieval systems that can help researchers to extract accurate and relevant information from the vast array of textual information.

# References

1. PDFBox - Java PDF library, The Apache Software Foundation, http://incubator.apache.org/pdfbox/index.html
2. Blaschke, C., Valencia, A.: The Frame-Based module of the SUISEKI information extraction system. J. Intell. Syst. **17**(2), 14–20 (2002)
3. Bunescu, R., Mooney, R., Ramani, A., Marcotte, E.: Integrating co-occurrence statistics with information extraction for robust retrieval of protein interactions from Medline. In: Workshop on Linking Natural Language Processing and Biology: Towards Deeper Biological Literature Analysis (BioNLP '06), pp. 49–56 (2006)
4. Bundschus, M., Dejori, M., Stetter, M., Tresp, V., Kriegel, H.P.: Extraction of semantic biomedical relations from text using conditional random fields. J. BMC Bioinformatics **9**(207), (2008)

5. Clark, J., Koprinska I., Poon J.: A neural network based approach to automated e-mail classification. In: IEEE/WIC International Conference on Web Intelligence, pp. 702–705 (2003)
6. Cohen, A., Hersch, W.: A survey of current work in biomedical text mining. Briefings in Bioinformatics. **6**, 57–71 (2005)
7. Coulibaly, I., Page, G.P.: Bioinformatic tools for inferring functional information from plant microarray data II: analysis beyond single gene. Int. J. Plant Genomics. (2008)
8. Craven, M.: Learning to extract relations from medline. In: AAAI-99 Workshop on Machine Learning for Information Extraction (1999)
9. Tsuruoka, Y., Jun'ichi, T.: Bidirectional Inference with the Easiest-First Strategy for Tagging Sequence Data, Proceedings of HLT/EMNLP, pp. 467–474 (2005).
10. Hall, M., Frank, E., Holmes, G., Pfahringer, B., Reutemann, P.W.I.H.: The WEKA data mining software: An update. SIGKDD Explorations **11**(1), (2009)
11. KDD: The Eighth ACM SIGKDD International Conference on Knowledge Discovery and Data Mining, pp. 23–26 Edmonton, Alberta, CA http://www.sigkdd.org/kdd2002 (2002)
12. McCallum, A., Li, W.: Early results for named entity recognition with conditional random fields, feature induction and web-enhanced lexicons. In: 7th Conference on Natural Language Learning (CoNLL), pp: 188-191 (2003)
13. MEDLINE, National Center for Biotechnology Information, http://www.ncbi.nlm.nih.gov/pubs/facsheets/medline.html
14. Miyao, Y., Sagae, K., Saetre, R., Matsuzaki, T., Tsujii, J.: Evaluating contributions of natural language parsers to protein-protein interaction extraction. J. Bioinformatics. **25**(3), 394–400 (2009)
15. Miwa, M., Saetre, R., Miyao, Y., Tsujii, J.: A rich feature vector for protein-protein interaction extraction from multiple corpora. In: Conference on Empirical Methods in Natural Language Processing (EMNLP '09), pp. 121–130 (2009)
16. PubMed, National Center for Biotechnology Information, http://www.ncbi.nlm.nih.gov/pubmed
17. Peri, S., Navaroo, J.D., Kristiansen, T.Z., Amanchy, R., Surendranath, V., Muthusamy, B., Gandhi, T.K., Chandrika, K.N., Deshpande, N., Suresh, S.: Human protein referncee database as a discovery resource for proteomics: J. Nuclein Acids Res, (2004)
18. Seymore, K., McCallum, A., Rosenfeld, R.: Learning hidden markov model structure for information extraction. In: AAAI Workshop on Machine Learning for Information Extraction, pp. 37–42 (1999)
19. Sutton, C., McCallum, A.: An introduction to conditional random fields for relational learning. Introduction to Statistical Relational Learning, MIT Press (2006)
20. Sutton, C.: GRMM: GRaphical Models in Mallet, http://mallet.cs.umass.edu/grmm (2006)
21. Sutton, C., McCallum, A., Rohanimanesh, K.: Dynamic conditional random fields: Factorized probabilistic models for labeling and segmenting sequence data. J. Mach. Learn. Res. **8**, 693–723 (2004)
22. Rau, L.F., Jacobs, P.S., Zernik, U.: Information extraction and text summarization using linguistic knowledge acquisition. J. Inform. Process. Manag. **25**(4), 419–428 (1989)
23. Tiwari, R., Zhang, C., Solorio, T.: A Supervised Machine Learning Approach of Extracting Coexpression Relationship among Genes from Literature. In: 11th IEEE International Conference on Information Reuse and Integration, pp. 98–103 (2010)
24. Voorhees, E.M.: Variations in relevance judgments and the measurement of retrieval effectiveness. In: 21st Annual Int. ACM SIGIR, New York, NY, USA, pp. 315–323 (1998)
25. Wei, X., Croft, B., McCallum, A.: Table extraction for answer retrieval. J. Information Retrieval. **9**(5), 589–611 (2006)

# Chapter 15
# Music Artist Similarity Aggregation

**Brandeis Marshall**

## 15.1  Introduction

AM/FM radio is no longer the primary source for listening to music.Online radio can be accessed in one of two approaches: *subscription* such as Sirius Satellite Radio and XM Satellite Radio and *free* such as AOL Radio, Pandora, MySpace and YouTube.

These music-enabling technologies has its user interface limitations and advantages. Several online music listening websites disallow the song "replay" and/or previous song option, which follows the traditional radio music paradigm. However, online music listening websites have permitted the use of "skip" and "pause" as data streaming and network bandwidth capabilities have improved in recent years. On the contrary, *MySpace* and *YouTube* are flexible venues to allow music listeners to select specific songs, providing song replay and pause options, without a pre-defined or computer-generated playlist option. The flexibility of *MySpace* and *YouTube* also gives a platform to music artists without a recording label for sharing their talents.

Each online music listening website allows music listeners to create a user account in hopes of tracking music genre and artist preferences. In most cases, the user chooses a radio station with a programmed playlist. In contrast, *Pandora* only needs a single music artist to begin the customized user playlist. If the user would like to listen to different music genres, she must provide a sample music artist for *Pandora* to generate an appropriate playlist. To assist their user in locating similar music artists, these portals design music recommendation methods to determine artist similarity. Music recommendation makes use of genre and artist profile information supplied by users. Currently, a consensus of appropriate artist similarity does not exist since each online radio portals creates its own unique process.

Some challenges facing music recommendation are multiple genre artists, music artist collaborations and artist similarity identification. Many music artists can be

---

B. Marshall (✉)
Computer and Information Technology, Purdue University, West Lafayette, IN 47907, USA
e-mail: brandeis@purdue.edu

T. Özyer et al. (eds.), *Recent Trends in Information Reuse and Integration*,
DOI 10.1007/978-3-7091-0738-6_15, © Springer-Verlag/Wien 2012

classified in more than one genre due to the artists deciding to alter her sound or assessing the influences of one genre onto another genre. Music artist collaborations have become popular in which collaborations are within and across music genre. These collaborations may occur on more than one song or album. For two artist collaborations, a music listener may like the song collaboration but only enjoy the music from one of the artists. Artist similarity is primarily user-driven since likeness is highly subjective. These challenges, therefore, make capturing artist similarity difficult.

In this paper,[1] we are concerned with identifying similar artists, which serves as a precursor to how music recommendation can handle the more complex issues of multiple genre artists and artist collaborations. We consider the individual most similar artist ranking from three public-use Web APIs (Idiomag, Last.fm and Echo Nest) as different perspectives of artist similarity. Each online radio portals has a particular methodology in determining artist similarity as discussed in Sect. 15.1. We aggregate these three ranking using five rank fusion algorithms. These fusion methods vary in how the information is merged including which aspects are exploited or ignored. If we had access to each Web API's artist similarity algorithm, we could apply score-based fusion algorithms that can be more accurate than their rank-based counterparts. We evaluate the performance of the aggregate result lists with respect to a ground truth list and examine the level of overlap amongst these Web APIs through rank fusion methods. By understanding this overlap, we can more easily isolate the multiple genre artists and artist collaborations.

The specific contributions of this paper are:

- Examine rank fusion as a solution to artist similarity.
- Perform a quantitative study of artist similarity using five fusions algorithms including Average, Condorcet-fuse, CombMNZ, PageRank and Median, and
- Conduct a performance analysis of artist similarity considering both precision and reciprocal rank.

## 15.2 Related Work

Music information is considered part of multimedia information retrieval since music research lies at the intersection of three dimensions as shown in Fig. 15.1. The paradigm dimension represents music as content-based e.g. vocal features [8,10,14,17,20] or text-based e.g. song lyrics, style and mood labels [5,19,21]. The objective dimension pinpoints the machine learning methodology, which is either identification (unsupervised learning) [20] or classification (supervised learning) [6, 8, 10, 14, 17, 19]. The data dimension identifies the data source for the music

---

[1]This chapter is an expanded version of the original publication "Aggregating Music Recommendation Web APIs by Artist" in the Proceedings of the International Conference on Information Reuse and Integration 2010, pages 75–79.

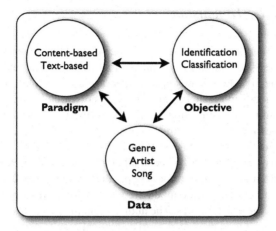

**Fig. 15.1** Dimensions of music research

research. This research uniquely centers on the objective dimension as determining the appropriate similar artists is a combination of accurate artist genre and style identification and classification.

## 15.2.1 Paradigm Dimension

A large section of music research focuses on content-based methods by processing the audio file in order to correctly determine a song's genre. However, content-based methods are computationally expensive, but has led to improvements in the music genre labeling, artist style classification and identifying song moods. These improvements have been applied to construct music recommendation systems [5, 13, 21] by using collaborative filtering methods to provide user-specific results using information from many users. However, prior work [5, 21] concentrates on the users' playlist through song properties including pitch, duration and loudness. While users' playlist are customized, artist similarity can assist in generating a playlist but has a wider appeal with greater song and artist diversity. Through collaborative filtering, artist similarity can also be assessed using text data.

In recommending music using text, music artists are typically classified into predefined or user-generated categories. Music recommendation varies from other Web 2.0 tools such as Web pages tagging (*Del.icio.us*) and image tagging (*Flickr*) due to its reliance on Type and Opinions tag classification taxonomy [2]. [2] emphasize that music listeners tend to label music and artist according to genre style and enjoy providing personalized opinions of the music while *Del.icio.us* and *Flickr* users classify by Topic, Time and Location. Due to the larger computational cost of content-based methods in music research, in this study, we rely on the text-based approaches using pre-defined expert reviewed labels.

## 15.2.2 Objective Dimension

The challenge of the objective dimension is deciding which data is valuable such as the artist, genre, song, album, style and/or mood. This data can be manipulated to semi-automatically identify or classify an artist into a set of genres, styles or provoke certain moods in music listeners. Previous research [7, 16, 19, 22] study the impact of artist similarity using supervised and unsupervised learning techniques.

[19] determines music artist similarity using pre-defined style and mood tags. The music style of a particular artist typically refers to a genre's subcategories. For example, music artist Usher has been labelled in the R&B genre, in which his music is categorized as urban, pop/rock, pop, contemporary, R&B, club/dance and electronic genre styles. The music mood are a set of adjectives (from a pre-defined list) that describes the artist's songs. The authors concluded that test based approach using style and mood tags produced nearly identical results using a content-based approach comparing 12 artists. We discover artist similarity using music style categories that leverage existing artist similarity rankings for several hundred artists across multiple music genres.

[16] show the similarity of human recommendation with automate music recommendation services provided by *Last.fm* and *Pandora*. Nevertheless, the results also note the limitations of human recommendation as some dependencies are not captured for an individuals musical taste. The landscape for finding new music has been vastly transformed thanks to the Internet [7] as it has helped new artists such as Taylor Swift (country) and Sean Kingston (reggae) find new listeners as well.

The work of [22] is the most similar work since five tagging approaches are compared including surveying, social tagging, gaming, web documentation and autotagging. The autotagging approach is the only method that does not require at least one song to be annotated by humans. The rank-based interleaving (RBI) approach determines appropriate tags, through fusion of social tagging, gaming, web documentation and autotagging approaches, resulting in higher precision. Hence, aggregation from imperfect methods can motivate the generation of more effective ground truth. We apply this methodology in determining artist similarity, rather than song genre classification.

## 15.2.3 Data Dimension

Music information has become more accessible online with *AllMusic*, *Pandora*, *AOLMusic*, *Last.fm*, *YouTube* and *SortMusic* websites and/or smartphone applications. *AllMusic* (formally AMG) contains a thorough repository of music artists; however, the similar artists are not ranked. *AllMusic* also lists "influenced by" and "followers" artists for a given music artist. The "influenced by" and "followers" artists are rarely labeled as similar artists resulting in a restrictive definition of similar artists.

Pandora is a music listening website, which automatically generates music playlists with a series of music genre stations, individual artist stations, personalized stations; however, *Pandora* requires a login and password and does not provide user access to the similar artists list through a Web API. *AOLMusic*, another online music portal, does not require login and password information and has only music genre stations. For each song, *AOLMusic* displays several recommended similar artists, but no Web API access to the complete ranked similar artist list.

The *Last.fm* music listening website has been used in music information retrieval in recent years [2, 7, 12, 16, 22]. The user-based tagging of *Last.fm* uncovered the frequency of tags by different users tend to be the most accurate in characterizing an artistõs style. Also, the authors propose a method in handling semantic similar tags, e.g. "hiphop", "hip hop", and "hip-hop", to remove duplicate tags due to spelling variations. Since we apply an aggregation approach, tag discrepancies are minimized.

## 15.3  Artist Aggregation

We discuss, in Sect. 15.1, each music recommendation tool's process in retrieving the most similar artists. Then, Sect. 15.2 describes different rank fusion methods. For aggregations, each Web API generates a ranked list of similar artists and serves as input to the rank fusion algorithms. Due to subjectivity of music preferences, ground truth in music similarity is difficult to determine. Hence, we consider a range in rank fusion methods to represent different perspectives of music listeners taste.

### 15.3.1  Music Recommendation Web APIs

To address the problem of inconsistent results among individual music recommendation Web APIs, we aggregate the artist query results from three music recommendation portal with Web APIs: *Idiomag, Last.fm* and *Echo Nest*. These three music recommendation Web APIs are selected since each is readily available for public use.

1. **Idiomag.**[2] *Idiomag* uniquely labels, or tags, a given artist by weighted genre names from a preset list of 144 acceptable genre names maintained by the company's staff. Then a manual weight is applied to each of the tags by Idiomag's expert musicians/music lovers. Lastly, the artists are ordered according to their labels' values.

---

[2]http://www.idiomag.com/api/

2. **Last.fm.**[3] *Last.fm* is highly user-centric by allowing any user to create self-defined tags. In contrast to *Idiomag*, *Last.fm* supports user-tagging which has led to a number of issues, including *duplicate tags* due to grammatical errors and *maliciously false tags* applied to artists. *Last.fm* combats this challenge by counting multiple occurrences of a single tag for a single artist as a vote for that artists tag. A tag's votes are reflected in how each tag is weighted. To determine musical similarity for an artist, *Last.fm* compares the tags of all artists in their database to the target artist.
3. **The Echo Nest.**[4] Due to intellectual property rights, *Echo Nest* has not unveiled their process of relating artists. Nevertheless, the company has revealed that artist information is generated in a number of ways such as the analyzation of the raw music, blogs, song lyrics and message board postings. While the exact method in rating artist similarity is unknown, test queries have shown that their artist database is multi-faceted with many musicians and genres.

## *15.3.2 Rank Fusion Methods*

Rank fusion, or rank aggregation, has been applied to disciplines such as sensor networks and the Web. However, in these disciplines, correct answers is presumed where relevancy of return results can be clearly measured with a general consensus. Music, on the other hand, is highly customizable as music listeners may enjoy songs and/or artists across music genres. In addition, music genres may overlap in style as one music genre is a descendent of multiple music genres and artists may bridge more than one music genre. We chose five existing rank fusion methods and discuss each below.

- **Average (Av).** Av [3] computes the average rank of artist amongst the three Web APIs and sorts these values to obtain the aggregate ranker. Average considers all rank information which may not be desired in the case of large amounts of incorrect information.
- **Median (Me).** Me [11] computes the median rank instead of the average rank of each artist. Median ignores all but one set of rank information that may be problematic when the median ranks are highly similar.
- **CombMNZ (MNZ).** MNZ [15] orders the information using a combination of the frequency of appearances and the ranks. MNZ relies on multiple appearances of the same artist in order to provide supporting evidence of similarity.
- **PageRank (Pg).** Pg [4] is the most popular rank fusion algorithm. It is an approximation to the Markov chain aggregator $MC_4$ [9], which computes the steady state probability through either navigation or randomly jumping.

---

[3]http://www.lastfm.com/api/

[4]http://developer.echonest.com/

- **Condorcet-fuse (Cfuse).** Cfuse [18] uses an unweighted directed graph in which an edge indicates the higher rank of one artist over another for each Web API. The magnitude of rank difference between one artist to another artist is omitted resulting in some information loss.

In the work of [1], information and robustness of six rank fusion methods and two optimization techniques were investigated by varying the correctness of input information on a statistical framework. The behavior of the rank fusion methods differ depending on the amount of noise, e.g. elevated rank position of a particular element, and misinformation, e.g. disagreement in the input information.

When noise and misinformation are low, the rank fusion methods produces the same or nearly identical lists and performs the same as compared to the ground truth. When misinformation remains low and noise increases, robustness becomes an important factor in which Pg provides the best performance. When noise is low but misinformation increases, there is an increasing need to incorporate as much of the input information as possible which leads to Av producing the best performance. When misinformation is bi-partisan, namely two polarized positions amongst the input information, Me has the best performance regardless of the noise level and Cfuse appears as a high performing method when the noise level is low.

When both misinformation and noise are at higher levels, method robustness and input information quality impact the which rank fusion method is the best performer. In this case, Av, MNZ and Pg are equally good with nearly identical performance as compared to a ground truth. As we will see through our experimental study, each genre has a distinct combination of noise and misinformation amongst the three Web APIs.

## 15.4 Experimental Study

To test artist similarity amongst the three Web APIs, we ran nearly 500 artist queries over ten popular music genres. We manually selected the query artists as to guarantee the return of at least ten similar artists from each music recommendation application. We present a sample of the artists queried by the rank fusion method in Table 15.1.

For each artist query, we examine the performance of each rank fusion method by calculating the precision. The precision $P$ is calculated by taking as input two ranked lists $m_i, m_j$ and finding the number of common elements in relation to the number of returned elements $k$. In our experiments, each Web API returns a ranked list of 10–15 similar artists. We chose $k = 10$. Formally, precision is defined as follows

$$P_k(m_i, m_j) = \frac{m_i \cap m_j}{k}$$

Precision is a commonly used measure to distinguish between relevance and non-relevance. However, precision does not indicate the degree of relevance, e.g. the

**Table 15.1** Sample artists

| | | |
|---|---|---|
| Alternative | 50 | Disturbed, Korn, Muse, Nickelback, Papa Roach, The Fray |
| Blues/Jazz | 48 | B.B. King, Nina Simone, Otis Redding, Ray Charles, Stevie Wonder, Doris Day |
| Country | 50 | Garth Brooks, Faith Hill, Toby Keith, Vince Gill, Willie Nelson |
| Electronic | 48 | Air, Daft Punk, Depeche Mode, Massive Attack, Zero 7 |
| Funk | 49 | Culture Club, Funkadelic, Musiq, Parliament, Sade |
| Hip-Hop | 48 | Janet Jackson, John Legend, TLC, Monica, Black Eyed Peas |
| Pop | 50 | Blackstreet Boys, Britney Spears, Coldplay, Justin Timberlake, Lady Gaga |
| Rap | 50 | Dr. Dre, Eminem, Lil' Wayne, Run DMC, Snoop Dogg, Young Jeezy |
| Reggae | 45 | Bob Marley, Black Uhuru, Matisyahu, Shaggy, Toots & The Maytals |
| Rock | 49 | Creed, Finger Eleven, Hinder, Rob Thomas, Train |

rank position of relevant data. To assess relevancy based on rank position, we use the reciprocal rank measure. Formally, reciprocal rank is defined as follows

$$RR_k(m_i, m_j) = \sum_{l=1,\ldots,k} \frac{1}{l} \; if \; m_i(l) = m_j(q)$$

where $l, q$ ($l = q$ or $l \neq q$) refer to a position in a ranking.

We investigate how each rank fusion method performs as the ground truth creating a symmetric matrix in which $P_k(m_i, m_j) = P_k(m_j, m_i)$. Since *Pandora* is the most popular music listening portal, we wanted to use Pandora's similar artist lists as the ground truth. However, Pandora does not provide a Web API to retrieve similar artists nor does Pandora provide any other means of accessing the rank-order list of similar artists.

We selected SortMusic[5] as the ground truth since it provides a comprehensive collection of various genre music artists. SortMusic is an entertainment portal of over 3,900 music artists founded in August 2003. For each artist, the SortMusic website displays biographies, discographies, music videos, music lyrics, similar artists and such. The similar artist tab on SortMusic provides a rank-order list for a given artist. However, SortMusic does not have a Web API so the gathering of similar artists must be done manually and not all artists are catalogued on their website.

As an example of the differences in the rank fusion methods, we display the output of the artist query for Usher in terms of the Web API rankings (Table 15.2) and the rank fusion methods results (Table 15.3). Since Ne-Yo appears highly ranked in two rankings, each rank fusion method gave it a rank of 1. Each Web API provides a different ordering of artists, but in general, we notice the higher overlap in artists between *Last.fm* and *Echo Nest*.

**SortMusic as Ground Truth.** To evaluate the relevancy of the rank fusion methods for artist similarity, we compute the precision and reciprocal rank using the

---

[5]http://www.sortmusic.com/

**Table 15.2** Web API rankings

| Rank | Idiomag | Last.fm | Echo Nest |
|---|---|---|---|
| 1 | Craig David | Ne-Yo | Mary J. Blige |
| 2 | Eamon | Mario | Ne-Yo |
| 3 | Sisqo | Chris Brown | Jagged Edge |
| 4 | Mario Winans | Bobby Valentino | Toni Braxton |
| 5 | R. Kelly | Trey Songz | Bobby Valentino |
| 6 | Baby Bash | Omarion | Trey Songz |
| 7 | Ciara | Marques Houston | Sammie |
| 8 | The Black Eyed Peas | Lloyd | Frankie J |
| 9 | Missy Elliott | Joe | Mariah Carey |
| 10 | Puff Daddy | Jagged Edge | Chris Brown |

**Table 15.3** Web API aggregate results

| Rank | Av | MNZ | Pg | Me | Cfuse |
|---|---|---|---|---|---|
| 1 | Ne-Yo | Ne-Yo | Ne-Yo | Ne-Yo | Ne-Yo |
| 2 | Bobby Valentino | Bobby Valentino | Bobby Valentino | Bobby Valentino | Jagged Edge |
| 3 | Trey Songz | Trey Songz | Trey Songz | Trey Songz | Toni Braxton |
| 4 | Craig David | Chris Brown | Jagged Edge | Chris Brown | Bobby Valentino |
| 5 | Mary J. Blige | Jagged Edge | Chris Brown | Jagged Edge | Trey Songz |
| 6 | Eamon | Craig David | Craig David | Craig David | Sisqo |
| 7 | Mario | Mary J. Blige | Mary J. Blige | Eamon | Sammie |
| 8 | Chris Brown | Mario | Eamon | Sisqo | R. Kelly |
| 9 | Jagged Edge | Eamon | Mario | Mario Winans | Baby Bash |
| 10 | Sisqo | Sisqo | Sisqo | R. Kelly | Ciara |

SortMusic artist similarity ranking and the result ranking from each rank fusion method. We display these performance values for each of the ten music genre in Table 15.4a-j and the overall performance in Table 15.4i. The best performance value is bolded for the precision and reciprocal rank.

MNZ provides the best precision performance in eight out of the ten music genres and 6 out of the ten music genres with respect to reciprocal rank. In addition, we notice that Pg performed nearly as well as MNZ in these cases. Based on the behavior of these rank fusion methods and the artist selected to represent a particular genre, the similar artist lists from the three Web API have higher levels of both noise and misinformation. Cfuse, with Me as a tie or close second, is consistently the worst performing rank fusion method when compared to the SortMusic artist similarity ranking. The country music genre is an anomaly due to Me producing the best performance by a signifiant margin. We conclude that the similar artist lists for country music are frequently bi-partisan (with differing opinions of rank positions) in which one Web API's rank contributes toward the resulting rank fusion list.

The music genres are quite varied in their precision and reciprocal rank values. Blues/Jazz, Funk and Rock have higher artist similarity than Pop and Country. Blues/Jazz, Funk and Rock are well-established music genres with very notable musicians who span generations of music listeners. However, Country is also

**Table 15.4** Performance by genre and overall

|  | $P_{10}$ | $RR_{10}$ |  | $P_{10}$ | $RR_{10}$ |  | $P_{10}$ | $RR_{10}$ |
|---|---|---|---|---|---|---|---|---|
| Av | 51.80% | **1.97** | Av | 60.41% | 2.27 | Av | 28.40% | 0.97 |
| Me | 46.40% | 1.54 | Me | 51.25% | 1.66 | Me | **38.20%** | **1.21** |
| MNZ | 53.40% | 1.96 | MNZ | **62.70%** | **2.31** | MNZ | 28.84% | 0.95 |
| Cfuse | 46.20% | 1.51 | Cfuse | 50.41% | 1.62 | Cfuse | 32.20% | 0.95 |
| Pg | **53.60%** | 1.94 | Pg | 62.29% | 2.30 | Pg | 27.40% | 0.93 |
| (a) Alternative | | | (b) Blues/Jazz | | | (c) Country | | |

|  | $P_{10}$ | $RR_{10}$ |  | $P_{10}$ | $RR_{10}$ |  | $P_{10}$ | $RR_{10}$ |
|---|---|---|---|---|---|---|---|---|
| Av | 53.90% | 2.16 | Av | 57.55% | 2.26 | Av | 58.33% | **2.30** |
| Me | 45.00% | 1.48 | Me | 51.63% | 1.78 | Me | 50.00% | 1.63 |
| MNZ | **58.12%** | **2.23** | MNZ | **60.40%** | **2.30** | MNZ | **59.79%** | 2.28 |
| Cfuse | 45.00% | 1.43 | Cfuse | 54.08% | 1.89 | Cfuse | 48.95% | 1.63 |
| Pg | 56.04% | 2.15 | Pg | 59.79% | 2.26 | Pg | 59.37% | 2.26 |
| (d) Electronic | | | (e) Funk | | | (f) Hip-Hop | | |

|  | $P_{10}$ | $RR_{10}$ |  | $P_{10}$ | $RR_{10}$ |  | $P_{10}$ | $RR_{10}$ |
|---|---|---|---|---|---|---|---|---|
| Av | 33.77% | **1.38** | Av | 51.40% | 2.07 | Av | 56.44% | 2.17 |
| Me | 29.55% | 0.89 | Me | 43.20% | 1.38 | Me | 49.33% | 1.66 |
| MNZ | **35.11%** | 1.36 | MNZ | **54.60%** | **2.13** | MNZ | **58.88%** | **2.18** |
| Cfuse | 28.44% | 0.88 | Cfuse | 42.80% | 1.38 | Cfuse | 48.66% | 1.60 |
| Pg | 34.88% | 1.33 | Pg | 53.40% | 2.07 | Pg | 57.77% | 2.14 |
| (g) Pop | | | (h) Rap | | | (i) Reggae | | |

|  | $P_{10}$ | $RR_{10}$ |  | $P_{10}$ | $RR_{10}$ |
|---|---|---|---|---|---|
| Av | 56.93% | 2.25 | Av | 50.91% | 1.98 |
| Me | 52.04% | 1.79 | Me | 45.72% | 1.51 |
| MNZ | **60.00%** | **2.28** | MNZ | **53.15%** | **2.00** |
| Cfuse | 50.40% | 1.70 | Cfuse | 44.77% | 1.46 |
| Pg | 58.36% | 2.21 | Pg | 52.30% | 1.96 |
| (j) Rock | | | (i) overall performance | | |

well-established but has very low artist similarity. The reasons for this discrepancy may be towfold: (1) SortMusic may be biased by having more collected information in Blues/Jazz, Funk and Rock music genre and (2) country music is flooded with new artists, thus making the similarity lists larger and resulting in less overlap. The Pop music genre, in contrast, is newer with more diverse set of artists. Pop artists may be included in other music genres Song collaborations have become more frequent and these collaborations may be classified as a "pop" song in which one of the artists is then classified as a pop artist.

**Precision Performance by Rank Fusion Method.** As any artist similarity ground truth will be biased to some degree, we discern how well relevant information can be retrieved from multiple sources. Hence we construct a rank fusion comparison matrix in which each rank fusion method serves as a ground truth. The precision relationship between each pair of rank fusion methods is calculated. For each row and column pair $(r_i, c_j)$, we present the precision value and if $i = j$, then

**Table 15.5**  Overall precision performance

|        | Av     | Me     | MNZ    | Cfuse  | Pg     |
|--------|--------|--------|--------|--------|--------|
| Av     | 100.0% | 68.79% | 93.63% | **65.45%** | 90.85% |
| Me     | –      | 100.0% | 70.75% | 77.11% | 70.66% |
| MNZ    | –      | –      | 100.0  | 65.97% | **95.29%** |
| Cfuse  | –      | –      | –      | 100.0% | 65.89% |

the precision is 100%. Since $P_k(m_i, m_j) = P_k(m_j, m_i)$, we only report the upper triangular matrix. In Table 15.5, we display the average precision across the ten genres. In bold, we highlight the highest precision, pair (MNZ, Pg) at 95%, and the lowest precision, pair (Av, Cfuse) at 65%.

The comparison of Av with the other rank fusion algorithms reveal a high precision of over 90% between (Av, MNZ) and (Av, Pg). Given the high correlation amongst Av, MNZ and Pg, these rank fusion methods tend to operate in a similar manner. Those rank fusion algorithms are highly dictated by the frequency of appearance. For Av, an artist ranked in the top-10 for all three APIs has a higher probability of producing as an average rank in the top-10. For MNZ, the method directly uses number of appearance and a weight measurement in its ranking scheme. In Pg, artists appearing in the top-10 of each API will have more incoming and outgoing links than those artists in one or two of the APIs. These links allows navigation of the graph more effective without cycles and the need to randomly jump to an unseen artist.

On the contrary, a lower precision of around 65% between rank fusion methods (Av, Me) and (Av, Cfuse) highlights the differences in the methods and increases the likelihood of finding (unconventional) partial matches such as new or cross-genre artists. Interestingly, (Me, Cfuse) are more similar to each other than to Av, MNZ and Pg with precision results of 77%. Both Me and Cfuse primarily focus on strict ordering of one artist over another artist. Since Me only considers the middle rank, artists are ranked in a localized environment. In Cfuse, the dominance of one artist in the ranking over another artist is based on all three APIs. When ties occur for either Me or Cfuse, the sort scheme is random, which can be a contributor to the performance variation between (Av, MNZ, Pg) and (Me, Cfuse).

As observed in Table 15.3, the contrast in artist ordering in (Av, MNZ, Pg) and (Me, Cfuse) is demonstrated with the inclusion of Toni Braxton, Mario Winans, R. Kelly, Baby Bash and Ciara in Me and Cfuse. The artist Mary J. Blige appears in Av, MNZ and Pg rankings but does not appear in either top-10 results for Me and Cfuse. The rank fusion results provide very similar rankings only when a high overlap amongst the Web APIs occurred. In the case of a large number of similar artists, the method behavior is emphasized providing different results.

Table 15.6 considers the music genre individually. The highest and lowest precision values are bolded for each music genre. The highest precision always includes the MNZ rank fusion method while the lowest precision always includes the Cfuse rank fusion method, which is consistent with the conclusion reached when SortMusic serves as the ground truth. MNZ and Cfuse vary in their rank order of

**Table 15.6** Genre precision performance

| | Av | Me | MNZ | Cfuse | Pg |
|---|---|---|---|---|---|
| Av | 100.0% | 73.80% | 94.00% | 69.40% | 91.19% |
| Me | – | 100.0% | 75.40% | 78.60% | 76.20% |
| MNZ | – | – | 100.0% | **69.20%** | **95.80%** |
| Cfuse | – | – | – | 100.0% | **69.20%** |

(a) Alternative

| | Av | Me | MNZ | Cfuse | Pg |
|---|---|---|---|---|---|
| Av | 100.0% | 69.79% | 92.29% | **67.91%** | 89.58% |
| Me | – | 100.0% | 72.91% | 76.25% | 71.87% |
| MNZ | – | – | 100.0% | 68.95% | **95.41%** |
| Cfuse | – | – | – | 100.0% | 68.75% |

(b) Blues/Jazz

| | Av | Me | MNZ | Cfuse | Pg |
|---|---|---|---|---|---|
| Av | 100.0% | 54.60% | 92.80% | 50.80% | 91.20% |
| Me | – | 100.0% | 53.80% | 74.00% | 52.60% |
| MNZ | – | – | 100.0% | 50.20% | **94.60%** |
| Cfuse | – | – | – | 100.0% | **49.00%** |

(c) Country

| | Av | Me | MNZ | Cfuse | Pg |
|---|---|---|---|---|---|
| Av | 100.0% | 70.83% | 93.54% | **65.00%** | 90.20% |
| Me | – | 100.0% | 73.12% | 75.83% | 72.91% |
| MNZ | – | – | 100.0% | 65.83% | **94.79%** |
| Cfuse | – | – | – | 100.0% | 66.66% |

(d) Electronic

| | Av | Me | MNZ | Cfuse | Pg |
|---|---|---|---|---|---|
| Av | 100.0% | 70.40% | 92.65% | **68.57%** | 90.20% |
| Me | – | 100.0% | 73.26% | 77.34% | 73.06% |
| MNZ | – | – | 100.0% | 70.20% | **96.12%** |
| Cfuse | – | – | – | 100.0% | 70.81% |

(e) Funk

| | Av | Me | MNZ | Cfuse | Pg |
|---|---|---|---|---|---|
| Av | 100.0% | 71.87% | 94.58% | **67.70%** | 92.29% |
| Me | – | 100.0% | 73.75% | 77.08% | 73.75% |
| MNZ | – | – | 100.0% | 68.33% | **96.25%** |
| Cfuse | – | – | – | 100.0% | 67.91% |

(f) Hip-Hop

| | Av | Me | MNZ | Cfuse | Pg |
|---|---|---|---|---|---|
| Av | 100.0% | 62.44% | 92.88% | **62.22%** | 90.88% |
| Me | – | 100.0% | 64.66% | 78.22% | 66.44% |
| MNZ | – | – | 100.0% | 63.55% | **95.55%** |
| Cfuse | – | – | – | 100.0% | 63.77% |

(g) Pop

| | Av | Me | MNZ | Cfuse | Pg |
|---|---|---|---|---|---|
| Av | 100.0% | 66.20% | 94.40% | **62.40%** | 90.80% |
| Me | – | 100.0% | 68.20% | 76.00% | 67.60% |
| MNZ | – | – | 100.0% | 63.20% | **94.40%** |
| Cfuse | – | – | – | 100.0% | 62.80% |

(h) Rap

| | Av | Me | MNZ | Cfuse | Pg |
|---|---|---|---|---|---|
| Av | 100.0% | 75.33% | 95.11% | 71.33% | 92.44% |
| Me | – | 100.0% | 76.22% | 78.22% | 76.22% |
| MNZ | – | – | 100.0% | 70.44% | **96.00%** |
| Cfuse | – | – | – | 100.0% | **69.11%** |

(h) Reegae

| | Av | Me | MNZ | Cfuse | Pg |
|---|---|---|---|---|---|
| Av | 100.0% | 73.06% | 94.08% | **69.80%** | 89.80% |
| Me | – | 100.0% | 76.53% | 79.80% | 76.53% |
| MNZ | – | – | 100.0% | 70.41% | **94.08%** |
| Cfuse | – | – | – | 100.0% | 71.43% |

(i) Rock

artists given the same input. The large pool of artist is accentuated by this large difference in the highest and lowest precision values. On the other hand, Rock artists are more concentrated with greater overlap (24.28% difference between highest and lowest precision).

**Reciprocal Rank Performance by Rank Fusion Method.** Through precision performance, we can only assess *how many* artists consistently matched across the Web APIs. We have no information about *which rank position* the artists appeared in the rank fusion results ranking. The reciprocal rank performance provides this evaluation. As with precision, we construct a rank fusion comparison matrix in which each rank fusion method serves as a ground truth. The reciprocal rank (RR) relationship between each pair of rank fusion methods is calculated. For each row and column pair $(r_i, c_j)$, we present the RR value and if $i = j$, then the RR is 2.93 $\left( = \sum_{l=1}^{k} 1/l \right)$. Since $RR_k(m_i, m_j) \neq RR_k(m_j, m_i)$, we report the RR value for every rank fusion pair.

In Table 15.7, we display the overall RR performance and bold the lowest and highest RR values. In parallel to Table 15.5, (Av, Cfuse) represent the least similar

**Table 15.7** Overall reciprocal rank performance

|        | Av   | Me   | MNZ  | Cfuse    | Pg       |
|--------|------|------|------|----------|----------|
| Av     | 2.93 | 2.45 | 2.86 | **2.22** | 2.82     |
| Me     | 2.50 | 2.93 | 2.56 | 2.46     | 2.57     |
| MNZ    | 2.82 | 2.51 | 2.93 | 2.31     | **2.88** |
| Cfuse  | 2.38 | 2.51 | 2.39 | 2.93     | 2.38     |
| Pg     | 2.75 | 2.48 | 2.85 | 2.30     | 2.93     |

**Table 15.8** Genre reciprocal rank performance

|        | Av   | Me   | MNZ  | Cfuse    | Pg       |
|--------|------|------|------|----------|----------|
| Av     | 2.93 | 2.57 | 2.86 | **2.38** | 2.83     |
| Me     | 2.59 | 2.93 | 2.63 | 2.53     | 2.65     |
| MNZ    | 2.86 | 2.63 | 2.93 | 2.45     | **2.88** |
| Cfuse  | 2.51 | 2.59 | 2.48 | 2.93     | 2.47     |
| Pg     | 2.83 | 2.66 | 2.87 | 2.48     | 2.93     |

(a) Alternative

|        | Av   | Me   | MNZ  | Cfuse    | Pg       |
|--------|------|------|------|----------|----------|
| Av     | 2.93 | 2.50 | 2.84 | **2.33** | 2.81     |
| Me     | 2.53 | 2.93 | 2.60 | 2.45     | 2.59     |
| MNZ    | 2.81 | 2.56 | 2.93 | 2.40     | **2.88** |
| Cfuse  | 2.46 | 2.53 | 2.48 | 2.93     | 2.47     |
| Pg     | 2.74 | 2.53 | 2.87 | 2.40     | 2.93     |

(b) Blues/Jazz

|        | Av   | Me   | MNZ  | Cfuse    | Pg       |
|--------|------|------|------|----------|----------|
| Av     | 2.93 | 2.05 | 2.85 | **1.55** | 2.83     |
| Me     | 2.22 | 2.93 | 2.30 | 2.17     | 2.27     |
| MNZ    | 2.74 | 2.08 | 2.93 | 1.72     | **2.87** |
| Cfuse  | 1.88 | 2.19 | 1.95 | 2.93     | 1.93     |
| Pg     | 2.53 | 1.91 | 2.77 | 1.64     | 2.93     |

(c) Country

|        | Av   | Me   | MNZ  | Cfuse    | Pg       |
|--------|------|------|------|----------|----------|
| Av     | 2.93 | 2.49 | 2.86 | **2.21** | 2.82     |
| Me     | 2.54 | 2.93 | 2.61 | 2.38     | 2.60     |
| MNZ    | 2.82 | 2.56 | 2.93 | 2.28     | **2.87** |
| Cfuse  | 2.41 | 2.51 | 2.42 | 2.93     | 2.43     |
| Pg     | 2.74 | 2.52 | 2.84 | 2.27     | 2.93     |

(d) Electronic

|        | Av   | Me   | MNZ  | Cfuse    | Pg       |
|--------|------|------|------|----------|----------|
| Av     | 2.93 | 2.50 | 2.85 | **2.41** | 2.82     |
| Me     | 2.52 | 2.93 | 2.59 | 2.61     | 2.61     |
| MNZ    | 2.81 | 2.57 | 2.93 | 2.49     | **2.88** |
| Cfuse  | 2.43 | 2.56 | 2.46 | 2.93     | 2.47     |
| Pg     | 2.77 | 2.57 | 2.87 | 2.51     | 2.93     |

(e) Funk

|        | Av   | Me   | MNZ  | Cfuse    | Pg       |
|--------|------|------|------|----------|----------|
| Av     | 2.93 | 2.53 | 2.87 | **2.31** | 2.84     |
| Me     | 2.57 | 2.93 | 2.60 | 2.49     | 2.61     |
| MNZ    | 2.87 | 2.59 | 2.93 | 2.38     | **2.89** |
| Cfuse  | 2.49 | 2.60 | 2.43 | 2.93     | 2.42     |
| Pg     | 2.81 | 2.59 | 2.87 | 2.39     | 2.93     |

(f) Hip-Hop

|        | Av   | Me   | MNZ  | Cfuse    | Pg       |
|--------|------|------|------|----------|----------|
| Av     | 2.93 | 2.31 | 2.85 | **2.10** | 2.82     |
| Me     | 2.39 | 2.93 | 2.46 | 2.44     | 2.49     |
| MNZ    | 2.80 | 2.36 | 2.93 | 2.17     | **2.88** |
| Cfuse  | 2.27 | 2.45 | 2.30 | 2.93     | 2.30     |
| Pg     | 2.73 | 2.33 | 2.85 | 2.18     | 2.93     |

(g) Pop

|        | Av   | Me   | MNZ  | Cfuse    | Pg       |
|--------|------|------|------|----------|----------|
| Av     | 2.93 | 2.40 | 2.86 | **2.12** | 2.82     |
| Me     | 2.45 | 2.93 | 2.54 | 2.42     | 2.54     |
| MNZ    | 2.81 | 2.49 | 2.93 | 2.26     | **2.87** |
| Cfuse  | 2.31 | 2.42 | 2.37 | 2.93     | 2.33     |
| Pg     | 2.75 | 2.47 | 2.85 | 2.26     | 2.93     |

(h) Rap

|        | Av   | Me   | MNZ  | Cfuse    | Pg       |
|--------|------|------|------|----------|----------|
| Av     | 2.93 | 2.61 | 2.87 | **2.40** | 2.84     |
| Me     | 2.63 | 2.93 | 2.65 | 2.52     | 2.65     |
| MNZ    | 2.88 | 2.65 | 2.93 | 2.45     | **2.88** |
| Cfuse  | 2.57 | 2.62 | 2.52 | 2.93     | 2.49     |
| Pg     | 2.83 | 2.64 | 2.88 | 2.42     | 2.93     |

(h) Reggae

|        | Av   | Me   | MNZ      | Cfuse    | Pg       |
|--------|------|------|----------|----------|----------|
| Av     | 2.93 | 2.56 | **2.90** | **2.40** | 2.82     |
| Me     | 2.59 | 2.93 | 2.66     | 2.54     | 2.66     |
| MNZ    | 2.84 | 2.60 | 2.93     | 2.49     | 2.87     |
| Cfuse  | 2.51 | 2.59 | 2.51     | 2.93     | 2.52     |
| Pg     | 2.78 | 2.64 | 2.85     | 2.50     | 2.93     |

(i) Rock

result rankings while (MNZ, Pg) represent the most similar result rankings. With 60–70% precision (Table 15.5) with >2.0 RR value for the (Av, Cfuse) pair, we can conclude that many of the artist similarity matches appear near the top of the result rankings, most likely including rank positions 1–4 $\left(\sum_{l=1}^{4} 1/l = 2.08\right)$ in order to obtain the average RR of 2.22. For the (MNZ, Pg) pair, the 2.88 average RR value is obtained by a combination of result rankings: (1) a large set of all rank positions match and (2) a small set of nine rank positions match with one bottom-rank position artist did not match between the two result rankings.

In Table 15.8, the RR values for each music genre is given, in which the lowest and highest RR values are in bold. In contrast to Table 15.6, the lowest RR value is always the (Av, Cfuse) pair. Interestingly, when (Av, Cfuse) pair does not have the lowest precision, it is the second lowest precision with about 1% difference from the lowest precision value. When comparing the lowest RR amongst the music genres, they vary due to the differences within each genre. Most notably, Country has the lowest precision (about 50%) and reciprocal rank (1.55), which indicates that most likely rank position 1 was not a similarity match but rank position 2–6 e.g., $\left(\sum_{l=2}^{6} 1/l = 1.45\right)$. Given the other genre's lowest RR values, rank position 1 seems to consistently be a similarity match. The highest RR value is typically the (MNZ, Pg) pair, except for the Rock music genre, in which (Av, MNZ) has a higher RR value. The individual music genre's highest RR value is nearly identical with the highest RR value of 2.88 (Table 15.7). Thus there is a strong correlation between MNZ and Pg.

The discovery of two ranking groups (Av, MNZ, Pg) and (Me, Cfuse) can provide music listeners a choice of aggregate music recommendation. The need for strongly linked similarity amongst the top artist result can be resolved by applying either Av, MNZ or Pg. To include partial matches and loose similarity, Me or Cfuse can be used for aggregation.

## 15.5    Conclusion

As music listening has transitioned from solely FM radio frequencies to online (customizable) radio, music artist similarity becomes hard to classify. On the Web, music listeners have a variation of online music portals including satellite radio venues, AOL Music, YouTube, Pandora, Last.fm, Idiomag and Echo Nest. The online music portals engage listeners with profile preferences options and skip song functionality to more accurately assess artist similarity. However, the variation in developing relationships amongst artists has led to different artist similarity rankings.

We address the problem of inconsistent similar artist results through aggregation of three online music portals (*Idiomag, Last.fm* and *Echo Nest*) with music recommendation APIs. Using five rank fusion methods, we test the relevancy of results from nearly 500 artist queries. Given a ground truth ranking of SortMusic, CombMNZ consistently performed better than the other rank fusion methods in terms of both precision and reciprocal rank.

Given the subjectivity of music ground truth ranking, we also examined the behavior of artist similarity through the rank fusion methods' perspective. We show that the Average, CombMNZ and PageRank rank fusion algorithms produced very similar precision results. Median and Condorcet-fuse are similar to each other but vary greatly with the other three ranking schemes. We observe highly similar artists can be retrieved using Average, CombMNZ or PageRank while a more relaxed similarity can be captured using either Median or Condorcet-fuse rank fusion method.

In future work, we would like to address semi-duplicate results in which music artists collaborate on more than one album and/or song. A music data repository of artists with collaboration will assist in reducing semi-duplicate noise. Also, a music collaboration structure can prove beneficial for assessing artist similarity quality where collaborating with other like-minded artists is becoming more commonplace.

# References

1. Adalı, S., Hill, B., Magdon-Ismail, M.: Information vs. robustness in rank aggregation: Models, algorithms and a statistical framework for evaluation. Special Issue on Web Information Retrieval, J. Digit. Inform. Manage. 5(5), 292–307 (2007)
2. Bischoff, K., Firan, C., Nejdl, W., Paiu, T.: Can all tags be used for search. In: Proceedings of ACM CIKM, pp. 203–212 (2008)
3. Borda, J.C.: Mémoire sur les élections au scrutin. In Histoire de l'Académie Royale des Sciences (1781)
4. Brin, S., Page, L.: The anatomy of a large-scale hypertextual web search engine. In: Proceedings of ACM WWW, pp. 107–117 (1998)
5. Chen, H.-C., and Chen, A.L.P.: A music recommendation system based on music data grouping and user interests. In: Proceedings of ACM CIKM, pp. 231–238 (2001)
6. Chen, L., Wright, P., Nejdl, W.: Improving music genre classification using collaborative tagging data. In: Proceedings of the ACM International Conference on Web Search and Data Mining, pp. 84–93 (2009)
7. Cunningham, S., Bainbridge, D., McKay, D.: Finding new music: a diary study of everyday encounters with novel songs. In International Conference on Music Information Retrieval (ISMIR) pp. 83–88 (2007)
8. Dikerson, K.B., Ventura, D.: Music recommendation and query-by-content using self organizing maps. In: Proceedings of the International Joint Conference on Neural Networks pp. 705–710 (2009)
9. Dwork, C., Kumar, R., Naor, M., Sivakumar, D.: Rank aggregation methods for the web. In: Proceedings of ACM WWW pp. 613–622 (2001)
10. Ellis, D.P.W., Whitman, B., Berenzweig, A., Lawrence, S.: The quest for ground truth in musical artist similarity. In: Proceedings of the International Symposium on Music Information Retrieval pp. 170–177 (2002)
11. Fagin, R., Kumar, R., Sivakumar, D.: Efficient similarity search and classification via rank aggregation. In: Proceedings of ACM SIGMOD pp. 301–312 (2003)
12. Geleijnse, G., Schedl, M., Knees, P.: The quest for ground truth in musical artist tagging in the social web era. In: International Conference on Music Information Retrieval(ISMIR) pp. 525–530 (2007)
13. Im, I., Hars, A.: Does a one-size recommendation system fit all? the effectiveness of collaborative filtering based recommendation systems across different domains and search modes. ACM Trans. Inform. Syst. 26(1), (2007)

14. Knees, P., Pampalk, E., Widmer, G.: Artist classification with web-based data. In: Proceedings of the International Symposium on Music Information Retrieval pp. 517–524 (2004)
15. Lee, J.H.: Analyses of multiple evidence combination. In: Proceedings of ACM SIGIR pp. 267–276 (1997)
16. Magno, T., Sable, C.: A comparison of signal-based music recommendation to genre labels, collaborative filtering, musicological analysis, human recommendation, and random baseline. In: International Conference on Music Information Retrieval(ISMIR) pp. 161–166 (2008)
17. McKay, C., Fuhinaga, I.: Improving automatic music classification performance by extracting features from different types of data. In: Proceedings of the ACM SIGMM International Conference on Multimedia Information Retrieval pp. 257–266 (2010)
18. Montague, M., Aslam, J.A.: Condorcet fusion for improved retrieval. In: Proceedings of ACM CIKM pp. 538–548 (2002)
19. Shao, B., Li, T., Ogihara, M.: Quantity music artist similarity based on style and mood. In: Proceedings of the ACM International Workshop on Web Information and Data Management pp. 119–124 (2008)
20. Shen, J., Shepard, J., Cui, B., Tan, K.L.: A novel framework for efficient automated singer identification in large music databases. ACM Trans. Inform. Syst. **27**(3), (2009)
21. Slaney, M., White, W.: Measuring playlist diversity for recommendation systems. In: Proceedings of the ACM Workshop on Audio and Music Computing Multimedia pp. 77–82 (2006)
22. Turnbull, D., Barrington, L., Lanckreit, G.: Five approaches to collecting tags for music. In: International Conference on Music Information Retrieval (ISMIR) pp. 225–230 (2008)

# Chapter 16
# Tag Grid: Supporting Multidimensional Queries of Tagged Datasets

**Ken Q Pu and Russell Cheung**

## 16.1 Introduction

Tag cloud is a phenomenon that has gained tremendous popularity as a method of collaborative organization of online information. With the ever growing number of pieces of information in the forms of Weblogs, forum posts, emails, news articles, online photographs, movie reviews, etc., users can easily be lost in the sea of hyperlinks and index pages. Tag cloud provides a community driven and visual technique to organize an evolving collection of data with ad-hoc created *tags*. Figure 16.1 is a sample tag cloud consisting of the most popular tags assigned to Flickr photos.[1]

Tag clouds are formed by allowing users to create short strings (e.g. *Java, Food recipe, Homework solution, comedy* ...), and associate one or more of them to data items. For example, the tag *sci-fi* and *action* may be assigned to the movie **Star Wars**. Formally, we can define the organization induced by a tag cloud as follows.

**Definition 1 (Tag cloud).** Given a collection of items, $X$. A tag cloud of $X$ is a set of tags $T$ and a tagging function $\theta : T \rightarrow 2^X$, assigning each tag $t \in T$ to a subset of items $\theta(t) \subseteq X$. The frequency of a tag is defined as $f(t) = |\theta(t)|$.

Visualizing tag clouds received a great deal of attention. By changing the font sizes of the tags, one can reflex the frequency $f(t)$ of the tags, as shown in Fig. 16.1. Different algorithms [7, 8] have been proposed to layout the tags with different font sizes in the visual layout of the tag cloud. The purpose of the visual tag cloud is that

---

[1]The tag cloud is obtained from http://en.wikipedia.org/wiki/Tag_cloud.

K.Q. Pu (✉) · R. Cheung
University of Ontario Institute of Technology, 2000 Simcoe Street N. Oshawa ON, Canada
e-mail: ken.pu@uoit.ca@mycampus.uoit.ca; cheugn@mycampus.uoit.ca

T. Özyer et al. (eds.), *Recent Trends in Information Reuse and Integration*,
DOI 10.1007/978-3-7091-0738-6_16, © Springer-Verlag/Wien 2012

06 africa amsterdam animals **architecture** art august australia autumn baby
barcelona **beach** berlin **birthday** black blackandwhite blue boston bw
**california** cameraphone camping canada canon car **cat** cats
chicago china **christmas** church city clouds color **concert** d50 day
dc december dog england europe fall **family** festival film florida
flower flowers food **france friends** fun garden geotagged
**germany** girl graffiti **green** halloween hawaii hiking **holiday** home
honeymoon hongkong **house** india ireland island **italy japan** july june kids la
**lake** landscape light live **london** losangeles macro **me** mexico mountain
mountains museum **music nature** new newyork newyorkcity newzealand
**night** nikon **nyc** ocean **paris** park **party** people portrait red
river roadtrip rock rome san **sanfrancisco** scotland sea seattle show **sky**
**snow** spain spring **street** summer sun sunset sydney taiwan texas
thailand tokyo toronto **travel** tree trees trip uk urban **usa**
**vacation** vancouver washington water **wedding** white winter
yellow york zoo

**Fig. 16.1** Tag cloud of frequent tags at Flickr

users of the Web database can easily understand the relevant topics, and drill across to other related data items by clicking on individual tags in the cloud. It is assumed that tags with higher frequency are more likely to be chosen, and hence are rendered with larger font sizes.

In addition to serve as a visual Web widget for navigation purposes, tag clouds also has the potential of supporting complex ad-hoc analytic queries on the underlying data items. Compared to the traditional relational data warehouses [14] and Online Analytic Processing (OLAP) [9], tag clouds are analogous to hierarchical dimensions of data cube [5]. How ever, associated with data analytics using tag clouds, we have some challenging issues:

- Tags are created by the user community in an ad-hoc fashion without any well-defined hierarchical or multidimensional organization.
- Tag names are decided by users, so they are non-standard and noisy.

In this paper, we focus on the problem of incorporating the tag cloud to data analytics and search. By introducing three new tag cloud query operators: multidimensional querying, horizontal tag expansion and drill-down tag expansion, users can treat with a tag cloud as a data cube, and explore and query the tagged data items *as well as* the tags themselves in an ad-hoc fashion. We have implemented the proposed methods in a Web interface we call *tag grid*. Through examples and performance evaluations, we demonstrate that tag grid is very efficient in runtime, and it offers very effective way of exploring large tag clouds such as the *Internet Movie Database* (IMDB) data set.

## 16.2   Related Work

Tag cloud has received much attention. Much of the work treat tag clouds as a navigational widget. [7] proposed methods on improving the usage of tag clouds as a visual information retrieval interface. [8] provided a mathematical measure of the visual compactness of the visual layout of a tag cloud and provided algorithms to reduce wasted space. Compared to the previous efforts on the visual representation of a tag cloud, our work focuses on the support of ad-hoc queries of tag clouds.

More recently, the connection between Online Analytic Processing (OLAP) and tag clouds was proposed by [1, 2]. Their contribution was a frame to visualize a sliced OLAP view as a tag cloud, thus treating data cubes as tag clouds. In constract, our interest is to mimic tag clouds as data cubes by providing users a richer set of multidimensional query constructs when interacting with tag clouds.

Ding et al proposed a system *Text Cube* [11] which utilizes OLAP queries to compute information retrieval measures for multidimensional text databases. In their work, they do not consider tag clouds of the underlying text database. Similarly, Lauw et al proposed TUBE [10] that discovers associations among documents and multidimensional text databases. Our work can be used to complement these systems as an ad-hoc querying system.

A preliminary version [13] of this paper appeared at the 11th IEEE International Conference on Information Reuse and Integration (IRI 2010).

## 16.3   Supporting Fuzzy and Multidimensional Queries Using Tag Cloud

Tag clouds have been very successful as a collaborative summarization tool and a visual navigation widget, but the potential to support more complex ad-hoc user queries has been largely ignored. Consider the Internet movies database (http://www.imdb.com)[2] which contains a large collection of movie titles and their details. Users are free to create and apply keywords to movies. In this scenario the data items are the movies, and the tags are the keywords. For instance, the movie *Star Wars: Episode VI - Return of the Jedi* is tagged by 142 keywords such as *alien, computer, desert, space-battle*, etc.

Multidimensional queries:   Users looking for a movie of interest can utilize the tags. For instance, fans of *Star Wars* may wish to find other movies that are tagged by both *computer* and *space-battle*. This requires a simple multidimensional query (treating *computer* and *space-battle* as dimensional elements, and movies as facts of a multidimensional data cube). The user may wish to generalize the simple multidimensional query of *computer* × *space-battle* to a more general grid of

---

[2]The data is available for research purposes from http://www.imdb.com/interfaces

$$(computer, technology) \times (space\text{-}battle, space)$$

This should result in a two dimensional table of possible movies:

|              | Computer | Technology |
|--------------|----------|------------|
| Space        | 38       | 30         |
| Space-battle | 12       | 17         |

Thus, it reveals that there are more choices of *space* and *computer* related films available.

Tag similarity: The user may wish to explore unknown tags in the process of searching for an interesting movie. Without having a particular movie in mind, the user may wish to find other tags that are similar to *computer*. The similarity is based on the movies that they tag. This type of tag level connection allows the user to navigate from one tag to another. For instance, using our system, the user can find that the *computer* tag is closely related to *internet, technology, cell-phone* etc., and can expand his search criteria if necessary.

Tag break down: In the IMDB database, there are 375 movies tagged by *technology* and only 6 tagged by *cryogenic*. Just by the count alone, one can see that *technology* is a fairly general tag, and the user may be interested at a granular break down of the *technology* movies in terms of a small number of highly representative components. The tag cloud can be used to find and represent these components. Using our system, one can break down *technology* movies into the following leading components: *computer, science, robot, engineering* etc. Using one or more of the break down components, the user may be able to zoom in the the movies that are of higher interests.

Fuzzy tag search: A unique feature of tag clouds is that it is community driven, so it is often the case that there is no standardization of the tags. Different users may choose to use slightly different tags for the same topic. For example, in the IMDB dataset, some movies are tagged by the keyword *happy* while others are tagged by *happiness*. Clearly, from these two tags should be considered similar, if not the same, in semantics. Thus, an important feature of analytic queries of tag clouds is the support of fuzzy matching of tag names.

Sections 16.3.1–16.3.3 describes the algorithms for performing multidimensional queries, tag similarity queries and tag break down queries – all supporting fuzzy tag matching.

## 16.3.1 Fuzzy Multidimensional Queries

In subsequent sections, we continue with the notation and symbols of Definition 1. Without loss of generality, we focus on two-dimensional queries.

**Definition 2 (Multidimensional Query).** Given two sets of tags $T_1$ and $T_2$, we denote the two dimensional query as $Q(T_1 \times T_2)$. The query returns a two-dimensional matrix of the form:

$$q : T_1 \times T_2 \to 2^X : (t_1, t_2) \mapsto \theta(t_1) \cap \theta(t_2)$$

mapping pairs of tags $(t_1, t_2)$ to a list of data items that are tagged by both $t_1$ and $t_2$.

Multidimensional queries as defined in Definition 2 are easy to evaluate, but they do not support fuzzy tag matching. For example, for the IMDB dataset, the two queries $Q(\{happy\} \times \{love\})$ and $Q(\{happiness\} \times \{love\})$ are completely disjoint! In fact, $|Q\{happy\} \times \{love\})| = 1$, whereas, $|Q(\{happiness\} \times \{love\})| = 65$. The reason is that most contributor to IMDB choose *happiness* as a tag, but one contributed used *happy* instead. Certainly, for a user of the IMDB dataset, it can be rather misleading to see that there only one movie on the topic of *happy* and *love*. To rectify the situation, we extend multidimensional queries to fuzzy multidimensional queries.

Let $\sigma : T \times T \to \mathbb{R}$ be a string similarity measure.[3] If $\sigma(t_1, t_2) = 1$, then the names of $t_1$ and $t_2$ are considered to be equivalent, and if $\sigma(t_1, t_2) = 0$, then their names are completely different.

**Definition 3 (Fuzzy Multidimensional Query).** Denote $\theta^{-1} : X \to 2^T$ as the inverse function of $\theta$. Namely $\theta^{-1}$ maps a data set $x$ to the associated tags $\theta^{-1}(x) \subseteq T$. Given two sets of tags $T_1, T_2$ and a similarity threshold $0 < c \leq 1$, the fuzzy multidimensional query $\tilde{Q}_c(T_1, T_2)$ is a query that returns a two dimensional matrix:

$$\tilde{q} : T_1 \times T_2 \to 2^X$$
$$: (t_1, t_2) \to \{x \in X : \exists t_i' \in T, \ \sigma(t_i, t_i') \geq c$$
$$\text{and } t_i' \in \theta^{-1}(x) \text{ for both } i = 1, 2\}$$

Intuitively, $q(t_1, t_2)$ contains data items $x$ whose tags $\theta^{-1}(x)$ contains at least one tag $t_1'$ similar to $t_1$ and one tag $t_2'$ similar to $t_2$.

There are many known string similarity measures [3, 6]. We choose to use the Jaccard similarity of $q$-grams of tag names.

**Definition 4 (Jaccard Similarity of $q$-Grams).** The $q$-gram of a string $s$ is the bag containing all substrings of $s$ with length $q$. We denote the $q$-gram of $s$ as $G_q(s)$. The Jaccard similarity of the $q$-grams of two tags $t_1$ and $t_2$ is defined as:

$$\sigma_J(t_1, t_2) = \frac{G_q(t_1) \cap G_q(t_2)}{G_q(t_1) \cup G_q(t_2)}$$

---

[3]Our fuzzy multidimensional query works with generic string similarity measures.

As an illustration, consider the tags *happy* and *happiness*. The 3-grams of the tags are:

- $G_3(happy) = \{hap, app, ppy\}$.
- $G_3(happiness) = \{hap, app, ppi, pin, ine, nes, ess\}$.

By the definition, we have $\sigma_J(happy, happiness) = 3/8$, which does not look all that similar. The solution to improve similarity between tags such as *happy* and *happiness* is to normalize tag names to a common root by means of *stemming* [12]. Stemming is a commonly used technique in natural language processing which maps a dictionary word to its root. Thus stem(happy) = stem(happiness) = happi. We choose to incorporate stemming to our Jaccard based similarity measure.

$$\sigma_{SJ}(t_1, t_2) = \sigma_J(\text{stem}(t_1), \text{stem}(t_2))$$

This similarity measure is quite powerful to compensate for the noisy tag names in tag clouds. For example:

- $\sigma_{SJ}(happy, happiness) = 1$.
- With 3-grams, $\sigma_{SJ}(happppy, happiness) = 3/4$ because stem(happppy) = happpi and stem(happiness) = happi. The 3-grams of the stemmed words only differ in one gram.

As defined in Definition 2, a multidimensional query is parameterized by a tuple of tags, and is evaluated to a collection of data set consisting of data times that are tagged by all the tags in the tuple. Due to the collaborative nature of the tag cloud, fuzzy multidimensional querying extends Definition 2 to allow the user to specify a threshold for the similarity as in Definition 3.

Using the similarity measure of $\sigma_{SJ}$, and a threshold $c = 0.75$, the fuzzy query $Q(happy, love)$ returns 66 movies, as the query result includes the movies tagged by *happy* and *happiness*.

In this section, we presented how to perform two-dimensional queries for the sake of brevity of notation. It is straight-forward to generalize the definitions to any finite dimensional queries involving multiple tag sets.

## 16.3.2 Horizontal Expansion: Finding Relevant Tags

Section 16.3.1 describes how to perform multidimensional queries that allow users to tabulate the data items along different dimensions formed by given tag sets. This is to assume that the user has a clear understanding of the available tags and their semantic meanings. Recall that a distinct property of tag clouds is that tags are created in a decentralized and uncontrolled fashion. Thus, we must also provide querying facility for users to search for relevant and interesting tags which can then be used as part of a multidimensional query for data item retrieval. Given that the tags are not immediately connected in any way, it is not possible for users to navigate and discover similar tags from known tags without additional query support. In this section, we present a query operation called *horizontal expansion* of tags to support inter-tag navigation. The name *"horizontal"* refers to the fact that we visualize the expanded similar tags in the *same* dimension as the initial tags.

The horizontal expansion operator, expand : $T \rightarrow 2^T$, maps tags to a set of "similar" tags expand($T$). Unlike fuzzy tag name matching in Sect. 16.3.1, the similarity measure of two tags $t_1$ and $t_2$ is measured by the data sets tagged by $t_1$ and $t_2$. We utilize a simplified version of the classic term frequency-inverse document frequency similarity to define the similarity measure.

**Definition 5 (Tag Similarity).** Given a data item $x \in X$, define the inverse frequency of $x$ as:

$$\text{idf}(x) = \log \left( \frac{|T|}{|\theta^{-1}(x)|} \right)$$

The similarity of two tags, $t_1$ and $t_2$, is defined as:

$$\text{sim}(t_1, t_2) = \frac{\sum_{x \in \theta(t_1) \cap \theta(t_2)} \text{idf}(x)^2}{\sqrt{\sum_{x \in \theta(t_1)} \text{idf}(x)^2 \cdot \sum_{x \in \theta(t_2)} \text{idf}(x)^2}} \qquad (16.1)$$

The inverse frequency idf($x$) of a data item $x$ is such that it penalizes $x$ for being tagged by too many tags. In the extreme case, if $x$ is tagged by all tags in $T$, then intuitively, we see that it contributes no information to the similarity of two tags, and thus, idf($x$) $= 0$ for such data item. For more details on scoring functions, refer to [3]. Equation 16.1 is a simplification of the well-known TFIDF measure [3] by assuming that every tag $t$ can only tag a data item at most once. It is simply the cosine similarity, $\frac{v_1 \cdot v_2}{\|v_1\| \cdot \|v_2\|}$, of the vectors $v_1$ and $v_2$ where

$$v_i(x) = \begin{cases} \text{idf}(x) & \text{if } x \in \theta(t_i) \\ 0 & \text{otherwise.} \end{cases}$$

The advantage of 16.1 is that it *only* depends on idf($x$), but not on any statistics of the tags, so we can very efficiently compute the similarity measures. More details on our implementation is presented in Sect. 16.4.

Now, we can define the horizontal expansion operator.

**Definition 6 (Horizontal Expansion).** Given a similarity threshold $0 < c \leq 1$, the horizontal expansion operator $\text{expand}_c : T \rightarrow 2^T$ is defined as:

$$\text{expand}_c(t) = \{t' \in T : \text{sim}(t, t') \geq c\}$$

Applying the expansion operator to the IMDB dataset, we get:

$$\text{expand}_{0.9}(\text{love}) = \{\text{male-female, friendship}\}$$
$$\text{expand}_{0.7}(\text{love}) = \{\text{male-female, friendship, death, marriage}\}$$

The property of expand($t$) is that tags $t' \in$ expand($t$) tend to have similar semantic meaning to $t$, so in the multidimensional query grid, they should all belong to the same dimension, hence the same "horizontal" expansion.

### 16.3.3   Drill-down Expansion: Finding Granular Tags

Horizontal expansion allows users to navigate from tag to tag to explore relevant but unknown tags in the tag cloud. Recall that multidimensional queries support multiple dimensions, and one limitation (or feature) of horizontal expansion is that it is designed to expand tags to other tags in the same dimension. How do we explore other potential dimensions? Given that tag clouds do not have prescribe dimensional meta-data, we must introduce yet another query operator which supports intra-dimensional navigation and exploration.

In OLAP databases, multidimensional views are to illustrate how a total sum is broken down into smaller components. For instance, we may be interested at how annual sales of cities are broken down into monthly sales over time. This would result in a two dimensional cube of geography v.s. time. We adopt the same philosophy in the design of the *drill-down expansion operator*: how do we break down the data items tagged by one tag to intersections with other tags?

The drill-down expansion operator, drill : $T \rightarrow 2^T$, maps tags $t$ to a tag set drill($t$) $\subseteq T$ which is best used as an alternate dimension.

**Definition 7 (Drill-down Expansion).** Given an integer $n > 0$, the drill-down operator is defined as:

$$\text{drill}(t) = \text{argmax}\{|\theta(S) \cap \theta(t)| : S \subset T, |S| \leq n, t \notin S\}$$

Namely, drill($t$) is the *optimal* tag set $S$ has the most common data items with $t$ while constrained by:

- $S$ cannot contain $t$.
- $S$ cannot have more than $n$ elements.

**Proposition 1.** *Computing the exact query result for* drill : $T \rightarrow 2^T$ *is NP-complete.*

The intractability result in Proposition 1 is not surprising because by the definition of $\text{drill}_n(t)$ is immediately equivalent to the maximum set cover problem [4]. The reduction is immediate: treat tags as covers and data items as set elements.

We present a greedy algorithm to approximate $\text{drill}_n(t)$ in Fig. 16.2.

The drill-down expansion is particularly useful to inspect very general movie tags such as *love* to more specific and granular tags. There are 6621 movies tagged by *love*. To get more specific break down of the *love* related movies, we can apply $\text{drill}_4$(love), and we get the following result:

---

**Algorithm: approximating drill-down expansion**

```
define drill_n(t):
    X = θ(t), data items to be covered
    S = ∅, tags in drill_n(t)
    S̄ = {t}, tags to be excluded
    while X ≠ ∅ and |S| ≤ n:
        t* = argmax{|θ(t') ∩ X| : t' ∈ T - S̄}
        S = S ∪ {t*}
        S̄ = S̄ ∪ {t*}
        X = X - θ(t*)
    end while
    return S
```

---

**Fig. 16.2** A greedy approximation algorithm to compute $\text{drill}_n$

| | Male-female | Death | Friendship | Soap-opera |
|---|---|---|---|---|
| Love | 2111 | 913 | 878 | 435 |

These are the top four tags that best divides the *love* movies. In this case, drill(love) resembles expand(love), but they do not agree exactly, and can differ significantly depending on the tag cloud.

If we apply $\text{drill}_4$ to *technology*, we get: {computer, science, engineering, robot}. So *technology* movies are mainly distributed among these tags.

*Remark.* We note that drill-down expansion can be extended to a set of tags $t_1, t_2, \ldots$. The definition of $\text{drill}_n(t_1, t_2, \ldots, t_m)$ is $n$ tags that best approximate $\bigcup_{i=1}^{m} \theta(t_i)$. We omit the details due to space limitation. Our implementation (Sect. 16.4) supports drill-down on multiple tags.

## 16.4  Implementation and Performance

We have implemented all the query operators in Sect. 16.3. The operators are integrated into a Web front-end which we refer to as *tag grid*. Tag grid offers a Web UI that supports two dimensional queries, horizontal and drill-down expansions. It constructs a matrix of movie counts, and the user can visualize the break-down of any row or column by means of a pie chart. The user can also click on any numeric count to see the actual list of movie titles.

### 16.4.1  The Interface

The overall user interface is shown in Fig. 16.3. Tag grid allows the user to search for an initial set of tags using a keyword search box (see right side of Fig. 16.3). Two dimensional queries are supported by dragging tags from the search box to cells

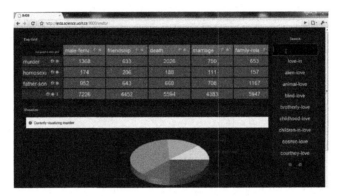

**Fig. 16.3** Tag grid: The web front-end for querying tag clouds

**Fig. 16.4** Horizontal and drill-down expansion buttons

| Tag Grid | love | | male-fem: | friendship | death | marriage |
|---|---|---|---|---|---|---|
| bar graph \| reset grid | 6621 | | 7226 | 4452 | 5594 | 4383 |

(a) Horizontal expansion applied

| Tag Grid | love | male-fem: | friendship | death | marriage |
|---|---|---|---|---|---|
| murder | 809 | 1368 | 633 | 2026 | 750 |
| homosexu | 631 | 174 | 206 | 188 | 111 |
| father-son | 600 | 952 | 643 | 660 | 708 |
| bar graph \| reset grid | 6621 | 7226 | 4452 | 5594 | 4383 |

(b) Drill-down expansion applied

**Fig. 16.5** Expanding *love* using horizontal and then drill-down expansions. (**a**) Horizontal expansion applied (**b**) Drill-down expansion applied

in either the top leading row or left leading column. Users can arbitrarily construct the tag matrix. Horizontal expansion is supported by the horizontal expansion button shown in Fig. 16.4, the right arrow circled on the top right. The drill-down expansion is supported by the drill-down button, the downward button circled on the left bottom in Fig. 16.4.

Figure 16.5 shows the sequence of performing navigation using horizontal and then drill-down expansions on the tag *love*.

## 16.4.2   Implementation Details

The tag cloud query evaluation is implemented in Java using Apache Lucene[4] keyword search engine to support approximate tag name search. Lucene is also used as a component of the horizontal and drill-down query operators. We first index the tags as documents, and the data items as terms of documents. Then, horizontal and drill-down expansion queries can be compiled into a sequence of keyword search queries, which are very efficiently evaluated by the Lucene search engine. Due to the lack of space, we omit the details of the query compilation and optimization. The Web front-end is implemented using Python Django[5] Web application framework. Javascript and AJAX are used extensively to provide a rich and interactive user experience with tag grid.

## 16.4.3   Performance Evaluation

The tag cloud queries are very efficient to evaluate. The following table summarizes some performance benchmark of our implementation when applied to the IMDB data set.

```
Total number of tags (keywords):        81,357 Total number of
data items (movies): 1,484,285
```

We ran a set of randomly selected two-dimensional queries, horizontal and drill-down expansion queries. The following table includes the average time of the 100 randomly chosen queries.

```
 two-dimensional query with single
cell: 0.002 sec fuzzy two-dim. query with single cell: 0.014 sec
horizontal expansion:                0.026 sec drill-down
expansion (n=5):          1.201 sec
```

The reason that drill-down query being the most expensive is that it needs to access the keyword index repeatedly.

Our experimental evaluation demonstrates that the fuzzy queries are very efficient to evaluation using modern text index structures. However, horizontal and drill-down expensions of more expensive because of the repeated access to the text index. In the future, we would like to investigate specialized text indices to improve the efficiency of drill-down expansions.

---

[4] See http://lucene.apache.org.
[5] See http://djangoproject.com.

## 16.5 Conclusion

We have presented a new way of interacting with tag clouds that offers flexible query constructs. Using the three proposed query operators (multidimensional query, horizontal expansion, drill-down expansion), users can treat a tag cloud as a multidimensional data cube, and navigate to interesting views of the underlying data items in an ad-hoc mention. Our implementation, *tag grid*, supports the three query operators via a Web interface. We have tested tag grid on the IMDB data set, and have shown that the operators can be very efficiently implemented on a reasonably large tag cloud. In our experience, tag grip offers interesting ways of interacting with the IMDB data.

## References

1. Aouiche, K., Lemire, D., Godin, R.: Collaborative OLAP with tag clouds: Web 2.0 OLAP formalism and experimental evaluation. In: Proceedings of WEBIST,08 (2008)
2. Aouiche, K., Lemire, D., Godin, R.: Web 2.0 OLAP: From data cubes to tag clouds. Lecture Notes in Business Information Processing. **18**, 51–64 (2008)
3. Flakes, W.B., Baeza-Yates, R.: Information Retrieval: Data Structures and Algorithms. Prentice Hall (1992)
4. Garey, M.R., Johnson, D.S.: Computers and Intractability: A guide to the theory of NP-completeness. Freeman (1979)
5. Gray, J., Chaudhuri, S., Bosworth, A., Layman, A., Reichart, D., Venkatrao, M., Pellow, F., Pirahesh, H.: Data cube: A relational aggregation operator generalizing group-by, cross-tab, and sub-totals. Data Min. Knowl. Discov. **1**(1), 29–53 (2004)
6. Gusfield, D.: Algorithms on strings, trees and sequences. Cambridge University Press (1997)
7. Hassan-montero, Y., Herrero-solana, V.: Improving tag-clouds as visual information retrieval interfaces. In: Proceedings of InSciT2006 (2006)
8. Kaser, O., Lemire, D.: Tag-cloud drawing: Algorithms for cloud visualization. In: Proceedings of Tagging and Metadata for Social Information Organization (2007)
9. Kimball, R.: The data warehouse toolkit: practical techniques for building dimensional data warehouses. Wiley (1996)
10. Hady, W., Lauw, Ee.-P.L., Hwee Hwa, P.: Tube (text-cube) for discovering documentary evidence of associations among entities. In: Proceedings of 2007 ACM symposium of applied computing (SAC'07) (2007)
11. Lin, C.X., Ding, B., Han, J., Zhu, F., Zhao, B.: Text cube: Computing ir measures for multidimensional text database analysis. In: Proceedings of 8th IEEE International Conference on Data Mining, ICDM'08 (2008)
12. Porter, M.F.: An algorithm for suffix stripping. Program, **14**(3), 130–137 (1980)
13. Pu, K.Q., Cheung, R.: Tag grid: supporting collaborative and fuzzy multidimensional queries of tagged datasets. In IRI'10: IEEE Conference on Information Reuse and Integration, pp. 364–367 (2010)
14. Ullman, J., Garcia-Molina, H., Widom, J.: Database Systems: The complete book. Prentice Hall (2001)

# Chapter 17
# A Novel Approach to Product Lifecycle Management based on Service Hierarchies

Stefan Silcher, Jorge Minguez, and Bernhard Mitschang

## 17.1 Introduction

Nowadays, one of the main challenges for companies is the effective management of IT-systems. In times when requirements and companies change steadily, the IT-infrastructures have to adapt to these changes as well: new systems have to be integrated or existing adapted. Even worse, these systems work together to support business processes of a company and thus the IT infrastructures become complex and difficult to manage.

### 17.1.1 Requirements in Product Lifecycle Management

The purpose of Product Lifecycle Management (PLM) is a continuous improvement of products and processes [24]. PLM covers all processes ranging from creating the first idea of a product over the production to the disposal of it. An important factor to achieve the desired improvements through PLM is to integrate the applications used in the product lifecycle. PLM identifies three main challenges [23]: Firstly, the concept, terms and acronyms within companies are not clearly defined. Secondly, the formats to save and record information between the different applications vary. Thirdly, the completeness and consistency of information cannot be guaranteed.

To solve these problems, the used software applications have to be integrated by providing a common platform in order to gain synergistic effects. By integrating all

S. Silcher (✉) · J. Minguez · B. Mitschang
Graduate School for advanced Manufacturing Engineering, GSaME, Institute for Parallel and Distributed Systems, IPVS, University of Stuttgart, Universitaetsstrasse 38, 70569 Stuttgart, Germany
e-mail: stefan.silcher@ipvs.uni-stuttgart.de; jorge.minguez@ipvs.uni-stuttgart.de; bernhard.mitschang@ipvs.uni-stuttgart.de

T. Özyer et al. (eds.), *Recent Trends in Information Reuse and Integration*, DOI 10.1007/978-3-7091-0738-6_17, © Springer-Verlag/Wien 2012

software applications across the product lifecycle, the exchange of data between these applications is mandatory and entails many benefits. Most advantageous is a faster product development. In particular, all involved software tools may collaborate with each other by simultaneously working on shared data, which is called Simultaneous Engineering [16]. Further benefits are an automated and thus faster and less error-prone data exchange, a minimized number of redundant data with the result of a reduced execution of redundant data management tasks, and a better data availability across the whole product lifecycle. Overall, this leads to reduced costs, a faster product development, an increased product quality and thus to a shorter time-to market.

### 17.1.2 Current Solutions for Application Integration in PLM

In [24], the author describes many reasons how to successfully realize PLM in an enterprise organization. However, the realization of an appropriate IT architecture even entails further issues. In most cases, the connection between two applications is realized via a direct point-to-point interface. Typically, heterogeneous data formats, e.g., flat files, XML files, or databases are used to exchange data between applications. This kind of realization leads to conversion problems, to loss of information along the chain of data propagation and to implementations of interfaces which are hardly to reproduce [9]. To overcome these problems, a canonical data format can be used to unify the data exchange format and thus reduce the problems of heterogeneous interfaces.

The major problem when integrating an increasing number of applications is the growing complexity of managing interfaces. When using only point-to-point interfaces, their number is raising exponentially. This leads to enormous efforts and costs for maintenance in the same order of magnitude. To reduce these disadvantages, the point-to-point architecture should be replaced by a hub-and-spoke architecture, which was introduced for Enterprise Application Integration (EAI) [9]. EAI integrates the applications by introducing a central Message Broker. The Message Broker can reduce the number of application interfaces, but each interface is still implemented individually.

### 17.1.3 Service-Based Integration Approach

This article describes another way of integrating applications by using a service-based approach. The benefits of SOA compared to the architectures above are a higher flexibility and interoperability. The principles of reusability and loosely-coupled services have driven Service Oriented Architecture (SOA) [8] to become the predominant paradigm for software design. Using services to implement the interfaces of the applications leads to consistent connections and reduced maintenance effort, in case of realizing the SOA with standardized technologies, e.g. Web services [27]. Another benefit is the possibility to assemble the implemented services to complex workflows, which are then available as new value-added services. The middleware of a SOA can be realized with an Enterprise Service Bus (ESB) [2], which offers additional functionality such as message queuing capabilities or monitoring service interactions. An ESB is typically a part of architectures for workflow management systems which offer administration and execution capabilities for workflows.

The approach in this article describes a solution based on an ESB that provides data provisioning services as well as an integration approach for applications [15]. We extend this previous work with a service hierarchy for the entire product lifecycle. This service hierarchy enables changes of atomic services without affecting the value-added services. The separation between data and business process integration additionally leads to a higher transparency. Furthermore, when changing business processes all dependent services can be easily identified and adapted accordingly.

### 17.1.4   Structure of this Document

The remainder of this article is structured as follows: Section 17.2 gives a motivation for a service-based PLM solution, pointing out the challenges of PLM and the accordant need for IT support. Furthermore, we introduce the concept of PLM application integration and analyze the main concepts of a SOA. Finally, we describe a common SOA implementation with Web services and the ESB. In Sect. 17.3, we present the benefits of using an ESB as middleware for Product Lifecycle Management via selected scenarios. Section 17.4 presents a detailed description of the service hierarchy concept and shows the benefits of using the hierarchy in one of the scenarios of Sect. 17.3. Afterwards, Sect. 17.5 presents an overview of related work, followed by the conclusion in Sect. 17.6.

## 17.2   Motivation for SOA in PLM

Usually, the implementations of IT infrastructures that support the Product Lifecycle Management become very complex and hard to maintain over the time due to continuously changing requirements. The systems are very inflexible and changes can only be performed with great efforts and overhead.

In this Section, we first present the requirements of PLM regarding the IT infrastructure (Sect. 2.1). Then, we describe the challenges of application integration within the PLM domain in Sect. 2.2. Finally, Sect. 2.3 introduces SOA and the concept of an ESB to address these challenges.

## 17.2.1 Product Lifecycle Management

The first approach of managing information in the product lifecycle came with Product Data Management (PDM) [24]. The data generated in the product development process are stored, managed and provided with PDM for other processes of the product development, but not for the whole product lifecycle. First, PLM extended the approach of PDM to use the product development data in all subsequent operations of the product lifecycle.

PLM embraces the whole chain of production from the conception of a product over its production to the recycling phase, and manages as well as distributes the product data over the whole lifecycle. Herein, the diversity of the used software applications is very large. Figure 17.1 presents a segmentation of a product lifecycle chain [23, 24].

The first phase of the presented product lifecycle chain is characterized by research and the generation of new product ideas. It deals with highly heterogeneous, semi-structured and unstructured data. Often, web tools like bulletin boards or blogs are used to capture this data which has to be structured to store it in an effective manner.

The product design and development (phase 2) uses many tools from the CAx area, mainly CAD (Computer Aided Design) tools. During this phase, a huge amount of data is generated. Most of this data is structured, but nevertheless it is a big challenge to manage this amount of data in an efficient manner to repeatedly retrieve it whenever necessary.

Phase 3, the production planning, is characterized by the Digital Factory [12], which is responsible for planning an optimized production using product, process and resource (PPR) data. Product information from the product development and resource information from the factory planning are linked to build a process plan for the production. To get an optimized production, the customer orders from the Enterprise Resource Planning (ERP) system are added and correlated with the PPR data within the simulation environment. When all data is loaded into the simulation, it runs with different configurations of the production line to identify the best configuration.

Once the production plan has been optimized in the digital environment, the production can start (phase 4). First, the scheduled production orders are sent to the Manufacturing Execution System (MES), which is responsible for the management of the assembly line. During the production, high amounts of data have to be stored, such as logging or sensor information. The same holds for information that describes faulty part series or warranty issues of assembled parts.

| Phase 1 | Phase 2 | Phase 3 | Phase 4 | Phase 5 | Phase 6 |
|---------|---------|---------|---------|---------|---------|
| Concept | Design & Development | Production-planning | Production | Maintain & Support | Retire & Dispose |

**Fig. 17.1** Product lifecycle chain

The completed products can then be sold. For maintenance and support reasons (phase 5), it is useful to link the customer data with the corresponding product data. In cases of complaints, guarantee issues and callbacks of faulty products, this information is valuable as well.

The last phase of the product lifecycle, namely the retire-and-dispose phase, covers the recycling of the products. For the disposal of a product, it is important to know which assemblies or parts could be used as spare parts or from which parts valuable materials could be extracted.

This short overview is just a very rough glance on data usage in the product lifecycle. Altogether, a huge amount of data is generated along the product lifecycle and this data shows a high heterogeneity regarding the data formats. As a consequence, an integrated data management and flexible information integration, in order to integrate CAD, ERP or Digital Factory applications, are fundamental requirements for any PLM approach.

## 17.2.2  Challenges in PLM Application Integration

In order to implement a successfully integrated PLM solution, there are some challenges and unresolved issues to be dealt with. From the IT perspective, an integration infrastructure for PLM should focus on:

- Enabling interoperability
- Loosely coupling of systems
- Integration of new PLM components and applications

This integration should not affect existing information systems in any phase of the product lifecycle, such as PDM, MES or ERP systems.

In the past 20 years, several initiatives have been started to develop standards and technologies capable of supporting data management and information provisioning in the entire production process [4, 25]. In order to face the interoperability challenges in PLM, some lessons learned from the adoption of Computer Integrated Manufacturing (CIM) need to be taken into account. The concept CIM tries to come up with an integrated production process by integrating all its phases. A major requirement is the integration of highly heterogeneous information systems. Two important conclusions derive from the adoption of CIM solutions and are essential for future PLM information systems to be successful:

1. Legacy systems cannot be easily replaced by future platforms with which they are not suitable to be integrated.
2. It is extremely difficult, complex and expensive to combine legacy, self-developed or newly acquired software components in a continuous CIM integration environment.

For these two reasons, the architectural principles, which drive integrated PLM frameworks, need to provide the necessary means to ease the integration and reuse of legacy systems. On this note, manufacturing companies have to face with three major challenges regarding their PLM infrastructure:

• Standardization
• Business process level integration
• Flexibility

First, the lack of standards can lead to vendor lock-in situations. When building up an integrated IT infrastructure for multiple phases of the product lifecycle, companies use proprietary technologies that claim to be open and extensible. However, quite the contrary, they install opaque information systems that are very difficult to extend or integrate with other applications. Therefore, the implementation of a PLM integration framework must also adopt standard communication protocols and a product data format standardized by industry consortia.

Second of all, application integration within the product lifecycle needs to be carried out at different levels: data, application and business level. At the data level, PLM data management has to be supported by a technology enabling the integration of data among multiple distributed and heterogeneous sources. At the application level, the integrated PLM infrastructure needs to exchange product, process, resource and production data and to provide standard communication protocols and mediation mechanisms that support and ease data exchange between diverse applications. At the business process level, all applications involved in the product lifecycle should be seamlessly accessible by all business processes that run the PLM information flows. Thus, a PLM integration infrastructure needs to provide the means to support the orchestration of different functions across diverse PLM components and applications.

Finally, one further important aspect an integrated PLM data management infrastructure needs to take into account is flexibility. This aspect can be achieved through the adoption of loosely coupled integration of PLM components and applications. An integration approach based on point-to-point interfaces is ubiquitous in heterogeneous manufacturing environments. However, it's a strategy that should be avoided as it leads to complex, inflexible, unsustainable and unreliable architectures with prohibitively high costs.

## 17.2.3 Service-Based Integration

The challenges explained in the former sections can be met by the design principles of the Service Oriented Architecture (SOA) and its adoption in Enterprise Application Integration (EAI). In order to describe how SOA can address these challenges, we now explain some basic concepts of service architectures.

**Service Oriented Architecture.** SOA is a paradigm of designing business applications by using – or reusing – services which are self-contained, platform-independent and discoverable. Two of the distinguishing principles of SOA are reusability of existing assets, such as legacy systems, and the loose coupling of services.

Web services can implement a SOA [27]. The use of XML-based and open standards has a great responsibility for the success of Web services. Widely adopted standards permit to easily define interfaces and message exchanges in Web services.

The basic communication within a SOA follows the "find, bind, invoke" pattern. The service provider publishes a service description. The service consumer looks it up in a service registry, accepts the interface contract and invokes the service. This communication method allows the client to delegate the dispatch of the request to a broker that is capable of finding a server that hosts the service. In Service Oriented Architectures, this mediation layer is known as Enterprise Service Bus.

**Enterprise Service Bus.** The concept of ESB is a paradigm middleware that combines messaging, services, data transformation and service orchestration to connect distributed applications across an enterprise while assuring reliability and transactional integrity [2]. The implementation of an ESB is a standard-based integration platform that supports different communication protocols and provides a message routing infrastructure. The ESB integration pattern keeps centralized control over configuration while supporting physical distribution via bus infrastructure services, such as message routing, mediation or addressing. This pattern is especially useful for extending ESB capabilities by deploying new services without affecting the existing infrastructure. These architectural principles and the standardization efforts enable PLM manufacturers to extend their systems without falling into an unpleasant vendor lock-in situation.

The approach presented in this article adopts the described service-oriented integration pattern in a PLM environment. In order to achieve a consistent and flexible integration between all PLM components, applications and manufacturing systems, we analyze the needed components in a SOA-based manufacturing integration platform.

In the next Section, we discuss the collection of service components that enhance an ESB by adapting the bus infrastructure to a PLM environment. Furthermore, we show how this extended ESB addresses the challenges of product lifecycle data management across all phases as introduced in Sect. 2.1 and describe the gain of using an ESB for PLM compared to other approaches.

## 17.3   Enterprise Service Bus for Product Lifecycle Management

The individual phases of the product lifecycle (with the same segmentation as introduced in Fig. 17.1) connected through the ESB by using Web services in a loosely coupled manner are presented Fig. 17.2. Some services provided by the ESB are illustrated in gray boxes within the ESB.

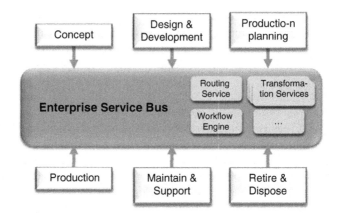

**Fig. 17.2** Product lifecycle chain managed via an enterprise service bus

The ESB has many benefits compared to other approaches. The integrated routing functionality and transformation services provide a loose coupling of the applications tied to the bus. Furthermore, workflows, which represent business processes, can be orchestrated and executed by the integrated workflow engine of the ESB.

Most realizations of ESB's are based on standards like XML and Web services. This provides a high flexibility, a high reusability and less effort to maintain than an integration strategy based on individual point-to-point interfaces.

The desired flexibility of the presented approach is achieved by mainly two components. Firstly, the description of Web service interfaces in the Web Services Description Language (WSDL) [27] allows humans and computers to read the properties of an interface. Therefore, a computer can automatically match a service request to available Web services, which fulfill the requirements of the requestor. Secondly, the routing service can be realized as a Content-Based Router (CBR) [9], which can route a message according to its content. Hence, the sender doesn't need to explicitly specify the recipient of a message, the recipient will be dynamically determined by the CBR. Therefore, the message can entail e.g. the name of the destination system, or the CBR can contain rules, which are selected depending of different variables, which are contained in the message. The message endpoint can be determined by a lookup in a database, where all message endpoints are stored. Using Web services and a CBR enables the desired loose coupling of the applications to the ESB.

We now present three scenarios that focus on processes between the different phases of the product lifecycle. They show how the usage of an ESB can considerably improve the data management during this lifecycle by regarding the applications and their interactions.

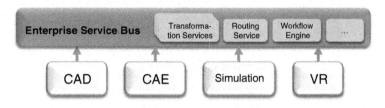

**Fig. 17.3** Support of simultaneous engineering by using an ESB

### 17.3.1 Improving Simultaneous Engineering by Using an Enterprise Service Bus

During the product development process, many different applications are simultaneously used to design the various properties of the product. Applications such as CAE (Computer Aided Engineering), simulation tools and Virtual Reality (VR) applications make use of the geometry data that are generated with CAD tools. An ESB can support this simultaneous engineering and the necessary data integration. The mentioned applications and the services to support the cooperation are illustrated in Fig. 17.3.

The benefits of using an ESB are the integrated transformation and routing services. Any application can call a defined service, e.g., to retrieve the needed product data without knowing the concrete data storage format of the involved CAD tools, the location of the service within the network or its accessibility.

The combination of workflow orchestration and data distribution within the ESB can be used to reduce the amount of exchanged data by sharing pointers where practicable instead of exchanging data by value [28]. Due to the fact that e.g. CAD programs produce hundreds of megabytes or gigabytes of data, using references can considerably reduce the amount of data to be transferred. Additionally, the processes can be executed faster, due to time savings in data handling.

### 17.3.2 Using Customer Relationship Management Information for Product Improvement

This scenario describes how the product improvement process can be enriched with Customer Relationship Management (CRM) information in a loosely coupled manner.

The product improvement process benefits from product data enrichment through semantic annotations that are based on CRM information [13]. The CRM application stores all customer complaints, suggestions, requests and relevant remarks to a product or to its parts. The challenge is to correlate these annotations with the specific version of the corresponding product or part in the used CAD tool. Currently, the feature technology [26] is used to aggregate elements within the CAD

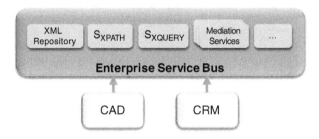

**Fig. 17.4** Integrating CAD and CRM by using an ESB

and CRM applications and provides the possibility for adding semantic information to these elements. This solution can define and re-use expert knowledge from the customers and integrate them with the computer-aided process chains.

By using an ESB, any CAD and CRM tool could be easily integrated together and enhanced with the described semantic annotations. This provides the desired loosely coupling. The integration of these applications and involved services to the ESB is shown in Fig. 17.4.

In this scenario, customer feedback is stored as annotations in a semantic repository based on XML languages such as RDF (Resource Description Framework http://www.w3.org/RDF/) or OWL (Web Ontology Language http://www.w3.org/TR/owl-features/). A mediation service transforms these annotations into the XML-Repository format. References to the annotations are sent over the bus to the CAD tool. The CAD tool, as well as other consumers, retrieves these annotations by making use of query services. For example, XQuery or XPath (http://www.w3.org/TR/xpath-functions/) are powerful XML query languages that permit query services to retrieve the desired data.

### 17.3.3 Improving the Interaction Between the ERP-System, the Digital Factory and the MES

The last scenario describes how the usage of an ESB can improve the interchange of data between the ERP system, the Digital Factory and the MES. We assume that the product, process and resource (PPR) data are held by the Digital Factory [12], the customer orders are managed by the ERP system and production orders are created in the Digital Factory by using customer order information and PPR data. Before executing the production orders, they are optimized towards the current assembly line by a simulation tool.

Using an ESB to integrate the applications shown in Fig. 17.5 improves flexibility by loosely coupled systems. The "find, bind, invoke" pattern can be used to interchange data between the applications without previous knowledge of the interfaces or locations. This pattern is supported by the routing service of the ESB. Transformation services are used to convert the data from the source format

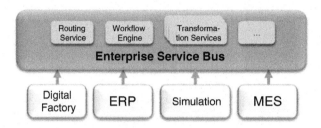

**Fig. 17.5** Improving collaboration by integrating various PLM applications via an ESB

to the destination format automatically. There are several alternatives for data transformation, like Champagne [14], WebSphere [33] or BizTalk Server [3].

An ESB may also contain a canonical format for messages that are exchanged over the bus. The definitions of XML templates for modeling events can support efficient data exchange between the mentioned applications. The XML canonical format to describe events reduces the number of necessary service adaptors that transform event formats, and it assures a higher extensibility of the model. Transformation services could be implemented with XSLT (Extensible Stylesheet Language Transformation) [32], which allow transforming proprietary event formats into the canonical XML format of the bus and vice versa. The sequence of data exchanges can be automated by a workflow. The simulation event can be repeatedly executed by a workflow until a predefined threshold is reached. Afterwards the optimized schedule of the production plan is transmitted to the MES. Execution and administration of the workflow is governed by the workflow engine of the ESB.

Additionally, we can use the approach of [28] in order to not affect the performance of the bus by overloading its capacity with large amounts of data. The concept of call by reference instead of call by value reduces the amount of transferred data in this scenario, too.

## 17.4   Hierarchy of Services

To integrate the whole product lifecycle with a SOA, many services are needed on different levels of granularity. We introduce a hierarchy of services in order to organize the different kinds of services according to the level of abstraction. The hierarchy consists of three levels:

1. Atomic service level, which is divided into application and mediation services
2. Integration service level
3. Value-added service level

The application services of the first level provide access to the data and functionality of single applications through respective interfaces. The mediation services are used for data transformation or message routing within the bus to enable data exchange

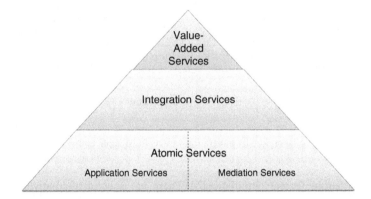

**Fig. 17.6** The service hierarchy

between the heterogeneous applications. On the second level, data integration is carried out. At the third level, the integration of services at the business process level is organized. The service hierarchy is shown in Fig. 17.6 and is described below in more detail.

### 17.4.1 Atomic Services

On the lowest level, the atomic services reside. These services are divided into two groups, the application and the mediation services.

The application services represent the interfaces of applications and expose their main functionality as encapsulated services, which can be executed or orchestrated in a complex workflow. The OMG standard PLM Services defines classes and interfaces for this kind of services [19]. Additionally, it describes different use cases such as "export/import of assembly data", "authorization", "authentication" or "change notification" to make different applications accessible in a standardized way. The PLM Services standard is based on other standards, e.g. STEP [25] and the "OMG Model Driven Architecture" standard [18]. IBM developed an implementation of the OMG PLM Services to build a framework for PLM integration based on these standards [6].

The mediation services carry out data transformation and routing tasks. They are deployed and executed in the ESB. The routing functionality can be realized with a Contend Based Router (CBR) [9], which processes incoming messages to determine their destination. Data transformation can be carried out by XSLT services. For example, an ERP system and the Digital Factory may rely on different data formats for describing customer orders. So, we need mediation services to convert these formats into each other when both systems exchange customer orders.

## 17.4.2   Integration Services

The second level of the service hierarchy describes the integration services. The integration services can be compared with an ETL (Extract – Transform – Load) process [11]. ETL is commonly used in data warehousing environments, but in both cases data is retrieved from a source system, converted into the destination data format if necessary and finally sent to the destination system.

Integration services build upon atomic services: Application services are used to retrieve the data from the source and to send it to the destination system, respectively. Mediation services are responsible for the correct routing of the messages as well as for the transformation of the data.

Combining different atomic services, namely the application and mediation services, into new services, enables the integration of applications within the product lifecycle on the data level. The integration services can be modeled in the workflow language Business Process Execution Language (BPEL) [21] and thus deployed, executed and managed within the workflow engine of the ESB.

## 17.4.3   Value-Added Services

The highest level of the service hierarchy comprises value-added services, which can be compared with workflows. These services combine different integration services to support the business processes of the enterprise. To model business processes, BPEL can be used. BPEL has become the de-facto standard to model and run business processes based on the orchestration of service interactions. The main benefits of BPEL are its modular design, its flexibility regarding generic XML data types and late binding of services as well as the sophisticated fault, compensation, and event handling capabilities [1]. In addition, many BPEL engines offer further useful capabilities, such as user interactions, workflow auditing and monitoring, or recovery of workflows.

## 17.4.4   Mapping Business Process to Data Integration Processes

The service hierarchy introduces a mapping of business processes to data integration services. This mapping supports a consistent business process and service management. Two main benefits of this mapping can be identified.

The first benefit is the identification of critical services. Thanks to the service hierarchy, critical business processes can be mapped to a number of integration and atomic services. We can identify these services as critical for the execution of the corresponding business process. To ensure the operability of critical business processes, the availability of the depending critical services has to be guaranteed and is therefore more important as the availability of non-critical services. To reach a higher service quality, development efforts can be increased for the critical services, especially a higher number and more detailed tests can decrease the risk of service failures.

On the other side, the knowledge about the mapping between business processes, integration and atomic services is important if a service faults. In this case, the affected business processes can be easily indentified and the corresponding workflows can be quickly adapted to solve the problem.

The second advantage of the mapping is related to the change of business processes. In this case, we can easily check for all depending integration and atomic services whether they need to be adjusted or not. Even when changing services, the dependent services and business processes can be identified and the effects on them can be determined.

The middle level with the integration services has a special role within the hierarchy. This level should make the changes of atomic service invisible for business processes. If an atomic service is changed, only the affected integration services have to be adapted as well as other atomic services. Therefore, changes of business processes only have to be performed when the business tasks change, thus isolating these from changes in the IT systems. One of the major benefits of the separation between business services and IT services that the presented service hierarchy represents is the management of independent lifecycles for business processes and data integration processes. Hence, capturing requirements, redesigning and adapting processes can be done separately on the business domain and on the IT service domain thanks to the different level of abstraction. Furthermore, the mapping between both domains allows identifying the relation between atomic services and value-added services thanks to the integration middle layer, namely the ESB. These relations are of great importance for an efficient adaptation of business processes.

Introducing the service hierarchy reduces the effort of managing business processes and services by increasing the transparency of service compositions. In case of errors or planned changes in all levels of the hierarchy, the affected business processes or services can be quickly identified and adjusted accordingly. This increases the adaptability of our infrastructure and the flexibility in business processes. The mapping introduced by the service hierarchy is illustrated in Fig. 17.7 where the integration of two PLM phases is described.

Each of the PLM phases contain many application services ($AS_{i1}$, $AS_{i2}$, etc.). These atomic services and different mediation services, e.g. $MS_2$, are combined together into one or more service compositions. We define these compositions as integration services, e.g. $IS_1$. Usually, an integration service executes an ETL process that propagates data from one atomic service to others, such as from $AS_{i3}$ to $AS_{j2}$. Integration services use mediation services to convert data from the source to the destination format. The tasks performed by such integration services are to integrate data from different PLM phases.

As already mentioned, the value-added services directly support the execution of business processes. In Fig. 17.7, the value-added service $VAS_1$ combines the integration services $IS_1$, $IS_2$ and $IS_3$ in a workflow that allows coordinating the data flow between the involved applications.

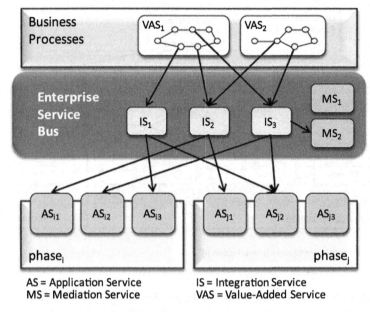

**Fig. 17.7**   Mapping business process to data integration processes

Assume that $VAS_1$ is a critical business process and that $VAS_2$ is a non-critical business process. This implies that all integration services and thus application services are critical for the execution of $VAS_1$. When changing $VAS_1$ by deleting $IS_2$ from the workflow, the importance of $IS_2$ decreases because this integration service is not critical any more. Consequently, the importance of the application services $AS_{i1}$ and $AS_{j1}$ also decreases and therefore their quality of service can be reduced.

### 17.4.5   Example of the Service Hierarchy

This section describes the above discussed benefits of the service hierarchy using the example of Sect. 3.3. This scenario introduces the automatic optimization and update of the production orders after receiving a new customer order.

First, the different applications need to get interfaces to become accessible. These interfaces are realized by atomic services as illustrated at the bottom of Fig. 17.8. The ERP system needs a service to retrieve the customer orders. The Digital Factory requires services to retrieve the product, process and resource information, as well as services to receive customer orders and to retrieve production orders. The simulation tool needs interfaces to receive and retrieve various data, e.g. PPR data. Finally, the MES has to receive the optimized schedule of the production orders.

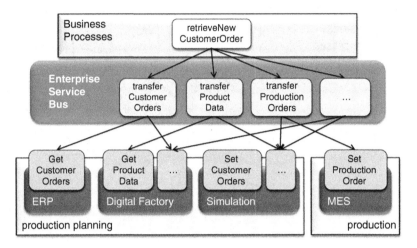

**Fig. 17.8** Example using the service hierarchy

All integration services are shown in the middle of Fig. 17.8 within the ESB. The first integration service calls the atomic service of the ERP system to receive the customer orders. Then, it sends these customer orders to the Digital Factory that converts them into production orders. If necessary, the data can be converted to an appropriate format for the consuming service by using XSLT services, which are left out in Fig. 17.8 for the sake of clarity. The same also applies to all other integration services.

A second integration service loads the product data from the Digital Factory into the simulation tool. Further integration services perform similar data provisioning processes for the process and resource data as well as for the production orders. A third integration service retrieves the optimized production orders from the simulation tool and sends them, if necessary transformed by a mediation service, to the MES.

To realize the business process, the described integration services can be integrated into the value-added service to realize the data movement. The example in Fig. 17.8 describes the workflow, which is started when a new customer order is received. Afterwards, the customer orders are transferred to the Digital Factory by using the first integration service described above. With the second and next integration services, the PPR data and production orders are transferred to the simulation tool. After the completion of these services, the workflow starts the optimization runs to improve the production plan. When the production orders schedule is optimized or when a predefined threshold is reached, the workflow continues by executing the last integration service to transfer the optimized production orders to the MES. Here, the current production plan is replaced with the new and optimized one that now takes the new customer order into account.

To implement this workflow, BPEL can be used. The advantage of modeling workflows in BPEL is that a BPEL workflow can be directly executed. Furthermore, an extension to the BPEL specification allows the interaction of workflows with humans, which is called BPEL4People (http://download.boulder.ibm.com/ibmdl/ pub/software/dw/specs/ws-bpel4people/BPEL4People_v1.pdf). Having such long-running workflows as the one described above, it is useful to have human interactions within the workflow to control the correct data flow and to avoid problems in the production.

## 17.5   Related Work

A continuous integration of the product lifecycle entails many benefits. Especially from an IT point of view, many improvements are achieved. In the following, we describe the most important work related to this article.

### 17.5.1   SOA Approaches of Different PLM Vendors

In the past few years the main PLM vendors like Dassault Systèmes, PTC and UGS Corporation (by now bought by Siemens AG and called Siemens PLM Software) started to integrate their various applications with a service-oriented approach [5]. However, all of these vendors use a proprietary integration middleware. Dassault Systèmes developed a collaborative infrastructure called ENOVIA [7], PTC uses PTC Windchill Info*Engine [22] and Siemens PLM Software make use of its Teamcenter to connect their various applications [5]. These solutions don't provide the same capabilities compared to the presented ESB approach. They are partly lacking in integrating systems of other vendors, flexibility in business processes is missing and applications are not loosely coupled to the integration middleware. Furthermore, the interfaces are not open, so it is hard or impossible for other software vendors to connect their applications to these middleware systems. As a result, the adoption of such solutions provokes vendor lock-in situations, making the integration with other applications extremely difficult and expensive. In the end, most companies end up replacing existing applications of other vendors to get the full benefit of the PLM vendors' solution.

### 17.5.2   Product Lifecycle Management

In the domain of services for PLM, two related works have to be mentioned: In [19], the OMG defines a specification on "Product Lifecycle Management Services". This specification defines a Platform Independent Model (PIM) for PLM component services and a Platform Specific Model (PSM) applicable to a Web service implementation defined by a WSDL specification, an SOAP binding, and

UML (Unified Modeling Language) Profiles for XML Schema. Based on this standard, our approach defines new value-added services that support business processes management and integration. The integration of business processes with PLM Services increases the flexibility and adaptability of the IT infrastructure.

In [6], IBM describes the implementation of these OMG services to realize selected scenarios. Furthermore, they offer an end-to-end integration to business process management. For the realization of our approach, we can use the implemented component services to build new integration and value-added services upon them.

### 17.5.3 Service Oriented Architecture

Regarding the ESB, a basic description and possible fields of application for ESB's can be found in [2]. A focus on Service Oriented Architecture and ESB in manufacturing environments as well as specific services required in such environments is described in [17]. Our approach is based on this work and we extend it to embrace the whole product lifecycle.

In [15], the authors describe a reference architecture for service-based data integration in a manufacturing environment. This reference architecture is designed for the production phase of the product lifecycle and uses the so called Manufacturing Service Bus (MSB), a domain-specific extension of an ESB for the production phase, as integration middleware. However, in the approach presented in this article, we extend the model of [15] to support the whole product lifecycle and we additionally introduce the service hierarchy for offering value-added services.

The Daimler AG has developed another service-based integration solution for the product design-and-development phase [10]. Their integration solution is based on the so-called Engineering Service Bus, which extends an ESB with requirements of the engineering domain. This approach focuses, like the MSB solution, on a single phase of the product lifecycle. As already mentioned, all the PLM phases have to be integrated to get all the benefits described in this article.

## 17.6 Conclusions

An integrated product lifecycle based on point-to-point interfaces is very complex, time-consuming and costly. Additionally, after implementing the interfaces, the complexity and costs of maintenance remain very high.

With our approach and the described scenarios, we have clearly shown that a Service Oriented Architecture considerably reduces this effort of implementing and maintaining interfaces within PLM. Adopting an ESB to the heterogeneous environment of the product lifecycle entails many benefits. The first benefit of using an ESB is the possibility to reuse services. The loose coupling of applications by using integrated routing and mediation services enables flexible and dynamic service binding and service composition. The second benefit is the usage of workflows via the integrated workflow engine, which offer an integrated modeling and execution framework for business processes. Furthermore, the ESB and standards like XML, Web services, XPath, XQuery or XSLT considerably improve the transparency, the flexibility and adaptability, as well as the interoperability of applications in the product lifecycle. Furthermore, they lead to a reduction of maintenance costs.

Introducing the service hierarchy increases the transparency of services by mapping business processes to data integration services. In case of errors or planned changes in any level of the hierarchy, the affected business processes or services can quickly be identified and adjusted accordingly. Furthermore, changes of atomic services can be made transparently, without affecting the value-added services. Hence, additionally to the improvements of the ESB and the usage of standard technologies, the introduced service hierarchy increases the adaptability of our infrastructure and the flexibility in business processes.

Altogether, a service-oriented integration middleware, namely the ESB, and the presented service hierarchy define the basis for a new generation of PLM services. They enhance the PLM functionality with value-added services and enable a continuous integration of all PLM phases.

**Acknowledgements** The authors would like to thank Thorsten Scheibler, who contributed to a previous publication on this topic, and Peter Reimann, who carefully reviewed this article.

Furthermore, the authors would like to thank the German Research Foundation (DFG) for financial support of the projects within the Graduate School of Excellence advanced Manufacturing Engineering (GSaME) at the University of Stuttgart.

# References

1. Akram, A., Meredith, D., Allan, R.: Evaluation of BPEL to Scientific Workflows. In: Proceedings of the 6th International Symposium on Cluster Computing and the Grid, Singapore, Malaysia (2006)
2. Chappell, D.A.: Enterprise Service Bus, O'Reilly, Sebastopol (2004)
3. Chappell, D.: Understanding BizTalk Server 2006. Whitepaper, Microsoft Corp. (2005)
4. ESPRIT Consortium AMICE (ed.): CIMOSA Open System Architecture for CIM, Springer, Berlin (1993)
5. CIMdata: Service-Oriented Architecture for PLM – An Overview of UGS' SOA Approach. CIMdata, Inc. (2006)
6. Credle, R., et al.: SOA Approach to Enterprise Integration for Product Lifecycle Management. IBM International Technical Support Organization (2008)
7. Dassault Systèmes: ENOVIA V6R2011x – FACT SHEET. http://www.3ds.com/fileadmin/PRODUCTS/ENOVIA/PDF/Datasheets/V6R2011x/ENOVIA-V6R2011x-factsheet.pdf (2010)
8. Erl, T.: Service-Oriented Architecture: Concepts, Technology, and Design. Prentice Hall International (2005)
9. Hohpe, G., Woolf, B.: Enterprise Integration Patterns: Designing, Building, and Deploying Messaging Solutions. Addison-Wesley Longman, Amsterdam (2003)

10. Katzenbach, A.: Engineering goes SOA – Role of SOA in Advanced Engineering IT. ProSTEP iViP Symposium Berlin (2008)
11. Kimball, R., Caserta, J.: The Data Warehouse ETL Toolkit: Practical Techniques for Extracting, Cleaning, Conforming, and Delivering Data. Wiley (2004)
12. Kühn, W.: Digital Factory – Integration of simulation enhancing the product and production process towards operative control and optimization. Int. J. Simulat. **7**, 27–39 (2006)
13. Lasi, H., Baars, H., Kemper, H.G.: Integration of customer based features in digital mock-ups. The IEEE International Conference on Industrial Engineering and Engineering Management, pp. 2010–2014 (2007)
14. Minguez, J., Jakob, M., Heinkel, U., Mitschang, B.V.: A SOA-based approach for the integration of a data propagation system. In: Proceedings of the IEEE International Conference on Information Reuse and Integration, pp. 47–52 (2009)
15. Minguez, J., Lucke, D., Jakob, M., Constantinescu, C., Mitschang, B., Westkämper, E.: Introducing SOA into Production Environments – The Manufacturing Service Bus. 43rd CIRP International Conference on Manufacturing Systems, Vienna, Austria, pp. 1117–1124 (2010)
16. Molina, A., Al-Ashaab, A.H., Ellis, T.I.A., Young, R.I., Bell, R.: A review of computer-aided Simultaneous Engineering systems. Research in Engineering Design, vol. 7, Springer, pp. 38–63 (1995)
17. MESA International, IBM Corporation and Capgemini: SOA in Manufacturing Guidebook. (2008)
18. OMG: Model Driven Architecture (MDA). OMG (2001)
19. OMG: Product Lifecycle Management Services. V 2.0, OMG (2009)
20. OASIS WS-BPEL Extension for People (BPEL4People) TC: BPEL4People specification Version 1.0. http://download.boulder.ibm.com/ibmdl/pub/software/dw/specs/ws-bpel4people/BPEL4People_v1.pdf (2007)
21. OASIS WS-BPEL TC: Web Services Business Process Execution Language Version 2.0. OASIS Standard, http://docs.oasis-open.org/wsbpel/2.0/OS/wsbpel-v2.0-OS.html (2007)
22. PTC: Windchill® Info*Engine. http://de.ptc.com/WCMS/files/57292/de/5074_WC_InfoEngine_DS_DE.pdf (2009)
23. Saaksvuori, A., Immonen, A.: Product Lifecycle Management, Springer, Berlin (2008)
24. Stark, J.: Product Lifecycle Management: 21st Century Paradigm for Product Realisation (Decision Engineering). Springer, Berlin (2004)
25. ISO 10303-1:Industrial automation systems and integration Product data representation and exchange - Overview and Fundamental Principles, International Standard, ISO TC184/SC4 (1994)
26. VDI-Gesellschaft Produkt- und Prozessgestaltung: VDI-Richtlinie 2218 - Information technology in product development - Feature Technology. Beuth Verlag, Berlin (2003)
27. Weerawarana, S., Curbera, F., Leymann, F., Ferguson, D.F., Storey, T.: Web Services Platform Architecture: Soap, WSDL, WS-Policy, WS-Addressing, WS-Bpel, WS-Reliable Messaging and More. Prentice Hall International (2005)
28. Wieland, M., Görlach, K., Schumm, D., Leymann, F.: Towards reference passing in web service and workflow-based applications. Proceedings of the 13th IEEE international conference on Enterprise Distributed Object Computing, IEEE Press, pp. 89–98 (2009)
29. W3C: Resource Description Framework (RDF). http://www.w3.org/RDF/ (2004)
30. W3C: OWL Web Ontology Language. http://www.w3.org/TR/owl-features/ (2004)
31. W3C: XQuery 1.0 and XPath 2.0 Functions and Operators. http://www.w3.org/TR/xpath-functions/ (2010)
32. W3C: XSL Transformations (XSLT) Version 1.0. W3C Recommendation, http://www.w3.org/TR/xslt (1999)
33. Wylie, H., Jones, J., Edwards, P.: IA81 WebSphere Business Integration Message Broker and Web Services. Tech report, Version 3.0, IBM (2004)

# Chapter 18
# Retrieving Reusable Software Components Using Enhanced Representation of Domain Knowledge

**Awny Alnusair and Tian Zhao**

## 18.1  Introduction

Software reuse simply allows us to efficiently build, enhance, and evolve software systems using artifacts and knowledge from existing ones. A systematic approach for reuse enables developers to construct quality software more quickly, reduces cost and risk, and certainly support software maintenance tasks. Currently, the most common practices of software reuse is the utilization of libraries of reusable components. Failure modes analysis of the reuse process shows that in order to be reused, a library component must be findable and understandable [5]. On one hand, understandability is usually hampered by the lack of informative API documentation of library components. On the other hand, finding suitable components is still a significant barrier for exploiting systematic software reuse.

In this paper, we present an approach for describing, retrieving, and exploring the various relationships among source-code components in large object-oriented reuse libraries or frameworks. In order to provide a formal and precise representation of library code, our approach relies on exploiting ontology representation of software knowledge. Ontologies provide means to explicitly describe concepts, objects, properties and other entities in a given application domain, and to represent the relationships that hold among these concepts. Due to their solid formal and reasoning foundation, ontologies can play an important role in domain engineering reuse; they can be used to structure and build a source-code knowledge base that

A. Alnusair (✉)
Department of Science, Math & Informatics, Indiana University Kokomo, 2300 S. Kokomo, IN 46904, Washington
e-mail: aalnusai@iuk.edu

T. Zhao
Department of Computer Science, University of Wisconsin-Milwaukee, 2200 E. Kenwood Blvd, P.O. Box 413, Milwaukee, WI 53211, USA
e-mail: tzhao@uwm.edu

T. Özyer et al. (eds.), *Recent Trends in Information Reuse and Integration*,
DOI 10.1007/978-3-7091-0738-6_18, © Springer-Verlag/Wien 2012

can be used by software agents (e.g., search engine) and certainly serve as a basis for semantic queries [14].

Towards that end, we have developed an ontology model that includes an enhanced software representation ontology. This ontology is further extended with additional component-specific knowledge and automatically populated with ontological instances representing various program elements of a given library. These instances are further annotated with respect to concepts from a domain-specific ontology. Ontology-based component search is thus performed by the semantic matching of user's requests expressed using terms from the domain ontology with component descriptions found in the populated knowledge-base. Furthermore, the knowledge population and indexing mechanisms used in our approach still allow searching the knowledge base using the most familiar keyword search. This is particularly useful when domain ontologies or semantic component annotations are lacking or incomplete. Users are thus able to retrieve components using pure semantic-based queries, keyword queries, type-based queries, or a mixture of these mechanisms.

## 18.2  Ontology Model for Describing Components

At the core of our ontology model for object-oriented component retrieval is a Source Code Representation Ontology (referred to afterwards as SCRO). This ontology provides a base model for capturing the relationships and dependencies among source-code artifacts. It models major concepts and features of object-oriented programs, including encapsulation, class and interface inheritance, method overloading, method overriding, and method signature information. SCRO's knowledge is represented using the OWL-DL[1] ontology language. OWL is a web-based conceptual modeling language used for capturing relationship semantics among domain concepts, OWL-DL is a subset of OWL based on Description Logic (DL) and has desirable computational properties for automated reasoning systems. OWL-DL's reasoning support enables inferring additional knowledge and computing the classification hierarchy (subsumption reasoning). SCRO defines various OWL concepts that map directly to source-code elements and collectively represent the most important concepts found in object-oriented programs. Furthermore, SCRO defines various OWL object properties, datatype properties, and ontological axioms to represent the various relationships among ontological concepts. SCRO is precise, well documented, and designed with ontology reuse [16] in mind. The availability of SCRO online (http://www.cs.uwm.edu/~alnusair/compre), along with an extended discussion of its knowledge representation and population mechanisms, allow researchers to reuse or extend its representational components to support any semantic-based application that requires source-code knowledge.

---

[1]http://www.w3.org/2004/OWL/

## 18.2.1   Domain-Specific Ontology

Domain ontologies describe concepts and structures related to a particular domain (e.g. finance, shopping, medicine, graphics, or the object-oriented programs domain as specified in SCRO). Domain-specific ontologies have been widely recognized as effective means for representing domain concepts and their relationships such that concepts are understandable by both computers and developers [11].

In our approach to component retrieval, we use a domain ontology that conceptualizes each software library we need to reuse. This ontology provides a common vocabulary with unambiguous and conceptually sound terms that can be used to annotate software components. Annotations in this context serves two key purposes. Firstly, both software providers and users can communicate using a shared and common vocabulary provided by this ontology. Thus enabling a precise retrieval of API components. Secondly, annotations bring different perspectives to typical program comprehension tasks. Users can familiarize themselves with terminology and conceptual knowledge that is typically implicit in the problem domain.

To this end, we have developed a mini-ontology for data retrieval in the Semantic Web applications domain. This ontology (referred to afterwards as SWONTO) has been used during the evaluation of our approach (cf. Sect. 18.4) and serves as a proof of concept that component search can be significantly enhanced through the use of domain ontologies. A small fragment of the ontologys taxonomy is shown on the upper right pane of Fig. 18.1 and the complete ontology can be found online (http://www.cs.uwm.edu/~alnusair/compre).

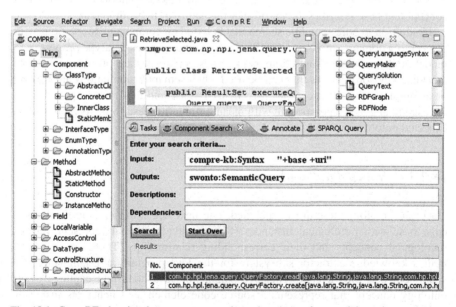

**Fig. 18.1**  CompRE: showing the component ontology, domain ontology, and the main search view

## 18.2.2   Component-Specific Ontology

In the context of component retrieval, we need a profound semantic description of the component's inner working structure and its interrelationships with other components in a given domain. Therefore, we extend SCRO's semantic representations of API structures and enrich it with additional component-specific descriptions that are required to uniquely identify and retrieve an API component. The result is a COMPonent REpresentation ontology (referred to afterwards as COMPRE). In addition to the concepts, axioms, and properties inherited from SCRO, COMPRE defines its own class hierarchy and relations for semantic component descriptions. For instance the `Component` concept represents software components in general and subsumes other OWL classes that represent Java-specific components such as `Method`, `ClassType`, and `InterfaceType`. A fragment of the ontologys taxonomy is shown on the left pane of Fig. 18.1 and the complete ontology, along with extended description can be found online (http://www.cs.uwm.edu/~alnusair/compre).

Moreover, COMPRE defines various ontological axioms and object properties that represent relationships among software components. These properties link components with their corresponding semantic descriptions specified in the domain-specific ontology. For example, `hasDomainInput` and `hasDomainOutput` and their corresponding inverse properties are used to annotate individual software components with domain terms representing the expected inputs and outputs of a given component. Moreover, COMPRE defines the `dependsOn` symmetric object property that defines dependency relationship between components. `describedBy` is also defined to link a component to a domain concept that best describes the purpose or the nature of the component.

In addition to pure semantic search that is based on component annotations with respect to a domain ontology, COMPRE also defines various datatype properties to model other metadata about components. These properties are provided to enable metadata keyword queries that can be used when semantic annotations are lacking or incomplete. For instance, the `hasInputTerms` and the `hasOutputTerms` are used to annotate a component with meaningful terms describing its input and its expected output.

## 18.2.3   Knowledge Population

Once the ontology structure is specified, one next needs to populate the knowledge base with ontological instances (OWL individuals) that represent various ontological concepts and their corresponding relationships. Therefore, we have built a knowledge extractor subsystem for the Java programming language. Our subsystem performs a comprehensive parsing of Java code and captures every ontology concept that represents a source code element and generates instances of all ontological properties defined in our ontologies for those program elements.

The generated semantic instances are serialized using RDF [2] triples. RDF is suitable for describing resources and provides an extensible data model for representing machine-processable semantics of data. For each application framework parsed, we thus generate a RDF ontology in Notation3 [3] format. This amounts to instantiating an OWL knowledge base for that framework. This knowledge base is managed by Jena [12], an open source Java framework for building Semantic Web applications.

The process of generating semantic instances for the concepts and relations specified in SCRO is completely automatic. However, the process of annotating components according to COMPRE's object properties is currently manual as it is the case for semantic annotations in general. Our tool though provides means for inserting these annotations directly into the knowledge-base, thus gradually building semantic descriptions for a particular API that can be shared, evolved, and reused by a community of users. On the other hand, metatdata modeled by COMPRE's datatype properties and other annotations provided by the Dublin Core Metadata Initiative (DCMI) [4], is generated automatically via direct parsing of the source-code. We thus capture and normalize method signatures, identifier names, source-code comments, and available Java annotations in order to obtain a meaningful term-based descriptions of components. These descriptions are lexically analyzed, stored, and indexed using the tokenization and indexing mechanisms provided by Apache Lucene [5], an open source, powerful, and full-featured text search engine.

Listing 18.1 shows a partial RDF description for a Jena API method. This method belongs to the `QueryFactory` class and usually used to create a `Query` object given the specified input. Notice that the fully qualified name of program elements is encoded but omitted in the listing for readability. In the next section, we show how this representation can be effectively used to retrieve API components. For an extended discussion of our ontologies, knowledge population, and complete samples of our knowledge extractor sub-system, we refer the reader to our ontologies website (http://www.cs.uwm.edu/~alnusair/compre).

## 18.3   Ontology-Based Search

The RDF description for the `create` method shown in Listing 18.1 clearly captures the component's metadata at the semantic and syntactic level. The underlying data structure of RDF is a labeled directed graph. Each node-arc-node in this graph represents a triple that consists of three parts, *subject*, *predicate* and *object*. Consider Listing 18.1 for example, the described method in this snippet, `create[..]`, is always the subject, ontology properties are predicates, and objects are either a

---

[2]http://www.w3.org/TR/rdf-primer
[3]http://www.w3.org/DesignIssues/Notation3
[4]http://dublincore.org/schemas/
[5]http://lucene.apache.org/

```
@base                 <http:.../ontologies/kb.n3>
PREFIX scro: <http:.../ontologies/scro.owl#>
PREFIX compre: <http:.../ontologies/compre.owl#>
PREFIX swonto: <http:.../ontologies/swonto.owl#>
PREFIX dc: <http://purl.org/dc/elements/1.1/>

<#QueryFactory.create[String,String,Syntax]>
 a scro:StaticMethod ;
 scro:hasInputType <#String> ;
 scro:hasInputType <#String> ;
 scro:hasInputType <#Syntax> ;
 scro:hasOutputType <#Query> ;
 scro:invokesMethod <#parse[Query,String,Syntax]>;
 scro:hasSignature "create[String,String,Syntax]";

 compre:describedBy    [ a swonto:QueryCreation];
 compre:hasDomainInput [ a swonto:QueryText];
 compre:hasDomainInput [ a swonto:URI];
 compre:hasDomainInput [ a swonto:QueryLanguageSy-
ntax];
 compre:hasDomainOutput [ a swonto:ExtendedQuery];

 compre:hasInputTerms "query string ...";
 compre:hasInputTerms "base URI ...";
 compre:hasInputTerms "query syntax URI ...";
 compre:hasOutputTerms "query ...";
 dc:description "create query ...";
                  ....
```

**Listing 18.1** Sample RDF descriptor for a Jena method in Notation3 syntax

resource, unlabeled node (blank node) or a literal value. For example, the first RDF triple below uses a property from SCRO to assert that the method has an input parameter of type String. The second triple associates this parameter with few terms describing its purpose. The third triple, however, tags the same input parameter with a meaningful concept (QueryText) from the domain ontology. Thus, giving the parameter an agreed-upon and meaningful description other than terms or the semantically vague String type.

1. create[..] scro:hasInputType String
2. create[..] compre:hasInputTerms "query string"
3. create[..] compre:hasDomainInput [a swonto:QueryText]

This multi-faceted description of components enables four different kinds of queries against the knowledge base: (*a*) Type or signature-based queries. (*b*) Metadata keyword queries. (*c*) Pure semantic-based queries; or (*d*) Blended queries of the previous three types.

Although supported by our tool, neither signature matching, nor keyword queries alone perform well in terms of recall and precision (cf. Sect. 18.5). Therefore, pure semantic-based queries that rely on domain-specific knowledge is our primary focus. In fact, it is extremely difficult for a user to precisely describe the requested component irrespective of knowledge that is specific to the application domain. Primarily, search techniques that rely on variations of keyword-based search suffer from synonymity and polysemic ambiguity that often lead to low recall and precision. On the other hand, signature matching techniques cannot distinguish between components that have the same signature but serve different purposes, e.g. using Jena API to *create* a new query vs *read* a query from a file. In semantic search, however, these limitations are completely dealt with since the semantics of each of the types in signatures are encoded and processed during search. Besides addressing knowledge representation effectively, semantic search offers extensible solutions to component retrieval. Since we are focusing on API usage and reuse, the descriptions shown in Listing 18.1 capture the component's interface and its relationships with other components. However, these description can be easily extended to capture other facets (e.g., component's environment) via introducing additional ontological properties.

Reasoning is one of the primary added benefits in semantic search. In addition to classifying and checking the consistency of our ontologies, a DL reasoner can also be used for inferring and thus enriching the knowledge base with additional knowledge that is not explicitly stated. Thus, playing a vital role in improving search precision and recall in comparison with other search techniques. DL Subsumption reasoning, for example, is typically used to establish subset inclusion relationships between different concepts and properties in the ontology. Consider the descriptions in Listing 18.1 for example, when pure semantic-based queries are used, users need only to provide domain concepts describing the component's interface. Therefore, if the user provides `SemanticQuery` as a domain output of the requested component, the method shown in the listing would still match this request since `SemanticQuery` subsumes `ExtendedQuery` as specified in the subsumption hierarchy of our domain ontology. Thus, automatically enabling an implicit form of query expansion.

It is notable that semantic-based retrieval alleviates many problems typically faced by tools that rely on exact keyword or type matching. One of the strengths of our approach, however, is the ability to utilize our various ontologies in order to perform blended search against the knowledge base. In particular, this is helpful when components in the knowledge base are not completely annotated or when users are still in the process of becoming familiar with the ontology. Consider for example a user who wishes to find a component in which the component's domain output type (`SemanticQuery`) and one of the actual input types (`Syntax`) are known. Furthermore, since the user is not sure about the other input types, she can provide a few terms to filter out the results. This request can be expressed using the following query in DL-like syntax:

$$Query \equiv compre : Component \sqcap$$
$$(\exists compre : hasDomainOutput . swonto : SemanticQuery) \sqcap$$
$$(\exists scro : hasInputType . compre - kb : Syntax) \sqcap$$
$$(\exists compre : hasInputTerms value "base\ uri")$$

As expected, executing this query returns not only the method shown in Listing 18.1 but also other unrelated methods. It turns out that the input terms specified in the query are very popular and are used to describe API methods that are used to `create`, `read` or even `parse` a semantic query. Nevertheless, this search mechanism is extremely flexible since it allows a wide range of queries to run against the knowledge base. In fact, the expressive power provided by our ontologies allows users to express their queries in more details than would otherwise be expressed with any alternative method. For instance, assume that the user was able to obtain a `Query` object as described in the previous example. The next natural step is to find a component that can take this query as input, execute it, and return the required result set. Browsing the Jena API looking for such a component or even querying using typical keyword-based queries would not return an answer. This appears to be a dead end since there is an intermediate query execution object that must be obtained to complete the task. However, using semantic search, this request can be expressed fairly easily as follows:

$$Query \equiv compre : Component \sqcap$$
$$(\exists compre : hasDomainInput . swonto : SemanticQuery) \sqcap$$
$$(\exists compre : hasDomainOutput . \exists compre :$$
$$hasDomainOutput . swonto : ResultSet)$$

More naturally, this query expresses the fact that we are looking for a component that takes a semantic query as input and returns another component that returns a query solution. Thus, querying for multiple components at the same time. In summary, semantic search is extremely expressive and robust in terms of precision and recall when the formal semantics expressed in the ontologies are fully utilized. However, search should also be achievable even when such semantics are incomplete or do not exist.

### 18.3.1 Implementation and Ranking Mechanisms

We have implemented this approach in a tool called CompRE, conveniently named after the main ontology in our model. CompRE is deployed as a plug-in for the Eclipse Integrated Development Environment (IDE). Figure 18.1 shows a snapshot of CompRE's main views in the Eclipse workbench.

When loaded for the first time, CompRE processes the library code and the component ontology in order to generate the initial knowledge base as described in Sect. 18.2.3. This process is completely automatic and efficient (it only took 4.5 s for parsing and processing the Jena framework). CompRE also includes a module that allows users to tag components with semantic references that correspond to concepts from the domain ontology. These annotations entered via drag and drop mechanisms, captured by the storage module, and stored automatically in the knowledge base. Upon the conclusion of the knowledge population process, a knowledge repository is created and becomes ready for answering user requests.

CompRE provides two separate views for formulating queries. The first view is provided as a simple data entry form as shown in the figure. In each entry box, users need to provide search restrictions that are either prefixed with an ontology name or provided as plain keywords enclosed within quotes. As described in the previous section, using the `compre-kb` prefix tells the system that this is in fact an actual API type specified in the knowledge base. However, the `swonto` prefix refers to a concept from the currently active domain ontology. Since the domain ontology can be different for each API, its name is provided as an external configuration parameter. Free-text requirements in each query may optionally utilize all fuzzy extensions supported by the Lucene's query parser, thus allowing a full-featured keyword-based search.

Once the form is filled, CompRE collects the search requirements and automatically generates the necessary SPARQL query that is executed against the knowledge base. CompRE also provides a query answering view for editing and executing regular SPARQL queries. This feature can be used by advanced users who wish to have more flexibility and full control over various aspects of the component ontology. For instance, users may wish to specify that the desired component `extends` a particular component or perhaps `usedBy` a certain number of components as a measure of its popularity in the target library. Regardless of the data entry mechanism used, CompRE executes the query, ranks the retrieved instances, and presents the result in a viewer that enables further exploration of each recommended component.

Ranking and ordering the retrieved candidates according to their relevancy to the current programming task is extremely important. It enhances user experiences with the system and saves, in some cases, navigating through many retrieved instances. While allowing blended queries in our approach ensures flexibility and robustness, it however complicates the ranking process. When blended or pure syntactic queries are submitted, we initially rely on the traditional, however solidly proven, information retrieval scoring mechanisms which are supported by Lucene. In order to improve this initial ranking, we refine this initial order based on suitability measures that consider the current user context. We thus parse the code that is currently being developed and create a context profile that includes all visible types that are either declared by the programmer or inherited in the user's context. We further analyze each retrieved candidate's signature in terms of the new input types that this candidate will introduce into the current context if selected by the user. Naturally, candidates that introduce more types should be assigned a lower rank value. However, finding the newly introduced type in the context profile will not count against this candidate's score.

With the absence of keywords in user queries, we apply only context-based heuristics such that candidates with exact matches are put at the top of the list while other candidates are ranked based on the number and type of their input and output types. For example, consider a user who is trying to search for a component that requires two particular input types, namely I1 and I2. Assume that the repository contains three components, namely C1, C2, and C3. Lets also assume that C1 is an exact match, C2 has only one input type (I1), and C3 requires three input types

(I1, I2, and I3). The system then ranks C1 first, C2 second, and C3 is in fact the least desired since it will introduce a new type to the user context, it is thus included in the result set to improve recall, however, ranked last.

These heuristics are simple, easy to implement, and work surprisingly well. It is, however left as future work the investigation of using other approximation techniques that are based on, for example, Natural Language Processing or Machine Learning, to improve ranking by capturing complex user patterns.

## 18.4 Experiments and Results

In this section, we report on two primary experiments we have conducted in order to assess the benefits acquired by our approach. Due to the lack of independent and standard benchmark test data, the evaluation of component search tools is generally subject to one's interpretation of what makes a good experiment. However, we designed our experiments such that they increase our confidence level of a fair evaluation. We have selected the Jena framework for testing CompRE. The domain ontology described in Sect. 18.2.1 fits naturally in the Jena's application domain. Moreover, our familiarity with Jena allows us to inform the study with tasks that are designed to test different aspects of our approach. In the next subsections, we discuss our findings.

### 18.4.1 Searching for Components

This experiment is designed to reveal the overlap between various kinds of searches supported by CompRE. The fundamental guiding hypothesis we test in this experiment is that pure semantic-based representation and annotation of library components improve search precision when compared with other techniques. Precision is defined as the ratio of the number of *relevant* component instances that are recommended by the tool to the total number of recommended instances. A recommended instance is considered relevant if it can be conveniently used to complete a certain programming task. Recall is the other commonly used metric in evaluating search systems, it is defined as the ratio of the number of *relevant* component instances that are recommended to the total number of relevant components in the repository. However, in these experiments we fix recall since there is typically one component in our repository that is relevant to the current task at hand. In fact, since we are focused on API reuse where our repository is an actual representation of the library itself rather than a corpus of sample code, recall values do not add a value to the evaluation. i.e, the component we are searching for is either found or not found.

We have selected twelve programming tasks. Six of these tasks were carefully designed by us and the remaining tasks were collected from the Jena developers

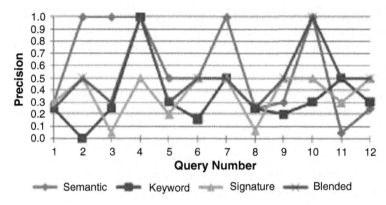

**Fig. 18.2** Precision of different kinds of searches supported by CompRE

forum[6]. Each of these tasks requires a query to be fired in order to search for a component that is required to complete the task. These tasks are diverse enough and cover various aspects of the problem domain. For space limitations, we do not include the tasks here, rather, an extended discussion of the tasks and results can be found in our website.

We then prepared the necessary coding environment and formulated four search queries for each task, i.e, one query for each kind of search supported by CompRE. Precision summery graph for running these queries is shown in Fig. 18.2. Results show a clear indication that pure semantic search yields better performance. In fact, precision values for semantic search could have been improved by utilizing all possible search properties provided by the component ontology. We have mainly focused on input/output properties in most cases.

As expected, semantic search tends to perform poorly when the components in the knowledge base are incompletely or incorrectly annotated. In Q11 for example, a required Jena method, `makeRewindable` was not completely annotated with the proper return type. Thus, search produced spurious components since only the input type was used during search. However, in most cases when proper tags exist, semantic search can precisely describe the needed component and improves overall precision values as seen in Fig. 18.2.

Metadata keyword-based search performs poorly due to the two well-known fundamental issues of polysemy and synonymy. Polysemy affects precision due to the diversity of meanings of some keywords. Synonymy, on the other hand, impacts recall since multiple keywords can have the same meaning, thus, increasing the chances of missing relevant components. These two problems become even more evident when searching for software components. This is due in part to inconsistent and often incomplete API descriptions of library code. Even when these descriptions exist, users who wishes to reuse the library are required to describe their requests

---

[6]http://tech.groups.yahoo.com/group/jena-dev/

using proper terms. What is needed to alleviate these problems is, in fact, a common vocabulary provided by ontologies to bridge the gap between the service provider and the service requester. Nevertheless, keyword search tends to return an exact match when a particular keyword is used to describe only a single component in the library (e.g., *clone* in Q4).

Signature based queries tend to yield low precision in cases where the component signature includes one or more semantically vague types such as the Java String type. The best example to illustrate this notion is Q3. In order to access a remote query service, one needs to provide, among other things, the URL of this service as well as a textual representation of the query; both of which are specified as String objects. Unless there is a clear semantic descriptions for such parameters, matchmaking would return many false positives (low precision).

As limited as it might be, blended search performed surprisingly well. We believe that it is often the case that the user is certain about a single API type that is used in the component's signature or a certain keyword that precisely describes some aspect of the component. These descriptions can also be coupled with semantic annotations to produce higher precision. These results show a clear indication that blended search needs to be investigated in more details.

Ranking as well as running time analysis have been computed as well. On average, semantic search achieved 1.75 rank over twelve queries, i.e., the desired component was ranked either at the first or at most at the second position. This ranking score outperforms other search schemes in which they achieved 2.27, 1.8, and 1.66 for keyword, signature, and blended search, respectively. However, the average observed response times were 15.5, 9, 2.5, for semantic, blended, and signature and keyword searches, respectively. All experiments were performed on a Windows XP machine with 1.8GHZ Intel processor and 1GB memory. This relatively lower performance for semantic queries is a result of having our queries run through the reasoner. In general, search time for all search methods is susceptible to increase as the size of the knowledge base increases; in the case of semantic search, speed is continuously improving as reasoners evolve. Achieving perfection in component search is near impossible, however, we believe that CompRE's internal mechanisms proved effective, and in the majority of cases, show a clear support for our hypothesis.

## 18.4.2 User Study

This experiment is designed to asses the usability of ontology-based search. Ontology creation is typically a subjective process and the people who create the domain ontology are usually not the end users of our system. Therefore, we designed this case study to understand the possible difficulties faced by end users in learning and using domain ontologies for successfully completing a particular search task against an unfamiliar API.

Six graduate-level MIS students from Northwestern University have voluntarily agreed to participate in this study. Although all students have at least seven months of Java experience and a good working knowledge of Semantic Web technologies, no student has been directly exposed to the Jena API. We delivered a one-hour tutorial that includes a brief introduction to the Jena API, brief introduction to our domain ontology, and a sample training task that explains CompRE's semantic search features. We then charged each student with other four independent Jena API programming tasks that vary in scope: *T1*) Data set creation and handling of multiple RDF gtaphs. *T2*) Query construction and execution over a given ontology model. *T3*) Result manipulation of query solutions; and *T4*) Access and treatment of remote services.

On average, two distinct components are needed to successfully complete a given task. An environment is setup for each task with a skeleton code, each student is then asked to finish three *consecutive* tasks using only CompRE's pure semantic search while the last task in the sequence must be completed using other alternative methods of student's choice. This last task was not included for the sake of comparison, instead, we wanted our volunteers to make a fair judgment in their feedback. For each participant, we recorded the actual time taken to complete each task, the final solution, and the queries used during their search. Finally, we asked the participants to fill a usability survey about their overall experience with semantic search and CompRE.

Table 18.1 shows task completion time measured from the time in which the task is presented to the student until the final correct answer is submitted. Numbers appeared in boldface represents tasks completed using semantic search while other numbers are underlined.

The most significant conclusion we can draw from these numbers is the correlation between the time taken by students to complete the first semantic task and the last one in the sequence. In most cases, there was a significant reduction in time as students became more familiar with the domain ontology and thus more able to construct more precise queries. This is confirmed by the responses we obtained upon the conclusion of the experiment. Four out of six students indicated that there is a small initial learning curve that was reduced fairly quickly as they became more

**Table 18.1** User study statistics

|  | Time (Minutes) | | | |
|---|---|---|---|---|
|  | T1 | T2 | T3 | T4 |
| S1 | 35 | 22 | 27 | 13 |
| S2 | 33 | 35 | 20 | 11 |
| S3 | 45 | 43 | 31 | 33 |
| S4 | 32 | 40 | 22 | 15 |
| S5 | 52 | 12 | 43 | 8 |
| S6 | 21 | 16 | 18 | 40 |
| *Avg (Semantic)* | 37.5 | 21.25 | 23.6 | 16 |
| *Avg (Alternative)* | 34 | 41.5 | 43 | 40 |

comfortable with the API vocabulary represented in the domain ontology. Since the last task has to be done without CompRE's assistance, five students argued that this task could have been completed faster had CompRE's assistance been allowed.

In the context of semantic search, a domain ontology provide a concise description of API content and vocabulary. This knowledge can be used by users not only to search for a particular component but also to successfully complete a coding task that may require more than one component. Consider task T2 for example, a programmer need to instantiate an intermediate object of type QueryExecution in the process of completing this task. We suspect that the coherent representation of ontology concepts and axioms aid users in arriving at such conclusions during the initial time invested in understanding and learning the taxonomy.

It is also natural to conclude that completing the last task (alternative task) should be relatively easier. After all, students have been using the same API, thus, the knowledge gained about this API after completing the first three tasks can be helpful. However, when examining the results, we did not see a dramatic improvement in response time for using alternative methods. We suspect that these alternative methods (e.g., exploring documentation, searching for code in the Web, etc.) do not provide a systematic and focused learning experience for programmers. In this study however, we did not intend to make a systematic comparison between semantic search mechanisms with usual search practices used by programmers. However, the obtained results clearly support our hypothesis and indicate that semantic search, in most cases, seems to yield better API learning experience and can certainly increase programmer's productivity.

Overall, nearly all participants provided positive comments about CompRE and semantic search. Five of the six students indicated that semantic search is effective and extremely precise. Some students suggested that it would be helpful if they had the chance to browse and explore the ontology in more details before started the coding tasks. Two students indicated that the SPARQL query view was indeed helpful and used in formulating more complex queries. However, these students requested a thorough integration of the domain ontology including its object properties into the CompRE's domain ontology view. Only one student reported a relative difficulty adapting to a new search approach after being familiar with other alternative methods. This user also requested a Tooltip feature such that when the user hovers over an ontology concept in the domain ontology view, a hover box appears with class description. Based on this sound and helpful feedback, we are currently adding new features as well as modifying CompRE's interface so it becomes more expressive by utilizing the full range of features provided by our component ontology.

## 18.5  Related Work

Due to the benefits acquired by systematic reuse, many researchers have tackled the reuse problem from various perspectives. Many approaches (e.g., Sourcerer [2], Assieme [7]) base their recommendation on analyzing a corpus of source-code structures of code downloaded from the Web. These tools employ traditional knowledge representation and variations of signature matching or keyword-based retrieval. Similar to CompRE, other tools (e.g., CodeBroker [17], Hipikat [3], Rascal [13]) leverage software understanding by being embedded in the development environment. However, unlike CompRE, these tools rely on a local repository of sample client code to search for components. CodeBroker [17] for example, use a combination of free-text and signature matching techniques. In order to retrieve appropriate matches, a CodeBroker's user must write high quality doc comments that precisely describe functionality. If the user comments did not retrieve satisfactory results, the system considers the signature of the method immediately following the comments. Finding a well documented code to populate the repository with is highly unlikely, especially in open source and legacy software.

The major problem with all previously mentioned IR-based systems is that precision usually suffers due to the knowledge representation and matching mechanisms used. All IR-based systems utilize traditional knowledge representations and hard-coded heuristics, a more flexible knowledge representation, e.g., based on Semantic Web technologies, will not only improve the possibilities to integrate and infer more knowledge, but it will also make the system's behavior more transparent by leveraging software understanding at higher levels [6].

Other component retrieval approaches (e.g., CodeGenie [10], Extreme Harvesting [8], and most recently, Code Conjurer [9]) apply automated testing techniques to analyze a corpus of client code harvested from the Web. Code Conjurer for example, helps agile development users in finding suitable components on the basis of unit test-cases. Therefore, users of the system has to write such test cases in order to invoke the system. Once invoked, the system contacts a remote server that finds suitable candidates based on the component's interface specified in the test case.

There are other approaches that base their retrieval on semantic techniques. However, the full potential of utilizing domain knowledge was not explored. Sugumaran and Storey [15] proposed an approach that utilizes domain models; a domain ontology is used mainly for term disambiguation and basic query refinement for keyword-based queries; these keywords are then mapped against the ontology to ensure that correct terms are being used. However, the ontology model used is rudimentary due to the lack of axioms and complex relationship modeling. Moreover, no semantic-based descriptions of the components have been used. Similarly, other proposals ([1] and [4]) employ ontologies to deal with the knowledge representation problem found in previous approaches. In [4], software assets are classified into domain categories (I/O, GUI, Security, etc.) and indexed with a domain field as well as other bookkeeping fields to facilitate free text search. Although the SRS [1] proposal uses similar indexing mechanism, it maintains two separate ontologies; an ontology for describing software assets as well as a domain ontology for classifying these assets. However, the structure of the source code assets and the semantic relationships between those assets via axioms and role restrictions were not fully utilized.

## 18.6 Conclusions and Future Work

We proposed an approach for component reuse. In addition to supporting pure semantic-based search, our approach also supports other kinds of search techniques. However, our studies showed evidence that pure semantic search that utilizes domain knowledge not only usable and achievable, but also improves precision of search results. Our results also showed that blended search has a great potential, we are currently conducting more case studies to asses the value of blended search. There are also two other future work directions. Firstly, ranking reused candidates has always been a challenge, therefore, we are currently investigating how could ranking be improved using semantic technologies. Secondly, we have not yet investigated how could one motivate library providers to ship domain ontologies with their software, or how could individually created ontologies be shared by a community of users.

## References

1. Antunes, B., Gomez, P., Seco, N.: SRS: A software Reuse System Based on the Semantic Web. In: 3rd International Workshop on Semantic Web Enabled Software Engineering (SWESE) (2007)
2. Bajracharya, S., Ossher, O., Lopes, C.: Sourcerer: An Internet-Scale Software Repository. In: First International Workshop on Search-driven Development: Users, Infrastructure, Tools and Evaluation (SUITE'09) (2009)
3. Cubranic, D., Murphy, G.C., Singer, J., Booth, K.S.: Hipikat: A project memory for software development. IEEE Trans. Software Eng. 31(6), 446–465 (2005)
4. Durao, F.A., Vanderlei, T.A., Almeida, E.S., Meira, S.R.: Applying a semantic layer in a source code search tool. In: 23rd ACM Symp. Appl. Comput. 1151–1157 (2008)
5. Frakes, W.B., Kang, K.: Software reuse research: Status and future. IEEE Trans. Software Eng. 31(7), 529–536 (2005)
6. Happel, H.J., Maalej, W.: Potentials of Recommendation Systems for Software Development. In: International workshop on Recommendation Systems for Software Engineering, pp. 11–15 (2008)
7. Hoffmann, R., Fogarty, J., Weld, D.S.: Assieme: Finding and Leveraging Implicit References in a Web Search Interface for Programmers. In: ACM Symposium on User Interface Software and Technology, pp. 13–22 (2007)
8. Hummel, O., Atkinson, C.: Extreme Harvesting: Test Driven Discovery and Reuse of Software Components. In: The IEEE International Conference on Information Reuse and Integration (IRI'04), IEEE Press, pp. 66–72 (2004)
9. Hummel, O., Janjic, W., Atkinson, C.: Code conjurer: Pulling reusable software out of thin air. IEEE Software. 25(5), 45–52 (2008)
10. Lemos, O., Bajracharya, S., Ossher, J., Morla, R., Masiero, P., Bald, P., Lopes, C.: CodeGenie: Using Test-Cases to Search and Reuse Source Code. In: Extended Abstract in the International Conference on Automated Software Engineering, pp. 525–526 (2007)
11. Musen, M.: Modern Architecture of Intelligent Systems: Reusable Ontologies and Problem-solving Methods. In: The American Medical Informatics Association annual symposium (AMIA'00) (2000)
12. McBride, B.: Jena: A semantic web toolkit. IEEE Internet Comput. 6(6), 55–59 (2002)

13. McCarey, F., Cinneide, M.O., Kushmerick, N.: Rascal: A recommender agent for agile reuse. Artif. Intell. Rev. **24**(3), 253–276 (2005)
14. Noy, F.N., McGuinness, D.L.: Ontology Development 101: A Guide to Creating Your First Ontology. Stanford Knowledge System Laboratory Technical Report KSL-01-05 and Stanford Medical Informatics Technical Report. SMI-2001-0880 (2001)
15. Sugumaran, V., Storey, V.C.: A semantic-based approach to component retrieval. ACM SIGMIS DATABASE. **34**(3), 8–24 (2003)
16. Uschold, M., Healy, M., Williamson, K., Clark, P., Woods, S.: Ontology Reuse and Application. In: The 1st International Conference on Formal Ontology in Information Systems (FOIS'98) pp. 179–192 (1998)
17. Ye, Y., Fischer, G.: Reuse-Conductive development environments. The Int. J. Automated Software Eng. **12**(2), 199–235 (2005)

# Chapter 19
# The Utility of Inconsistency in Information Security and Digital Forensics

Du Zhang

## 19.1 Introduction

Inconsistency in knowledge, information and data is ubiquitous. It turns out that not all inconsistencies are bad and useless, some are even desirable [27]. The utilities of inconsistency in information security and digital forensics are the focus of this research work. The point we want to drive home is that inconsistency can be a very effective tool in accomplishing the objectives in the aforementioned areas.

Inconsistency is an important, complex, and multi-dimensional phenomenon, and has been studied in the following individual areas of information security and digital forensics: firewalls, intrusion detection systems, deception based defensive mechanisms, and digital image forensics, in an unrelated fashion [2,4,10,12,14,16, 18,20,21,24,25]. There has not been a concerted effort to bring results obtained in those individual information security areas into a single context and setting to foster cross fertilization of ideas and approaches that will lead to more general and better solutions to the issues we have at hand. This paper, an extended version of [28], is meant toward that goal. Instead of addressing the technical details, we bring the utility of inconsistency in those areas into focus.

Inconsistencies manifest themselves in a variety of ways in information security and digital forensics. How they can be utilized as effective tools in accomplishing the objectives in these areas also differs significantly. In this paper, we first take a close look at how inconsistencies arise in firewalls and propose an algorithm to detect several types of firewall inconsistency. Compared with related work, our approach is based on decomposing a system-wide set of firewall rules into smaller and relevant groups of rules according to possible policy actions and traffic types to improve detection effectiveness. We then examine how different types of

Du Zhang (✉)
Department of Computer Science, California State University, Sacramento,
CA 95819-6021, USA
e-mail: zhangd@ecs.csus.edu

T. Özyer et al. (eds.), *Recent Trends in Information Reuse and Integration*,
DOI 10.1007/978-3-7091-0738-6_19, © Springer-Verlag/Wien 2012

inconsistency can be utilized in aiding the intrusion detection process. A special type of inconsistency called *setuid* inconsistency is defined and investigated next. Setuid inconsistency is very much a double-edged sword in operating system (OS) access control mechanisms in that it allows user processes to gain elevated privileges for OS services, but at the expenses of creating potential security vulnerabilities when the setuid programs are not properly developed. Inconsistent deceptions, an emerging area of research, add another defensive technique in our toolbox for information assurance and security. Finally we discuss inconsistency in digital forensics with emphasis on inconsistency in digital image forensics.

The rest of the paper is organized as follows. A brief overview of the related work is given in Sect. 19.2. Section 19.3 investigates the firewall inconsistency and describes a new approach in capturing inconsistent firewall rules. Section 19.4 examines several different types of inconsistencies in the context of intrusion detection systems. Section 19.5 defines and discusses the setuid inconsistency. Section 19.6 highlights a new area in information security, i.e., using inconsistent deceptions as an effective defensive technique to fend off attackers. In Sect. 19.7, we discuss how inconsistency can be used in digital forensics as an effective tool, in particular, in detecting forged digital images. Finally Sect. 19.8 concludes the paper with remark on future work.

## 19.2  Related Work

Because our intent is to bring results in several areas in information security and digital forensics into focus under the main theme of the utility of inconsistency, we will look at related work from several different perspectives.

There have been extensive studies on network security in general and firewalls in particular. Results in [8] highlighted the complexity and error-prone nature in configuring reliable and consistent network security policies due to the semantics in ACLs and interactions in network security devices. A comprehensive classification of network security policy conflicts was offered in [8]. A static analysis toolkit was described in [25] to detect firewall anomalies such as security policy violations, inconsistencies, and inefficiencies in both individual firewalls and distributed firewalls. The toolkit was implemented through modeling firewall rules with binary decision diagrams. The work in [2] described a formal correctness proof process for a conflict detection algorithm for firewall ACLs. The theorem proving tool called the Coq Proof Assistant was used in the correctness proof of the algorithm. A structured firewall design methodology was proposed in [7] that was based on designing a firewall using a formalism called the firewall decision diagram (FDD) first and then translating the FDD into a compact and functionally equivalent ACL. A heuristic process was proposed in [15] for detecting local inconsistent firewall rules. The process was broken down into a sequence of inconsistency detection and isolation, inconsistent rule identification, and inconsistency characterization, and had a quadratic time complexity.

In the area of the intrusion detection process, inconsistency has been utilized as one of the effective tools to identify intrusions. There are signature based and anomaly detection based approaches to intrusion detection systems (IDS). Numerous results in IDS have been reported in the literature. In this paper, we only focus our attention on a subset of the results to highlight the role inconsistency plays in the intrusion detection process. The results in [20] indicated the presence of inconsistency in IDS rules for signature based detection. Inconsistency in system calls can also be indicative in intrusions [17, 19]. Data flow inconsistency played a role in intrusion confinement in [5] once an intrusion was identified. Inconsistency in file systems is another important factor in IDS [9, 14].

How to write setuid programs in operating systems has been a contentious area almost since the concept was introduced in [16] in 1979. Over the past thirty years, issues regarding setuid programs have never ceased to challenge us [1, 3, 12, 18, 21]. In this paper, we define and introduce the concept of setuid inconsistency that we think will play an important role in developing secure system software. To the best of our knowledge, this is the first time that setuid inconsistency is defined.

Inconsistency in deception was a new concept initially proposed in [13]. Though still needing to be further developed, the promise of taking full advantage of inconsistency as a defense strategy in deceiving attackers can never be underestimated.

Finally in digital forensics, inconsistency has been playing an increasingly important role in helping detect digital forgeries. In this paper, we only focus our attention on inconsistency in digital images. The results in [4, 8] highlighted the lighting inconsistency in digital images. Inconsistency of noise level was used to identify digital image forgeries in [11]. The work in [23] described how inconsistency of blocking artifacts in JPEG images was the basis for detecting tampered images.

## 19.3   Inconsistency in Firewalls

Firewalls are a cornerstone to network security. A firewall consists of hardware and software components that collectively establish a controlled interface for traffic flows between an organization's internal network and the Internet at large [10]. In addition to the high-level security policy mechanisms through *blacklists* and *whitelists*, there are low-level *access control lists* (ACL) a firewall uses to define filtering patterns for authorized traffic flows to go through in both directions and fend off illegitimate attempts to access. Typically, an ACL contains a number of rules that have the following format:

| ID | Act | Protocol | Src_IP | Src_port | Des_IP | Des_port |

where IDs embody the priority, Protocol indicates the type of protocol involved, Src_IP and Src_port refer to the source IP address and port number respectively,

**Table 19.1** Sample ACL

| ID | Act | Protocol | Src_IP | Src_Port | Des_IP | Des_Port |
|----|-------|----------|-------------|----------|-------------|----------|
| 1 | deny | tcp | 192.168.1.1 | any | 10.37.2.* | 21 |
| 2 | allow | udp | *.*.*.* | any | 192.168.1.0 | 24 |
| 3 | deny | tcp | 192.168.1.1 | any | 10.37.2.* | 22 |
| 4 | deny | udp | 172.16.1.0 | 24 | 192.168.1.0 | 24 |
| 5 | allow | tcp | 192.168.1.1 | any | 10.37.2.3 | 21–22 |
| 6 | deny | udp | 10.1.1.0 | 24 | 192.168.0.0 | 16 |
| 7 | allow | udp | 172.16.1.0 | 24 | 192.168.*.* | <100 |

Des_IP and Des_port are the destination IP address and port number, and Act is the action taken by the firewall regarding the traffic: e.g., allow, or deny.

Though it is a common practice to have several rules target a specific traffic pattern, in reality only the first applicable rule from the top of the rule base matters because of the top-down priority and the first-matching semantics (the first matching rule matters) in the firewall execution mechanism, thus neutralizing the remaining applicable rules. There are firewalls that rely on the last-matching semantics (the last matching rule applies). In this paper, our focus is on inconsistencies arising from the first-matching ACLs. Inconsistency arises when firewall rules have conflicting policy actions with regard to the same traffic flow. For instance, in the ACL shown in Table 19.1, there are a number of conflicting circumstances as follows:

1. Rule 4 and rule 2 are conflicting because rule 4 denies UDP packets from 172.16.1.0 to 192.168.1.0, which is allowed by rule 2. This is problematic because rule 2's priority is higher than that of rule 4, hence UDP packets always come through from 172.16.1.0 to 192.168.1.0, completely neutralizing the effect of rule 4 which attempts to block a specific illegitimate traffic. In this case, a more specific rule should be placed before a more general rule.
2. On the other hand, rule 5 conflicts with the combination of rule 1 and rule 3.
3. Since the scope of rule 7 is a generalization of that of rule 4 (or rule 4 represents an exception to rule 7), and the two have antagonistic actions, hence the two are inconsistent.[1]
4. There exists an intersection between the address spaces in rule 2 and rule 6 for UDP traffic with inconsistent actions, thus presenting yet another type of inconsistency.

The results in [7] defined the following types of firewall rule inconsistency: *shadowing* inconsistency (cases 1 and 2), *exception* inconsistency (case 3), and *correlation* inconsistency (case 4). These types of inconsistency can occur in both a single-firewall (intra-firewall or intra-policy) and multiple-firewall (inter-firewall or inter-policy) settings [7, 24] (Table 19.2).

---

[1] According to [7], exceptions to general filtering decisions may not be errors. This type of inconsistency thus serves as a warning and needs to be highlighted to firewall administrators.

**Table 19.2** Types of firewall inconsistency

| | | Single firewall | Multiple firewalls |
|---|---|:---:|:---:|
| Inconsistency | Shadowing | ✓ | ✓ |
| | Exception | ✓ | |
| | Correlation | ✓ | |
| | Spuriousness | | ✓ |

How to devise network security policies and design firewall ACLs in accordance to the policies, and how to properly arrange rules in a firewall ACL so as to avoid cases of inconsistent rules are important issues. These issues, however, are beyond the scope of this paper as our attention is focused on how to detect inconsistent rules once they get introduced into a firewall ACL.

Since the actions such as allow and deny in a firewall ACL are syntactically different and semantically opposite of each other, they can be represented as a set of mutually exclusive and jointly exhaustive predicates [25, 26].

**Definition 1.** For a firewall rule that has its id (priority) $i$ and that allows or denies traffic type $p$ from $sip/sp$ to $dip/dp$, we can introduce the following first order atomic formula with predicates allow/deny to represent it[2]:

allow($i$, $p$, $sip$, $sp$, $dip$, $dp$)

deny($i$, $p$, $sip$, $sp$, $dip$, $dp$)                                                    □

Hence, the essence of a rule is captured by its corresponding first order formula where the predicate denotes rule action and terms represent firewall rule id, protocol, source IP, source port, destination IP, and destination port, respectively. The assertion of allow($i$, $p$, $sip$, $sp$, $dip$, $dp$) implies the negation of deny($j$, $p$, $sip$, $sp$, $dip$, $dp$) and vice versa. Given a firewall and its ACL, we can rewrite rules in ACL in terms of atomic formulas through the allow and deny predicates and use $\Omega_{ACL}$ to denote the set of formulas for the firewall.

**Definition 2.** For two rules $R_j$ and $R_k$ with source and destination patterns $SD_j = <sip_j, sp_j, dip_j, dp_j>$ and $SD_k = <sip_k, sp_k, dip_k, dp_k>$, assuming $\alpha \in SD_j$ and $\beta \in SD_k$, we use $\alpha \leftrightarrow \beta$ to denote that $\alpha$ and $\beta$ are corresponding terms in $SD_j$ and $SD_k$, respectively. When $(\alpha \not\subseteq \beta) \wedge (\beta \not\subseteq \alpha)$, we use $\alpha \not\equiv \beta$ to denote the disjointedness between the two. We can define the following pattern matching scenarios according to [7]:

• *Exact matching*, denoted $R_j(SD_j) = R_k(SD_k)$, if we have:

$$(sip_j = sip_k) \wedge (sp_j = sp_k) \wedge (dip_j = dip_k) \wedge (dp_j = dp_k)$$

---

[2]The set of actions different firewall devices have may differ. Typical action sets include: {permit, deny}, {accept, deny}, or {discard, protect, bypass}. Different predicates can be utilized accordingly.

- *Inclusive matching*, denoted $R_j(SD_j) \subset R_k(SD_k)$, when we have:

$$[(\text{sip}_j \subseteq \text{sip}_k) \wedge (\text{sp}_j \subseteq \text{sp}_k) \wedge (\text{dip}_j \subseteq \text{dip}_k) \wedge (\text{dp}_j \subseteq \text{dp}_k)]$$

$$\wedge [\exists \alpha \in SD_j \exists \beta \in SD_k ((\alpha \leftrightarrow \beta) \wedge (\alpha \subset \beta))]$$

- *Correlating matching*, denoted as $R_j(SD_j) \bowtie R_k(SD_k)$, if we have:

$$[\exists \alpha \in SD_j \exists \beta \in SD_k ((\alpha \leftrightarrow \beta) \wedge (\alpha \neq \beta) \wedge (\alpha \cap \beta \neq \emptyset))]$$

$$\wedge [\forall \mu \in (SD_j - \alpha) \forall \lambda \in (SD_k - \beta)(\mu \subseteq \lambda)]$$

- *Disjoint*, denoted $R_j(SD_j) \not\equiv R_k(SD_k)$, when we have:

$$\forall \alpha \in SD_j \forall \beta \in SD_k [(\alpha \leftrightarrow \beta) \wedge (\alpha \neq \beta)] \qquad \square$$

**Definition 3.** Given $\Omega_{ACL}$ for a firewall, we can decompose $\Omega_{ACL}$ into the following two subsets:

- $D = \{\text{deny}(\ldots)|\text{deny}(\ldots) \in \Omega_{ACL}\}$
- $A = \{\text{allow}(\ldots)|\text{allow}(\ldots) \in \Omega_{ACL}\}$

We further decompose D and A into the following subsets according to the protocol involved (assuming that there are $m$ different types of protocols):

- $D = D_{p1} \cup \ldots \cup D_{pm}$
- $A = A_{p1} \cup \ldots \cup A_{pm}$, where

$$D_{pi} = \{\text{deny}(\_, p_i, \ldots)|\text{deny}(\_, p_i, \ldots) \in \Omega_{ACL}\} \quad \text{and}$$

$$A_{pi} = \{\text{allow}(\_, p_i, \ldots)|\text{allow}(\_, p_i, \ldots) \in \Omega_{ACL}\} \qquad \square$$

**Definition 4.** The domains for firewall inconsistencies with regard to a particular type of traffic flow (protocol) can be defined as follows, where $m$ is the number of protocols:

- $\DJ(p_1) = D_{p1} \times A_{p1} \cup A_{p1} \times D_{p1}$

. . . . . . . . .

- $\DJ(p_m) = D_{pm} \times A_{pm} \cup A_{pm} \times D_{pm}$ $\qquad \square$

Algorithm 19.1 describes the process that captures intra-policy inconsistencies in firewall rules. The algorithm can be used to detect and identify shadowing, correlation, and exception inconsistency as defined in [7]. Using the algorithm, we

Input:              $\Omega_{ACL}$, **D**, **A**, and $m$ protocols involved in firewall rules;
Output:             Conflict$_{ACL}$;     //total number of conflicting cases for ACL

Conflict$_{ACL}$ = Ø;
Conflict$_{shad}$ = Ø;
Conflict$_{excep}$ = Ø;
Conflict$_{corre}$ = Ø;

**for** (i=1; i<= $m$; i++) {
 $\forall d \in D_{pi} \exists a \in A_{pi}$ {
  **if** [[(d = deny($j, p_i, sip_j, sp_j, dip_j, dp_j$) $\in D_{pi}$))
                 $\wedge$ (a = allow($k, p_i, sip_k, sp_k, dip_k, dp_k$) $\in A_{pi}$) $\wedge$ ($R_j(SD_j) = R_k(SD_k)$))]
      $\vee$ [(a = allow($j, p_i, sip_j, sp_j, dip_j, dp_j$) $\in A_{pi}$)
                 $\wedge$ (d = deny($k, p_i, sip_k, sp_k, dip_k, dp_k$) $\in D_{pi}$)) $\wedge$ (($R_j(SD_j) = R_k(SD_k)$))]]
  **then** { Conflict$_{shad}$ = Conflict$_{shad}$ $\cup$ {$R_j, R_k$} }

  **if** [[(d = deny($j, p_i, sip_j, sp_j, dip_j, dp_j$) $\in D_{pi}$))
                 $\wedge$ (a = allow($k, p_i, sip_k, sp_k, dip_k, dp_k$) $\in A_{pi}$) $\wedge$ ($R_j(SD_j) \bowtie CR_k(SD_k)$))]
      $\vee$ [(a = allow($j, p_i, sip_j, sp_j, dip_j, dp_j$) $\in A_{pi}$)
                 $\wedge$ (d = deny($k, p_i, sip_k, sp_k, dip_k, dp_k$) $\in D_{pi}$)) $\wedge$ ($R_j(SD_j) \bowtie R_k(SD_k)$))]]
  **then** { Conflict$_{corre}$ = Conflict$_{corre}$ $\cup$ {$R_j, R_k$} }

  **if** [[(d = deny($j, p_i, sip_j, sp_j, dip_j, dp_j$) $\in D_{pi}$))
                 $\wedge$ (a = allow($k, p_i, sip_k, sp_k, dip_k, dp_k$) $\in A_{pi}$) $\wedge$ ($R_j(SD_j) CR_k(SD_k)$))]
      $\vee$ [(a = allow($j, p_i, sip_j, sp_j, dip_j, dp_j$) $\in A_{pi}$)
                 $\wedge$ (d = deny($k, p_i, sip_k, sp_k, dip_k, dp_k$) $\in D_{pi}$))$\wedge$ ($R_j(SD_j) CR_k(SD_k)$))]]
  **then** { Conflict$_{excep}$ = Conflict$_{except}$ $\cup$ {$R_j, R_k$} }
 }
}

Conflict$_{ACL}$ = Conflict$_{shad}$ $\cup$ Conflict$_{corre}$ $\cup$ Conflict$_{excep}$
**return**(Conflict$_{ACL}$).                                                           □

**Algorithm 19.1:** Capturing inconsistent firewall rules

can identify the set of inconsistent cases for the sample ACL of Table 19.1 as shown
in Fig. 19.1 below.

Compared with the related work in [2, 15, 24], our approach has several salient
features. Because inconsistency arises only when two rules have antagonistic actions
(i.e., two rules both recommending "deny the traffic" or "allow the traffic" will
not be involved in an inconsistency), we decompose an entire set of firewall rules
into two subsets containing "allow" and "deny" recommendations, respectively.
Since inconsistency occurs when the same type of traffic is recommended with
two conflicting actions, we further divide "allow" and "deny" sets into smaller sets
with regard to different protocols involved. Then detection process only takes place
with relevant smaller groups of rules (see Fig. 19.2). For instance, an inconsistency
having a rule of "deny" action on a "tcp" traffic flow must involve another rule
of "allow" action with the same type (tcp) of traffic. Our proposed approach will
result in a smaller set of rules to work with, leading to a more effective detection
process.

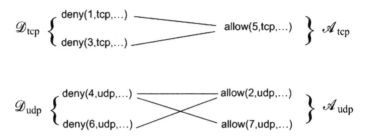

**Fig. 19.1** Cases of inconsistency for ACL in Table 19.1

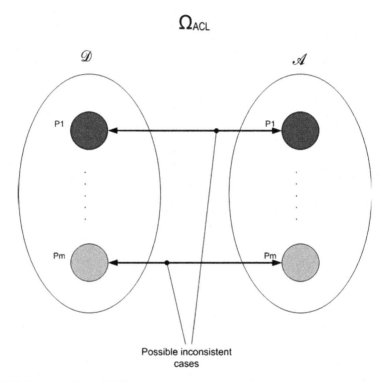

**Fig. 19.2** Decomposition of ACL.

Now we analyze the performance of the algorithm. Given the ACL for a firewall, assuming that it contains $n$ rules with two actions (allow and deny) and that rules with allow and rules with deny are equally divided in number ($\frac{n}{2}$). Assume that there are $m$ different protocols (or traffic patterns) for the ACL control and that the rules with allow and deny for controlling the $m$ protocols are equally split ($\frac{n}{2m}$) in both of sets $D$ and $A$. Fig.19.2 indicates the scenario.

The time complexities of checking for inconsistent ACL rules according to the patterns in Definition 2 can be examined in the following cases.

Without decomposing ACL rules according to actions and protocols first, the detection process is essentially a brute force comparison process that would take the following number of comparisons:

$$1 + 2 + 3 + \cdots + (n-1) = \frac{n(n-1)}{2} = \frac{n^2 - n}{2}$$

If we decompose the ACL rules based on just the possible rule actions (so as to take advantage of the necessary condition that two rules conflict when they have different rule actions), but without taking into consideration of different protocols involved, then the process takes the following number of comparisons, assuming that each of the sets $D$ and $A$ has $\frac{n}{2}$ rules:

$$\left(\frac{n}{2}\right)^2 = \frac{n^2}{4}$$

In our proposed algorithm, we recognize a second necessary condition for conflicting rules: two rules with different actions become inconsistent when they apply to the same protocol (traffic pattern). That is, inconsistent cases can only exist in subsets of rules with the same color in $D$ and $A$ (Fig. 19.2). Hence we decompose the ACL rules according to the scenario in Fig. 19.2 before we attempt to initiate the detection process. The time complexity of the proposed algorithm can be obtained as follows. For each of the $m$ protocols, it takes $\left(\frac{n}{2m}\right)^2$ comparisons to detect inconsistent rules. Since there are $m$ protocols, the total amount of effort is $m\left(\frac{n}{2m}\right)^2$ which translates into the following:

$$m\left(\frac{n}{2m}\right)^2 = \frac{n^2}{4m}$$

When the total number of actions involved in rules is more than two, the time complexities vary for the aforementioned cases. Table 19.3 summarizes the results.

**Table 19.3** Summary of time complexities

|  | $ACT = 2$ | $ACT = 3$ | $ACT = 4$ |
|---|---|---|---|
| No rule decomposition | $\dfrac{n^2 - n}{2}$ | $\dfrac{n^2 - n}{2}$ | $\dfrac{n^2 - n}{2}$ |
| Rule decomp. w.r.t ACT | $\dfrac{n^2}{4}$ | $\dfrac{n^2}{3}$ | $\dfrac{3n^2}{8}$ |
| Rule decomp. w.r.t ACT and PROTOCOL | $\dfrac{n^2}{4m}$ | $\dfrac{n^2}{3m}$ | $\dfrac{3n^2}{8m}$ |

## 19.4   Inconsistency in Intrusion Detection

Firewall ACLs offer packet filters that only allow legitimate traffic to pass. To identify various types of attacks, *deep* packet inspection is needed that goes beyond the header fields and into the actual contents that packets carry. Many IDSs developed so far are based on either *signature* identification or *anomaly* detection [10]. In this section, we offer an overview on the roles inconsistency plays in IDS.

In the signature-based IDS for known attacks, an attack signature is defined in terms of some detection rules. When inconsistencies arise in those rules, they introduce ambiguities into the depictions of host events and network traffic, compromise the IDS performance, and subject the host system to a vulnerable situation. IDS rules are different from firewall rules in at least the following areas [20]: (1) firewall rules often have a common format but IDS rules don't; (2) firewall rules usually have a small set of actions such as {allow, deny} whereas IDS rules may result in a whole host of possible responses; and (3) rule ordering plays a pivotal role in firewall rules, but not necessarily in IDS rules. The work in [20] reported two types of inconsistencies found in IDS rules: *inter-rule* inconsistencies and *intra-rule* inconsistencies. Inter-rule inconsistencies happen when network traffic enables two IDS rules that have conflicting responses or when an exception rule (a specific case) contradicts with a general rule (a general case). Intra-rule inconsistencies refer to conflicting circumstances where conditions of a single rule contradict with each other.

Intrusion detection can be carried out through an approach called *multi-variant execution* environment. The approach works through running several slightly different versions of the same program in lockstep, having a monitor comparing the behaviors of those variants at certain synchronization points (e.g., when a system call is made), and looking for inconsistencies which are indicative of attacks [17]. The monitor is implemented in user space, doing away with the need to change OS kernel [17].

In *host-based* IDS (as contrast to *network-based* IDS), traces of system calls have been utilized to detect intrusions. Anomalous subsequences of system calls not consistent with normal patterns will be evidence for malicious programs or unauthorized accesses [19]. The data are sequential in nature. The system call based detection approach relies on modeling the call sequences and establishing similarity measures between sequences.

Data flow inconsistency can be utilized as the basis for *intrusion confinement* [5]. Two data items $d_n$ and $d_m$ are said to be inconsistent if $d_n$ precedes $d_m$ and $d_m$ precedes $d_n$. Data inconsistency can be the consequence of malicious transactions. The detection process hinges on computing the precedence set for the data items involved. If $d_n$ appears in the precedence set of $d_m$ and $d_m$ in $d_n$'s precedence set, then a data inconsistency is found.

Detecting inconsistencies in file systems has been a hallmark of IDS since early nineties [9]. File inconsistencies arise when an intruder attempts to modify system utilities, alter file contents or attributes, and access unauthorized files [14]. There are tools that detect such file inconsistencies through periodic integrity checks. Some run in the user mode [9] while others in the kernel mode [14].

## 19.5   Setuid Inconsistency

A cornerstone to access control mechanisms for files, shared memory sections and semaphores in Unix or Linux systems is based on the notion of access permissions granted to individual users and groups of users. When a user logs in to a computer system, a user identification number (UID) and a group identification number (GID, or multiple GIDs) are generated for the user. UIDs establish the basis for accounting, restricted access to privileged kernel services, signal processing, file system access, and disk space allocation and GIDs also helps facilitate file system access and disk space allocation [12].

Processes initiated by users also have their respective UIDs and GIDs. Normally a process' UID and GID are determined either by its initiating user's UID and GID or by inheriting from its parent process. Similarly, a file also has its owner's UID and GID associated with it. If a process' UID matches the UID of a file, then the process has the *owner permissions* to the file. If the UID of a process does not match that of a file but the GID of the file matches one of the GIDs of the process, then the process has the *group permissions* to the file. If neither of the UID or GIDs of a process match those of a file, then the process has *others permissions* to the file.

In many circumstances, a user process may need to access files that require system privileges (e.g., password file when changing password, mail boxes when sending emails, print queues when submitting print jobs). To enable a user process to carry out such activities, a *setuid* mechanism has been developed to facilitate requests of user processes [1, 3, 16, 18, 21]. The mechanism works as follows:

- The kernel keeps two user identifiers for a process, the *effective UID* (euid) and the *real UID* (ruid). In normal circumstances, euid = ruid referring to the actual user's UID.
- Euid establishes the basis for the system to determine applicable file access permissions a user process is entitled with regard to a file.
- There is a set of programs called *setuid programs* belonging to users (possibly the user with UID = 0, i.e., the root) that have appropriate privileges in accessing pertinent files (e.g., the program that can change the password file or the program that can append an email to a user's inbox file).
- There is the *setuid bit* associated with a setuid program file. When the bit is set for a setuid program, a user process executing the setuid program (via one of the exec family of functions) will have its euid set to the UID of the owner of the setuid program file, resulting in ruid ≠ euid for the user process. The user process' ruid, however, remains to be its actual UID.
- During the time when ruid ≠ euid, the user process is running the particular setuid program to carry out some task (e.g., changing the password file as a result of password change) that requires elevated privileges. In other words, the setuid program is now being executed as part of the invoking user process which has an euid that allows an ordinary user process to enjoy some escalated

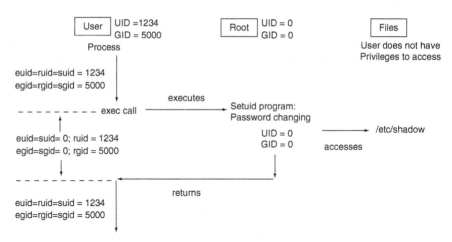

**Fig. 19.3** Temporal aspect of euid-ruid discrepancy

privileges in accessing an otherwise off-limit file. Figure 19.3 illustrates this circumstance.

- When the setuid program finishes executing, the (elevated) euid of the user process is reverted back to the value of its ruid, resulting in euid = ruid. So the user process returns to its normal privileges for file accessing (see Fig.19.3).

Now we are in a position to define *setuid inconsistency* as follows. Figure 19.4 highlights the condition under which ruid $\neq$ euid (suid in Fig. 19.4 refers to *saved uid* [12]).

**Definition 5.** Given euid and ruid for a process, setuid inconsistency refers to the circumstance when ruid $\neq$ euid for the process. We use the following to denote this type of inconsistency:

$$\Im_{setuid}(\rho, \mu, \phi, \Delta)$$

where $\rho$ is a setuid-bit enabled program, $\mu$ is a user process that (has the permission to execute and) is executing $\rho$, $\phi$ is the file that $\mu$ does not have privileges to access but can be manipulated by $\rho$, and $\Delta$ is the time period during which ruid $\neq$ euid for $\mu$.                                                                                                                                                             □

When Setuid inconsistency arises, there is a period ($\Delta$) during which a user process has acquired elevated privileges (ruid $\neq$ euid) in file accessing (Fig.19.3). Though setuid inconsistency allows a user process to have augmented privileges for OS services, it creates security vulnerabilities when the setuid programs are not properly developed. Setuid inconsistency related security vulnerabilities can stem from race conditions (e.g., lpr vulnerability), environment variables (e.g., PATH, IFS, LD_PRELOAD) related issues, use of publically writable directories (e.g., /tmp), resource exhaustion, buffer overflow, chsh, sendmail, or openSSH [1, 18, 22].

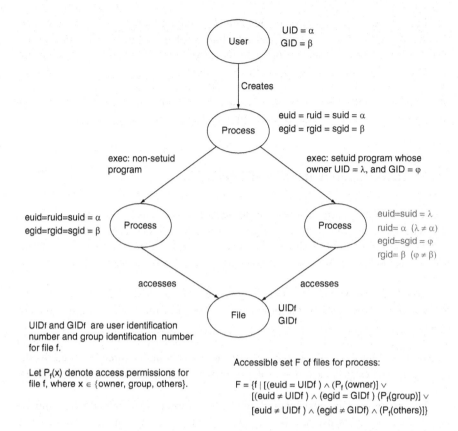

**Fig. 19.4** Setuid inconsistency

## 19.6   Inconsistency in Deception

In the field of information security, deception is one of the defensive techniques to fend off attackers. A deception is geared toward subjecting the attacker to perceiving a false reality known as a *fiction* [13]. There are consistent deceptions and inconsistent deceptions. A consistent deception constructs a fiction and tries to convince the attacker into believing the fiction. Examples include honeypots, honeynets, sandboxes, and so forth.

Inconsistent deceptions, on the other hand, use conflicting responses to the attacker's requests so as to confuse or disorient the attacker [13]. Inconsistent deceptions complement consistent deceptions as a useful defensive technique, and enjoy the benefit of not needing to keep track of the history of previous states or responses like the consistent deception approach does. It is generally the case that keeping a water-tight consistent deception, especially at a low level such as a raw device, is very challenging and costly.

There are *semantic* inconsistencies and *data* inconsistencies when inconsistent deceptions are deployed as a defensive technique [13]. Not only can data returned to the attacker be inconsistent, but also the response to the attacker's request can be semantically irrelevant and incoherent. Implementing this defensive measure may call for either *vertical* inconsistency (multiple methods to the same task return inconsistent results) or *horizontal* inconsistency (single method for the same task returns inconsistent results at different times) [13].

## 19.7   Inconsistency in Digital Forensics

The objective of digital forensics is to seek explanations to the state of digital artifacts. There are many areas in digital forensics: network and firewall forensics, database forensics, mobile device forensics, image forensics, and so forth. In this section, we take a look at how inconsistency can play an important role in image forensics.

Though digital watermarking and signature are two major techniques of image authentication, the state-of-the-practice today is that the bulk of digital images captured and stored do not come with either of the two. In the absence of such active measure, digital forgery detection has to rely on passive or blind approaches that work without prior information about the images. Usually the blind approaches take advantage of the presence of inconsistencies of different sorts.

The first type of inconsistency digital image forgery detection methods utilize is the *lighting* inconsistency [4, 8]. Digital image manipulations or tampering have become a common place. As a result, digital image forensics has in recent years emerged to offer a means to discern the fidelity of digital imagery [8]. In digital images, lighting inconsistency refers to a circumstance where shapes, colors, locations or directions, or number of sources of reflected light on objects are inconsistency. For instance, when specular highlights (small white specks of reflected light in people's eyes) in an image result in inconsistent lighting directions, number of sources, or shapes, it is evidence of image altering. So any presence of lighting inconsistency can be used to identify image tampering [4, 8].

The second type of inconsistency that digital image forgery detection methods can take advantage of is the *local noise level* inconsistency [11]. Normally in an authentic digital image, the noise is uniformly distributed across the entire image. Tools used to alter images often inject locally randomized noise to the forged regions so as to conceal the evidence of tampering, thus creating inconsistencies in the levels of noise in the image. The work in [11] described an approach that detects tampering through recognizing noise level inconsistencies in an image. The approach is based on decomposing an image into segments of various noise levels.

The third type of inconsistency that proves to be useful in digital image forgery detection is the *blocking artifact* inconsistencies [23]. In JPEG images, due to the fact that digital camera manufacturer and image processing software often utilize different JPEG quantization tables to balance between compression ratio and image quality, different blocking artifacts will be injected into the images as a result. When a digitally forged image is created from several sources, the resulting image invariably contains different sorts of compression artifacts. In addition, many image manipulation operations such as image splicing, re-sampling, or skin optimization, will generate differential blocking artifacts. These blocking artifact inconsistencies can serve as an effective tool for detecting tampered images.

## 19.8   Conclusion

In this paper we highlighted various types of inconsistencies in areas of information security and digital forensics and described how they can be utilized as effective tools in accomplishing the objectives in these areas. Our intent is not meant to address the technical issues in depth in all areas of information security and digital forensics. Rather, our goal is to bring these otherwise isolated results from different information security areas into a single context to raise the awareness of the role inconsistencies play in these areas, and to foster cross fertilization of ideas and approaches that hopefully can lead to better overall solutions to the security and forensics issues we have to contend with. The take-home message is that inconsistency is a very important and powerful phenomenon and its utilities can never be underestimated in our research endeavors.

The contributions of this research work are threefold. (1) We described an algorithm to detect several types of firewall inconsistency. Our proposed firewall inconsistency algorithm has some unique features in decomposing a system-wide set of firewall rules into smaller and relevant groups of rules according to possible policy actions and traffic types, and detecting inconsistencies with regard to relevant subsets of rules. Though empirical study is still forthcoming, the preliminary results based on qualitative comparison with related work indicate that our approach is more effective. (2) We defined the setuid inconsistency, and called attention to the setuid inconsistency related security vulnerabilities. (3) We highlighted the utilities of various types of inconsistencies in IDS, deception based defensive mechanism, and digital image forensics as having been reported in the literature.

Future work can be pursued in the following directions. (1) Implementing the firewall inconsistency detection algorithm and collecting its performance data. (2) Cataloging additional types of inconsistency toward a comprehensive taxonomy in the fields of information security and digital forensics.

**Acknowledgements** We would like to express our appreciation to anonymous reviewers whose comments help improve both the technical contents and the presentation of this paper.

# References

1. Bishop, M.: How to Write A Setuid Program. USENIX Login. **12**(1), 5–11 (1987)
2. Capretta, V., Stepien, B., Felty, A.: Formal Correctness of Conflict Detection for Firewalls. In: Proceedings of the Fifth ACM Workshop on Formal Methods in Security Engineering: From Specifications to Code pp. 22–30 (2007)
3. Chen, H., Wagner, D., Dean, D.: Setuid Demystified. In: Proceedings of Eleventh USENIX Security Symposium August 2002 pp. 171–190 (2002)
4. Farid, H.: Seeing Is Not Believing. IEEE Spectrum. **46**(8), 44–51 (2009)
5. Fayad, A., Jajodia, S., McCollum, C.D.: Application-Level Isolation Using Data Inconsistency Detection. In: Proceedings 15th Annual Computer Security Applications Conference (ACSAC) pp. 119–126 (1999)
6. Gouda, M.G., Liu, A.X.: Structured Firewall Design. Comput. Networks. **51**(4), 1106–1120 (2007)
7. Hamed, H., Al-Shaer, E.: Taxonomy of Conflicts in Network Security Policies. IEEE Comm. Mag. **44**(3), 134–141 (2006)
8. Johnson, M.K., Farid, H.: Exposing Digital Forgeries by Detecting Inconsistencies in Lighting. In: Proceedings of ACM Multimedia and Security Workshop pp. 1–10 (2005)
9. Kim, G.H., Spafford, E.H.: The Design and Implementation of Tripwire: A File System Integrity Check. In: Proceedings of the 2nd ACM Conference on Computer and Communications Security November 2–4, 1994, pp. 18–29 (2010)
10. Kurose, J., Ross, K.: Computer Networking, Addison Wesley (2010)
11. Mahdian, B., Saic, S.: Using Noise Inconsistencies for Blind Image Forensics. Image and Vision Computing Vol. 27, pp. 1497–1503 (2009)
12. McKusick, M.K., Bostic, K., Karels, M.J.: The Design and Implementation of the 4.4 BSD Operating System Addison Wesley (1996)
13. Neagoe, V., Bishop, M.: Inconsistency in Deception for Defense. In: Proceedings of the 2006 Workshop on New Security Paradigms Schloss Dagstuhl, Germany, Sept. 19–22 pp. 31–38 (2006)
14. Patil, S., et al: I$^3$FS: An In-Kernel Integrity Checker and Intrusion Detection File System, In: Proceedings of Large Installation System Administration Conference pp. 67–78 (2004)
15. Pozo, S., Ceballos, R., Gasca, R.M.: A Heuristic Process for Local Inconsistency Diagnosis in Firewall Rule Sets. J. Networks. **4**(8), 698–710 (2009)
16. Ritchie, D.M.: Protection of Data File Contents, United States Patent#4,135,240, January 16, 1979 http://www.google.com/patent?vid=USPAT4135240 (2010)
17. Salamat, B., Jackson, T., Gal, A.: Orchestra: Intrusion Detection Using Parallel Execution and Monitoring of Program Variants in User-Space Proceedings of the 4th ACM European conference on Computer systems, pp. 33–46 (2009)
18. Setuid: Checklist for Security of Setuid Programs. http://www.homeport.org/$\sim$adam/setuid.7.html (2010)
19. Snyder, D.: On-line Intrusion Detection Using Sequences of System Calls, MS thesis, Department of Computer Science, Florida State University (2001)
20. Stakhanova, N., Li, Y., Ghorbani, A.A.: Classification and Discovery of Rule Misconfigurations in Intrusion Detection and Response Devices. In: Proceedings of the World Congress on Privacy, Security, Trust and the Management of e-Business pp. 29–37 (2009)
21. Tsafrir, D., Da Silva, D., Wagner, D.: The Murky Issue of Changing Process Identity: Revising 'Setuid Demystified', USENIX Login 33:3 pp. 55–66 (2008)
22. Vulnerability Note: Vulnerability Note VU#40327, OpenSSH UseLogin option allows remote execution of commands as root. http://www.kb.cert.org/vuls/id/40327 (2010)
23. Ye, S., Sun, Q., Chang, E.C.: Detecting Digital Image Forgeries by Measuring Inconsistencies of Blocking Artifacts, Proceedings of the IEEE International Conference on Multimedia and Expo pp. 12–15 (2007)

24. Yuan, L., et al.: FIREMAN: A Toolkit for Firewall Modeling and Analysis. In: Proceedings of the IEEE Symposium on Security and Privacy (2006)
25. Zhang, D.: Quantifying Knowledge Base Inconsistency via Fixpoint Semantics. Springer Transactions on Computational Science II LNCS 5150 pp. 145–160 (2008)
26. Zhang, D.: Taming Inconsistency in Value-Based Software Development. In: Proceedings of 21st International Conference on Software Engineering and Knowledge Engineering Boston Mass. pp. 450–455 (2009a)
27. Zhang, D.: Inconsistency: The Good, The Bad, and The Ugly. In: Proceedings of the 10th IEEE International Conference on Information Reuse and Integration pp. 182–187 (2009b)
28. Zhang, D.: Inconsistencies in Information Security and Digital Forensics. In: Proceedings of the 11th IEEE International Conference on Information Reuse and Integration pp. 141–146 (2010)